Lecture Notes in Artificial Intelligence 7132

Subseries of Lecture Notes in Computer Science

LNAI Series Editors

Randy Goebel
University of Alberta, Edmonton, Canada
Yuzuru Tanaka
Hokkaido University, Sapporo, Japan
Wolfgang Wahlster
DFKI and Saarland University, Saarbrücken, Germany

LNAI Founding Series Editor

Joerg Siekmann
DFKI and Saarland University, Saarbrücken, Germany

W0193052

Sanjay Modgil   Nir Oren   Francesca Toni (Eds.)

Theory and Applications of Formal Argumentation

First International Workshop, TAFA 2011
Barcelona, Spain, July 16-17, 2011
Revised Selected Papers

 Springer

Series Editors

Randy Goebel, University of Alberta, Edmonton, Canada
Jörg Siekmann, University of Saarland, Saarbrücken, Germany
Wolfgang Wahlster, DFKI and University of Saarland, Saarbrücken, Germany

Volume Editors

Sanjay Modgil
King's College London
Department of Informatics
Strand, London, WC2R 2LS, UK
E-mail: sanjay.modgil@kcl.ac.uk

Nir Oren
University of Aberdeen
Department of Computer Science
Aberdeen, AB24 3UE, Scotland, UK
E-mail: n.oren@abdn.ac.uk

Francesca Toni
Imperial College London
Department of Computing
South Kensington Campus, London SW7 2AZ, UK
E-mail: ft@imperial.ac.uk

ISSN 0302-9743   e-ISSN 1611-3349
ISBN 978-3-642-29183-8   ISBN 978-3-642-29184-5 (eBook)
DOI 10.1007/978-3-642-29184-5
Springer Heidelberg Dordrecht London New York

Library of Congress Control Number: Applied for

CR Subject Classification (1998): I.2, H.4, H.3, H.5, C.2, F.1, J.1

LNCS Sublibrary: SL 7 – Artificial Intelligence

Typesetting: Camera-ready by author, data conversion by Scientific Publishing Services, Chennai, India

Printed on acid-free paper

Springer is part of Springer Science+Business Media (www.springer.com)

Preface

Recent years have witnessed a rapid growth of interest in formal models of argumentation and their application in diverse sub-fields and domains of application of AI, including reasoning in the presence of inconsistency, non-monotonic reasoning, decision making, inter-agent communication, the Semantic Web, grid applications, ontologies, recommender systems, machine learning, neural networks, trust computing, normative systems, social choice theory, judgement aggregation and game theory, and law and medicine. Argumentation thus shows great promise as a theoretically grounded tool for a wide range of applications.

TAFA-11, the First International Workshop on Theory and Applications of Formal Argumentation, aimed at contributing to the realization of this promise, by promoting and fostering uptake of argumentation as a viable AI paradigm with wide-ranging application, and providing a forum for further development of ideas and the initiation of new and innovative collaborations.

We invited submission of papers on: formal theoretical models of argumentation and application of such models in (sub-fields of) AI; evaluation of models, both theoretical (in terms of formal properties of existing or new formal models) and practical (in concretely developed applications); theories and applications developed through inter-disciplinary collaborations. We received 32 submissions, of which we accepted 9 as full papers and 12 as short papers. Extended and improved versions of all full papers are included in these proceedings, as well as extended and improved versions of eight short papers that were re-reviewed after the workshop.

The papers included in these proceedings cover the following topics:

- Properties of formal models of argumentation
- Instantiations of abstract argumentation frameworks
- Relationships among different argumentation frameworks
- Practical applications of formal models of argumentation
- Argumentation and other artificial intelligence techniques
- Evaluation of formal models of argumentation
- Validation and evaluation of applications of argumentation

In addition to paper presentations, the workshop also included an extended panel session on the topic: "The future of argumentation: what is its added value and how we communicate this to researchers in the artificial intelligence community and beyond." The panel was conducted by three influential researchers in the area of formal argumentation: *Carlos Chesnevar* (Universidad Nacional del Sur, Argentina), *Martin Caminada* (Université du Luxembourg, Luxembourg), and *Stefan Woltran* (Vienna University of Technology, Austria). The panelists addressed and debated (with one another and the workshop participants) the following questions:

1. Which main challenges do we need to face for argumentation theory to have a real impact on applications?
2. Are any of the argumentation systems currently available ready for deployment?
3. Have we identified suitable "killer" applications already? If not, which direction should we look at for a "killer" application?
4. Do we need any further theoretical developments to pave the way toward applications and if so in which direction?
5. Which "industry" is most likely to be receptive to our methodologies/ techniques?
6. Would it be useful to "team up" with any other field (in AI, or computer science, or elsewhere) in order to have a higher impact/more powerful techniques?

The panel stirred a lively debate among the 25 or so workshop participants. Passions often ran high: a testament not to fundamental divisions within the community, but rather a desire to ensure that "we get things right" and so realize the promise of argumentation.

December 2011

Sanjay Modgil
Nir Oren
Francesca Toni

Organization

TAFA-11 took place at the Universitat de Barcelona, Barcelona, Catalonia (Spain) during July 16–17, 2011, as a workshop at IJCAI-11, the 22nd International Joint Conference on Artificial Intelligence.

Workshop Chairs

Sanjay Modgil	King's College London, UK
Nir Oren	University of Aberdeen, UK
Francesca Toni	Imperial College London, UK

Program Committee

Leila Amgoud	IRIT, Toulouse, France
Katie Atkinson	University of Liverpool, UK
Pietro Baroni	University of Brescia, Italy
Floris Bex	University of Dundee, UK
Elizabeth Black	Universiy of Utrecht, The Netherlands
Guido Boella	University of Turin, Italy
Ivan Bratko	University of Ljubljana, Slovenia
Gerhard Brewka	University of Leipzig, Germany
Martin Caminada	University of Luxembourg, Luxembourg
Carlos Chesnevar	Universidad Nacional del Sur, Argentina
Sylvie Doutre	University of Toulouse 1, France
Phan Minh Dung	Asian Institute of Technology, Thailand
Paul Dunne	University of Liverpool, UK
Dov Gabbay	King's College London, UK
Massimilliano Giacomin	University of Brescia, Italy
Tom Gordon	Fraunhofer FOKUS, Germany
Anthony Hunter	University College London, UK
Antonis Kakas	University of Cyprus, Cyprus
Nicolas Maudet	Universite Paris Dauphine, France
Peter McBurney	University of Liverpool, UK
Sanjay Modgil	King's College London, UK
Pavlos Moraitis	Paris Descartes University, France
Timothy J. Norman	University of Aberdeen, UK
Nir Oren	University of Aberdeen, UK
Simon Parsons	City University of New York, USA
Henry Prakken	Utrecht University and University of Groningen, The Netherlands

Iyad Rahwan   Masdar Institute, UAE and Massachusetts
                            Institute of Technology, USA
Chris Reed   University of Dundee, UK
Nicolas Rotstein   University of Aberdeen, UK
Guillermo Simari   Universidad Nacional del Sur, Argentina
Francesca Toni   Imperial College London, UK
Leon van der Torre   University of Luxembourg, Luxembourg
Serena Villata   University of Turin, Italy
Simon Wells   University of Dundee, UK
Stefan Woltran   Vienna University of Technology, Austria

Additional Referees

Mark Snaith

Sponsoring Institutions

TAFA-11 was endorsed by the Agreement Technologies COST action.

Table of Contents

Theory and Applications of Formal Argumentation

Probabilistic Argumentation Frameworks

Hengfei Li, Nir Oren, and Timothy J. Norman

Department of Computing Science, University of Aberdeen, Aberdeen, AB24 3UE,
Scotland
{h.li,n.oren,t.j.norman}@abdn.ac.uk

Abstract. In this paper, we extend Dung's seminal argument framework
to form a probabilistic argument framework by associating probabilities
with arguments and defeats. We then compute the likelihood of some set
of arguments appearing within an arbitrary argument framework induced
from this probabilistic framework. We show that the complexity of computing this likelihood precisely is exponential in the number of arguments
and defeats, and thus describe an approximate approach to computing
these likelihoods based on Monte-Carlo simulation. Evaluating the latter
approach against the exact approach shows significant computational sav-
ings. Our probabilistic argument framework is applicable to a number of
real world problems; we show its utility by applying it to the problem of
coalition formation.

1 Introduction

Likelihoods and probabilities form a cornerstone of reasoning in complex domains. When argumentation is used as a form of defeasible reasoning, uncertainty
can affect the decisions reached during the reasoning process [27]. Uncertainty
can also affect applications of argumentation technologies in other ways. For
example, in the context of a dialogue, uncertainty regarding the knowledge of
participants can affect both the dialogue outcome, and the utterances the participants choose to make. Furthermore, if uncertainty is viewed as a proxy for
argument strength, questions immediately arise regarding argument interaction
and the strength of conclusions given an argument system.

In this paper we examine the role of probabilities in an abstract argument
framework. Within such a framework, an argumentation semantics defines a
method by which a set of justified arguments can be deduced. As a reasoning
approach, a semantics takes an argumentation framework as its knowledge base
and produces a set of justified arguments as its output. The problem we address
thus involves identifying the effects of probabilities on argument justification.

At the intuitive level, our approach is relatively simple. Starting with Dung's
abstract argumentation framework[9] as its base[1], we assign probabilities to arguments and defeats. These probabilities represent the likelihood of existence of
a specific argument or defeat, and thus capture the uncertainties inherent in the

---

[1] Though as discussed in Section 6, our techniques are applicable to nearly any other
argumentation framework.

S. Modgil, N. Oren, and F. Toni (Eds.): TAFA 2011, LNAI 7132, pp. 1–16, 2012.

argument system. Within such a *probabilistic argument framework* (abbreviated PrAF), all possible arguments neither definitely exist, nor completely disappear. Instead, all elements of the framework have a different chance of existing. The semantics of such a framework then identify the likelihood of different sets of arguments being justified according to different types of extensions.

Now, since we are interested in the likelihood of a set of arguments being justified we are, in a sense, reversing the standard semantics of argumentation. Rather than identifying *which* arguments are in some sense compatible, we are instead identifying a set of arguments and asking what their likelihood of being compatible is (with respect to the other arguments, defeats and probabilities which make up the framework). Answering this type of question has a number of real world applications, including to the domains of trust and reputation [32] and coalition formation [28].

As we show, a naïve approach to computing the likelihood of some set of arguments being justified within a probabilistic argumentation framework based on the standard laws of probability has exponential computational complexity with respect to the number of arguments even in situations where the underlying semantics has linear complexity. Given that this is impractical for most real-life scenarios we propose, and evaluate, an approximation method based on the idea of Monte-Carlo simulation for calculating the likelihood of a set of arguments being justified.

The remainder of this paper is structured as follows. In the next section, we describe and formally define probabilistic argumentation frameworks, and explain the naïve method for performing computations over such PrAFs. Section 3 then details the Monte-Carlo simulation based approximation method. In Section 4, we empirically evaluate the performance of both of our techniques. An illustrative application for which PrAFs are particularly applicable is detailed in Section 5, following which Section 6 provides a more general discussion together with suggestions for future work. We then summarise our results and conclude the paper in Section 7.

2   Probabilistic Argumentation Frameworks

In this section, we extend Dung's argumentation framework to include uncertainty with respect to arguments and defeats. Essentially, we assign a probability to all elements of the argument framework, namely to every argument and element of the defeat relation. It should be noted that our approach can be easily extended to other frameworks such as the bipolar [7], evidential [20] and value based argumentation frameworks [5] as probabilities can be also assigned to the additional elements of these frameworks (e.g. to the members of the support relation in the case of bipolar frameworks). We begin this section by briefly describing Dung's system, following which we discuss our extensions and methods for reasoning about probabilistic frameworks.

Definition 1. *(**Dung Argumentation Framework**) A Dung argumentation framework DAF is a pair (Arg, Def) where Arg is a set of arguments, and Def ⊆ Arg × Arg is a defeats relation.*

A set of arguments S is conflict-free *if* $\nexists a, b \in S$ *such that* $(a,b) \in Def$. *An argument a is* acceptable *with respect to a set of arguments S iff* $\forall b \in Arg$ *such that* $(b,a) \in Def$, *$\exists c \in Arg$ such that* $(c,b) \in Def$. *A set of arguments S is* admissible *iff it is conflict free and all its arguments are acceptable with respect to S.*

From these definitions, different semantics have been defined [4]. These semantics identify sets of arguments which are, in some intuitive sense, compatible with each other. For example, the grounded semantics yield a single extension which is the least fixed point of the characteristic function $F_{AF}(S) = \{a | a \in Arg$ is acceptable w.r.t $S\}$. In the remainder of this paper, we will concentrate on the grounded semantics due to its computational tractability [11].

2.1   Formalising Probabilistic Argumentation Frameworks

A probabilistic argumentation framework extends Dung's argument framework by associating a likelihood with each argument and defeat in the original system. Intuitively, a PrAF represents an entire set of DAFs that exist *in potentia*. A specific DAF can then has a certain likelihood of being *induced* from the PrAF.

Definition 2. *(**Probabilistic Argumentation Framework**) A Probabilistic Argumentation framework PrAF is a tuple (A, P_A, D, P_D) where (A, D) is a DAF, $P_A : A \rightarrow (0 : 1]$ and $P_D : D \rightarrow (0 : 1]$.*

The functions P_A and P_D map individual arguments, and defeats to likelihood values. These represent the likelihood of existence of an argument within an arbitrary DAF induced from the PrAF. As discussed below, P_D is, implicitly, a conditional probability. It should be noted that the lower bound of these probabilities is not 0 (but approaches it in the limit). This requirement exists because any argument or defeat with a likelihood of 0 cannot ever appear within a DAF induced from the PrAF, and is thus redundant.

A PrAF represents the set of all DAFs that can potentially be created from it. We call this creation process the inducement of a DAF from the PrAF. All arguments and defeats with a likelihood of 1 will be found in the induced DAF, which can then contain additional arguments and defeats, as specified by the following definition.

Definition 3. *(**Inducing a DAF from a PrAF**) A Dung argument framework $AF = (Arg, Def)$ is said to be induced from a probabilistic argumentation framework $PrAF = (A, P_A, D, P_D)$ iff all of the following hold:*

- *$Arg \subseteq A$*
- *$Def \subseteq D \cap (Arg \times Arg)$*
- *$\forall a \in A$ such that $P_A(a) = 1$, $a \in Arg$*
- *$\forall (f, t) \in D$ such that $P_D((f, t)) = 1$ and $P_A(f) = P_A(t) = 1$, $(f, t) \in Def$*

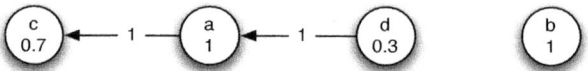

Fig. 1. A graphical depiction of a PrAF

We write $I(PrAF)$ to represent the set of all DAFs that can be induced from PrAF.

A DAF induced from a PrAF thus contains a subset of the arguments found in the PrAF, together with a subset of the defeats found in the PrAF, subject to these defeats containing only arguments found within the induced DAF. The process of inducing a DAF eliminates information regarding likelihoods found in the original PrAF.

Now, consider a situation where a number of entities are participating in a dialogue, and one of them (labelled α) would like to compute what conclusions might be drawn at the end of this interaction. Let us assume that α has arguments a and b in its knowledge base, and it believes that the other dialogue participants have arguments c and d in their knowledge base. This belief is however uncertain; c is believed to be known by the others with a likelihood of 0.7, and d with a likelihood of 0.3. Now let us assume that argument a defeats c and d defeats a. For simplicity, we assume that these defeat relations have no uncertainty associated with them (i.e. $P_D = 1$ for each of them). Formally, this can be represented by the PrAF following PrAF, illustrated in Figure 1.

$$(\{a,b,c,d\}, \{(a,1),(b,1),(c,0.7),(d,0.3)\}, \{(a,c),(d,a)\}, \{((a,c),1),((d,a),1)\})$$

Given this PrAF, we can induce the following DAFs:

$$(\{a,b\}, \{\}), \qquad (\{a,b,c\}, \{(a,c)\}),$$
$$(\{a,b,d\}, \{(d,a)\}), (\{a,b,c,d\}, \{(a,c),(d,a)\})$$

Clearly, b appears in the grounded extension of all of these DAFs, while a appears in the grounded extension of 3 out of 4 induced DAFs. Now, α might want to identify the likelihood of a being justified (i.e. in the grounded extension) at the end of the dialogue, perhaps to decide whether to advance it or not (assuming that advancing an argument has some associated utility cost [21]).

2.2   Probabilistic Justification

Our goal is to compute the likelihood that some set of arguments exists and is justified according to some semantics within the DAFs induced from a PrAF. This likelihood can be obtained from the basic laws of probability, and we detail this procedure next. We make one critical simplifying assumption, namely that the likelihood of one argument (defeat) appearing in an induced DAF is independent of the likelihood of some other argument (defeat) appearing. With this assumption in hand, we begin by computing the likelihood of some DAF being induced from the PrAF.

As mentioned earlier, the P_D relation associates a *conditional* probability with each possible defeat. That is, for some arguments a, b

$$P_D(a, b) = P((a, b) \in Def | a, b \in Arg) \text{ for the induced DAF } (Arg, Def)$$

Informally, the probability of some DAF AF being induced from a PrAF can be computed via the joint probabilities of the arguments and defeat relations appearing in AF. In order to formalise this concept compactly, we must identify the set of defeats that *may* appear in an induced DAF. We label this set as *DefA*. Given a DAF with arguments $Args$, and a *PrAF* containing defeats D

$$DefA = \{(a, b) | a, b \in Args \text{ and } (a, b) \in D\}$$

This allows us to compute the probability of some DAF AF being induced from a PrAF, written $P^I_{PrAF}(AF)$, by computing the joint probabilities of independent variables as follows:

$$P^I_{PrAF}(AF) = \prod_{a \in Arg} P_A(a) \prod_{a \in A \setminus Arg} (1 - P_A(a)) \prod_{d \in Def} P_D(d) \prod_{d \in DefA \setminus Def} (1 - P_D(d))$$

$$(1)$$

Applying this to our earlier example, $P^I_{PrAF}((\{a, b\}, \{\})) = 0.21$.

Proposition 1. *The sum of probabilities of all DAFs that can be induced from an arbitrary PrAF is 1. That is, $\sum_{a \in I(PrAF)} P^I_{PrAF}(a) = 1$.*

Now our goal is to identify the likelihood of some set of arguments being consistent with respect to some set of argumentation semantics. Such a semantics may return one or many extensions for a given argument framework, and we formalise our notion of consistency through the definition of a *semantic evaluation function*, $\xi^S(AF, X)$ which returns *true* if and only if the set of arguments X is deemed consistent using the semantics S when evaluated over the argument framework AF. Thus, for example $\xi^G(AF, X)$ could return true if the set of arguments X appears as a subset of the grounded extension of AF.

Then, following on from Proposition 1, given some $PrAF$, the likelihood of X being consistent according to the semantics S is defined as follows:

$$P_{PrAF}(X) = \sum_{AF \in I(PrAF)} P^I_{PrAF}(a) \text{ where } \xi^S(AF, X) = true \qquad (2)$$

Referring again to our earlier example, $P_{PrAF}(\{a, b\}) = 0.7$.

While we can utilise Equations 1 and 2 to compute the exact likelihood of a set of arguments being justified with regards to some semantics, the size of the set of possible DAFs which can be induced from a PrAF grows exponentially with regards to the number of arguments and defeats within the PrAF, resulting in exponential time complexity (not including the computational costs associated with computing the results of ξ^S). This is clearly impractical for a large set of arguments, and in the next section, we examine an approximate method for determining these likelihoods.

3   Approximate Solutions in Probabilistic Argumentation Frameworks

In this section we describe a Monte-Carlo simulation based approach to computing $P_{PrAF}(X)$ for an arbitrary set of arguments X. At an abstract level, a Monte-Carlo simulation operates by repeatedly sampling a distribution many times in order to approximate it. More specifically, such a simulation has three basic steps. First, given a possible set of inputs, a subset of these inputs is selected according to some probability distribution. Second, some computation is performed using the selected inputs. Finally, the results of repeating the first two steps multiple times is aggregated. Monte-Carlo simulation has a long history, and has been applied to a variety of computationally difficult problems including inference in Bayesian Networks [19], reinforcement learning [31] and computer game playing [8].

In this context of probabilistic argumentation frameworks, this process involves randomly inducing DAFs from a PrAF, with the likelihood of an arbitrary DAF being induced being dependant on the underlying probability distribution of its individual members. We thus sample the space of possible DAFs in a way that approximates the DAFs true distribution in the probability space.

The only source of uncertainty in Equation 2 lies in P_{PrAF}^{I} which in turn depends only on the probabilities found in the underlying PrAF. Therefore, in order to approximate $P_{PrAF}(X)$ we need only sample the space of arguments and defeats found in the PrAF. Algorithm 1 describes this process more precisely.

The algorithm samples N DAFs from the set of inducible DAFs. A single DAF is generated by randomly selecting arguments and defeats according to their likelihood of appearance (Lines 4-7 and 10-14 respectively). This resultant DAF is then evaluated for the presence of X through the ξ^S function (Line 16), and if this function holds, the DAF is counted. $P_{PrAF}(X)$ is finally approximated as the ratio of the total number of DAFs in which $\xi^S(X)$ holds to the number DAFs sampled (Line 20).

The following proposition states that as our number of trials increases, the error in our approximation of $P_{PrAF}(X)$ shrinks.

Proposition 2. *If we denote the output of Algorithm 1 as $P'_{PrAF}(X)$, then as $N \to \infty$, $P_{PrAF}(X) - P'_{PrAF}(X) \to 0$. More specifically, there is some $N \in \mathbb{Z}^+$ and $\epsilon \in \mathbb{R}^+$ such that for all $M > N$, if M trials are run, $|P_{PrAF}(X) - P'_{PrAF}(X)| < \epsilon$.*

This proposition means that our algorithm has an anytime property: it may be terminated at any time, and earlier terminations will still provide an approximation to the true probability, albeit with a greater error than would be provided from a later termination.

While this proposition provides some guarantees regarding the accuracy of our results given enough trials, it does not answer one critical question: how many trials must be run to ensure (with some level of confidence) that our approximation has only a small level of error?

Algorithm 1. An algorithm to approximate $P_{PrAF}(X)$

Require: A Probabilistic Argumentation Framework $PrAF = (A, P_A, D, P_D)$
Require: A set of arguments $X \subseteq A$
Require: A number of trials $N \in \mathbb{N}$
Require: A semantic evaluation function, ξ^S
 1: $Count = 0$
 2: **for** $I = 0$ to N **do**
 3:     $Arg = Def = \{\}$
 4:     **for all** $a \in A$ **do**
 5:         Generate a random number r such that $r \in [0, 1]$
 6:         **if** $P_A(a) \geq r$ **then**
 7:             $Arg = Arg \cup \{a\}$
 8:         **end if**
 9:     **end for**
10:     **for all** $(f, t) \in D$ such that $f, t \in Arg$ **do**
11:         Generate a random number r such that $r \in [0, 1]$
12:         **if** $P_D((f, t)) \geq r$ **then**
13:             $Def = Def \cup \{(f, t)\}$
14:         **end if**
15:     **end for**
16:     **if** $\xi^S((Arg, Def), X) = true$ **then**
17:         $Count = Count + 1$
18:     **end if**
19: **end for**
20: **return** $Count/N$

In order to answer this question, we note that the results of a Monte-Carlo simulation can be viewed as a normal distribution over possible values for $P_{PrAF}(X)$, and with $P'_{PrAF}(X)$ as its mean. Given this, we may make use of the notion of a *confidence interval* in order to answer our question. In statistics, a confidence level of l for a given a confidence interval CI and a mean p' can be read as stating that the true mean lies within $p' \pm CI$ with a likelihood of l. Such a confidence interval is dependant on the observed likelihood of an event and the number of trials used to make the observations. We can thus recast our problem to ask how many trials need to be run in order to ensure that the confidence interval around $P'_{PrAF}(X)$ (i.e. its error) is smaller than some value ϵ with some specific confidence level (e.g. 95%).

Probably the most common approach to computing such an interval is the normal approximation interval [18], which is defined as follows:

$$p' \pm z_{1-(\alpha/2)} \sqrt{\frac{p'(1 - p')}{n}} \tag{3}$$

Here, p' is the observed mean, n is the number of trials, and $z_{1-(\alpha/2)}$ the $1 - (\alpha/2)$ percentile of the normal distribution. In the experiments described in Section 4, we required a 95% confidence level, resulting in $z_{1-(\alpha/2)} = 1.96$. Then we get

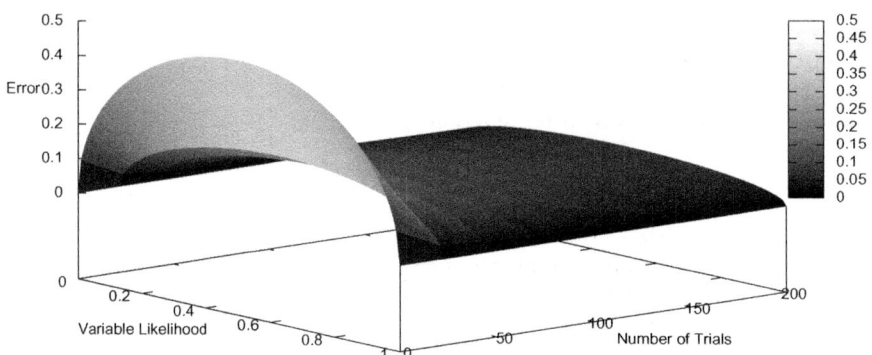

Fig. 2. The relationship between likelihood of a variable, the number of observations made and the error in the observed likelihood

the following equation to compute the number of trials required to achieve an error level below ϵ:

$$N > \frac{p'(1-p')}{\epsilon^2}(1.96)^2 \qquad (4)$$

However, this approximation is problematic in our situation as p' is either 0 or 1 after a single trail, which will break down the calculation. To overcome this problem, we utilise the Agresti-Coull interval [1] instead. The general form of this interval is the same as that of Equation 3. However, the values of n and p' are computed differently:

$$n = N + z_{1-\alpha/2}^2 \qquad p' = \frac{X + z_{1-\alpha/2}^2/2}{n}$$

Here, N is the number of trials and X is the number of "successes" observed. Intuitively, it adds two different trivial numbers to the number of trials and the number of "successes" respectively, which ensures p' will never be 0. Approximating $1 - \alpha/2$ with the value 2 yields the above new equation. This method guarantees the real "anytime" feature of our algorithm.

$$N > \frac{4p'(1-p')}{\epsilon^2} - 4 \qquad (5)$$

Figure 2 provides a plot of this function. As seen here, initially, as the number of trials increase, the error falls off rapidly. However, this shrinking of the error quickly ceases, and additional trials serve to reduce the error by only a small amount. It should also be noted that the likelihoods of variables with extreme values (i.e. near 0 or 1) can be approximated far more quickly than variables with values near 0.5.

Given a desired error level ϵ and confidence level, Equation 5 provides us with a new stopping condition for Algorithm 1. The **for** loop of Line 2 can be substituted for a `while` loop which computes whether the expected error level falls below ϵ given the number of iterations that have been run so far. If this is the case, the loop can end, and the algorithm will terminate.

4   Evaluation

We have described, given some PrAF, two approaches to computing the likelihood of a chosen set of arguments being justified with respect to some semantics. While it is clear that the exact approach is exponential in complexity, it is useful to identify the approximate number of arguments in a PrAF at which point this becomes impractical. Similarly, in order to use it in real world settings, the approximate running time of the Monte-Carlo based approach must also be evaluated.

We implemented both of the approaches described in the paper using SWI-Prolog[2]. For simplicity, we associated likelihood values only with arguments within the PrAF; all defeats had a likelihood of 1. The goal of our first experiment was to identify the effects of differently sized PrAFs on the runtimes of the exact approach, and of the Monte-Carlo based approach with different error tolerances ($\epsilon = 0.01$ and $\epsilon = 0.005$). In order to do so, we evaluated the approaches on identical PrAFs with each PrAF containing between 1 and 16 arguments. Our semantic evaluation function $\xi^S(X)$ computed whether X formed a subset of the grounded extension. We ran our experiment 20 times for each unique number of arguments, and Figure 3 shows our results. As expected, the time taken by the exact approach increases exponentially; the Monte-Carlo based approaches overtake the exact approach at around 13 (when $\epsilon = 0.01$) and 15 (when $\epsilon = 0.005$) arguments. The introduction of uncertainty into the defeats relation would increase the number of DAFs that can be induced from the PrAF meaning that our results, in a sense, represent the best case for the exact approach.

In order to more closely examine the effect of ϵ and the size of the PrAF on the performance of our approximate algorithm, Figure 4 compares the average number of iterations, and runtime, required to achieve the desired level of accuracy against the number of arguments found in the PrAF. As expected, an increase in the size of the PrAF has only a linear effect on the runtime of our algorithm. This increase occurs due to an increase in the time required to computing the membership of grounded extension (as computing this has linear complexity) rather than additional iterations. In other words, the complexity level of our algorithm depends on the complexity of computing membership under some semantics. This result can clearly be seen from Figure 2; the number of iterations required to obtain a certain error level do not depend on the number of arguments and defeats in the PrAF, but only on the joint probabilities obtained from the PrAF. Figure 2 also predicts another result clearly seen in Figure 4, namely that as the permitted error shrinks, the standard deviation of the number of iterations that must be executed grows. This is because the number of iterations required to obtain an error ϵ when the joint probability in

---

[2] http://www.swi-prolog.org

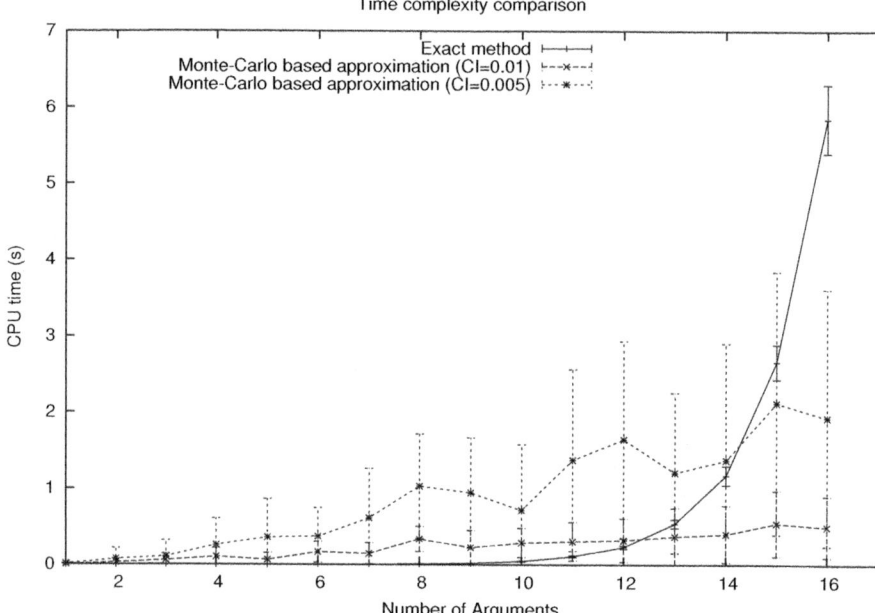

Fig. 3. Comparison of runtimes between the exact and Monte-Carlo based approaches. Error bars indicate 1 standard deviation.

question is close to 0 or 1 grows much more slowly than when the probability is close to 0.5.

Finally, it can also be seen that there exists some variability between the number of iterations required and the time to execute these iterations; this arises due to the underlying Prolog implementation, and the number of iterations is thus a better indicator of algorithm performance.

5   Applying PrAFs to Coalition Formation

In this section, we describe an application of our approach to a real world problem, namely coalition formation. According to [28], "Coalition formation is a fundamental form of interaction that allows the creation of coherent groupings of distinct, autonomous, agents in order to efficiently achieve their individual or collective goals". Coalition formation is applicable to both virtual domains such as e-commerce (where virtual organisations can form in order to satisfy a customer's requirements [24]), and physical domains where, for example, a search and rescue team must be composed of agents with specific capabilities in order to be able to undertake some mission [26].

Most approaches to coalition formation treat the problem as one of utility maximisation; agents will join a coalition if being in the coalition will yield a greater utility than not. Here, we show how to address the problem of coalition formation from a very different perspective. This different perspective allows us

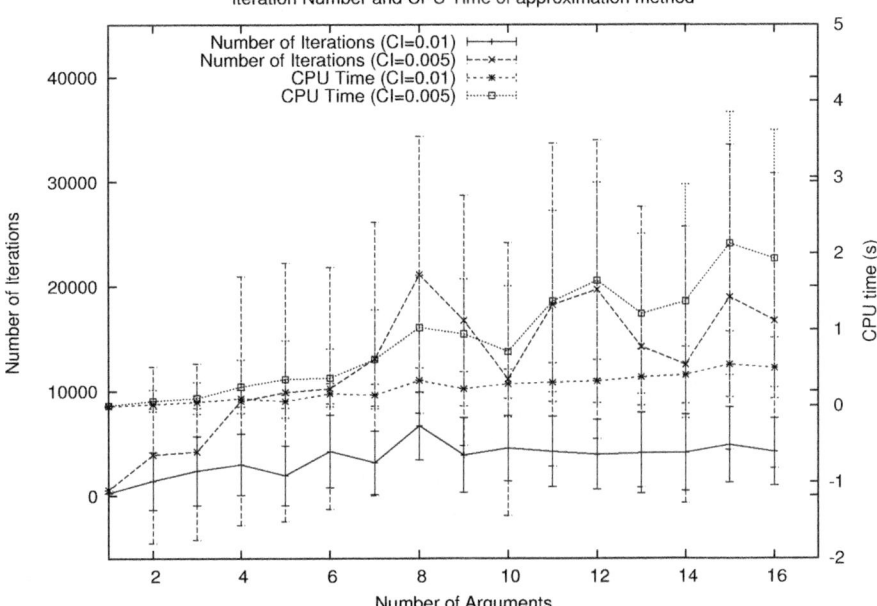

Fig. 4. Comparison of runtimes and number of iterations between the Monte-Carlo based approaches with different ϵ values. Error bars indicate 1 standard deviation.

to explore an aspect of the social dimensions involved in coalition formation; i.e. the notion of whether or not an individual's presence in a coalition may influence another's membership. More specifically, we model a system containing agents with different capabilities, each of which has a prior probability of joining the coalition, and a probability of preventing other potential coalition members from joining the coalition. We would then like to determine what the probability of a coalition forming which is capable of achieving some task.

Translating this problem into a PrAF is trivial. Each agent can be represented as an argument within the PrAF, an associated P_A equal to its prior probability of joining the coalition. Defeats then represent the likelihood of the presence of one member in the coalition preventing another member from joining. Computing the likelihood of a coalition containing specific members can then be computed by computing P_{PrAF}.

As an illustrative example[3], consider a small mercenary team consisting of a leader h, a pilot m, a mechanic b and an expert in persuasion f. Now assume that the presence of the pilot cannot be tolerated by the mechanic, and that f is generally disliked by other team members (to varying degrees); f's presence in a coalition will increase the risk that others will not join. Finally, assume that both f and h are often busy, and occasionally cannot join the team. This situation can be represented by the PrAF shown in Figure 5.

---

[3] This example is based on the characters from a 1980's television series.

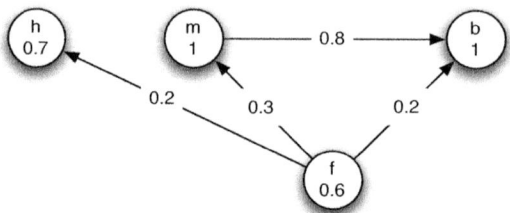

Fig. 5. A PrAF representing the coalition formation example

A system user could use the techniques presented in this paper to compute the likelihood of a specific team being formed, for example consisting of h, m and b (this would be 0.037632+0.056=0.093632. The first value is the likelihood of the full team forming, and the second, the probability of the team forming without f). Given this likelihood, the user might decide to change their goals, or add new agents to the system to increase the chances of success.

The discussion thus far has concentrated on determining whether a coalition can be formed containing some specific set of agents. However, in the context of coalition formation, the goal is often to form a coalition consisting of agents taking on some set of specific roles (e.g. a coalition requires two mechanics and a pilot). One approach to determining the likelihood of forming such a coalition involves identifying all possible ways in which such a coalition can form, and combining the probabilities of each individual coalition to obtain an aggregate probability. However, this approach does not scale well as the size of the system increases. We intend to investigate techniques for dealing with this issue in future work, and discuss it further in the next section.

6   Discussion and Future Work

The use of likelihood in different facets of argumentation for modelling strength or uncertainty of arguments has a long and rich history. Most commonly, such likelihood measures have served as a proxy for argument strength [27,14], or causal strength between arguments [23]. Argument framework incorporating uncertainty about arguments have utilised probabilities to compute the likelihood of some conclusion holding using a variety of different methods. For example, [17,16] consider probability in the construction of argument, deriving a probability of argument from the probabilities of its premises by several different methods. In the context of abstract argumentation frameworks, some approaches for modelling uncertainties such as assigning a numerical values [10] or preference ordering [2] to attacks have been developed. Another approach to strengths of argument involves counting the number of subsets which meet the requirements of some (multiple status) semantics [3], and in which the argument under question appears. The ratio of this number to the total number of extensions then serves to act as a measure of strength for the argument. Our approach is similar in spirit to this latter work as we compute the likelihood of some subset of arguments appearing. However, the introduction of probabilities, through the definition of a PrAF, makes the approach applicable to both single and multiple status semantics, with the distinct advantage of the former's tractability. Another similar work is by Janssen et al [15] in which they do not use probability but also a value between 0 and 1 to model the strength of arguments and attack relations. They borrow the idea from fuzzy logic and these values capture the concept of *degree*; they define their semantics based on fuzzy logic operations.

Within the machine learning and knowledge representation communities, Bayesian Networks [25] form a popular approach to modelling and reasoning about uncertainty. Such networks allow one to reason about the posterior probability of the value of variables within the network given observations about some other parts of the network. The idea of representing a PrAF as a Bayesian Network is intuitively appealing, but not trivial. The DAF induced from a PrAF captures the probability that an argument exists, while the PrAF based semantics computes the likelihood that an argument (or set of arguments) is present in an extension. It is not clear how both of these values can be encoded in, or computed from, a single Bayesian Network, and as future work, we intend to investigate whether, and how, a Bayesian Network representing a PrAF can be constructed.

In Section 5, we discussed one possible application for PrAFs, namely answering questions about the likelihood of a coalition with certain characteristics being formed. We discussed one shortcoming, namely the inability of the basic approach to deal with the notion of roles in coalition formation, and suggested one method for overcoming this shortcoming. Another more nuanced approach involves the use of resource bounded argumentation frameworks [30], which would allow us to place requirements on team composition via constraints, and thus also allow for more nuanced team formation. Another shortcoming involves our underlying Dung based model wherein only defeats between arguments are modelled. Constructing a PrAF on top of a bipolar framework (e.g. [7,20]) would allow us to cater for situations where one agent is more likely to enter into a coalition if some other agent will be present. Another way of achieving this would be to lift the independence assumption regarding the likelihood of argument and defeat relation likelihoods, and all of these form enticing possibilities for future work.

PrAFs and the techniques described in this paper can be applied to other argument frameworks and domains. For example, a value based argumentation framework (VAF) [5] provides a model of determining whether some set of arguments will be accepted by audiences containing agents with different preferences over the defeat relation. Constructing a PrAF on top of such a VAF can allow us to answer questions such as "what is the likelihood of all members in the audience accepting this argument". Clear applications of this include opponent modelling [21] and heuristics for argument [22,29,12]. Another interesting possibility lies in associating a probability distribution with the preferences of the audience within the VAF.

Apart from the coalition formation and argument strategy domains, the ideas associated with constructing and evaluating PrAFs can also play a role in other domains where the notion of the strength of an argument is relevant. For example, in the area of trust and reputation [32], PrAFs can be used to associate reputation information with individual agents. Distrust relationships (following [13]) or biases in trust relationships (following [6]) can then be constructed through the defeats relation , and, by using a bipolar framework, trust relationships can be created through support links. The resultant PrAF can then be used to compute the likelihood of some set of agents considering one another trustworthy.

7   Conclusions

In this paper we introduced *probabilistic argumentation frameworks*. These frameworks add the notion of likelihood to all elements of an abstract argument framework (in this paper, we concentrated on Dung argument frameworks, and thus associated likelihoods with arguments and defeats), and are used to determine the likelihood of some subset of arguments appearing within an extension. The exact method for determining this likelihood has exponential complexity, and is thus impractical for use with anything other than a small argumentation system. To overcome this limitation, we introduced a Monte-Carlo simulation based approach to approximate the likelihood. This latter technique scales up well, providing good results in a reasonable period of time, and has anytime properties, making it ideal for use in almost all situations.

PrAFs have applications to a myriad of domains. In this paper, we focused on one such domain, namely coalition formation, and described how PrAFs can be used to assist a system designer. While we have touched on the applications of PrAFs to other domains, and suggested a number of extensions to their basic representation, we intend to further explore their potential applicability to additional argumentation frameworks and application domains.

References

1. Agresti, A., Coull, B.A.: Approximate is better than "exact" for interval estimation of binomial proportions. The American Statistician 52(2), 119–126 (1998)
2. Amgoud, L., Cayrol, C.: Inferring from inconsistency in preference-based argumentation frameworks. Journal of Automated Reasoning 29, 125–169 (2002)
3. Baroni, P., Dunne, P.E., Giacomin, M.: On extension counting problems in argumentation frameworks. In: Proceeding of the 2010 Conference on Computational Models of Argument: Proceedings of COMMA 2010, pp. 63–74. IOS Press, Amsterdam (2010)
4. Baroni, P., Giacomin, M.: Semantics of abstract argument systems. In: Simari, G., Rahwan, I. (eds.) Argumentation in Artificial Intelligence, pp. 25–44. Springer, US (2009)
5. Bench-Capon, T.: Value based argumentation frameworks. In: Proceedings of the 9th International Workshop on Nonmonotonic Reasoning, Toulouse, France, pp. 444–453 (2002)

6. Burnett, C., Norman, T.J., Sycara, K.: Stereotypical trust and bias in dynamic multi-agent systems. ACM Transactions on Intelligent Systems and Technology (in press)
7. Cayrol, C., Lagasquie-Schiex, M.C.: Bipolar Abstract Argumentation Systems. In: Rahwan, I., Simari, G. (eds.) Argumentation in Artificial Intelligence, ch. 4, pp. 65–84. Springer, Heidelberg (2009), http://www.springerlink.com
8. Coulom, R.: Efficient Selectivity and Backup Operators in Monte-Carlo Tree Search. In: van den Herik, H.J., Ciancarini, P., Donkers, H.H.L.M(J.) (eds.) CG 2006. LNCS, vol. 4630, pp. 72–83. Springer, Heidelberg (2007)
9. Dung, P.M.: On the acceptability of arguments and its fundamental role in non-monotonic reasoning, logic programming and n-person games. Artificial Intelligence 77(2), 321–357 (1995)
10. Dunne, P.E., Hunter, A., McBurney, P., Parsons, S., Wooldridge, M.: Weighted argument systems: Basic definitions, algorithms, and complexity results. Artificial Intelligence 175(2), 457–486 (2011)
11. Dunne, P.E., Wooldridge, M.: Complexity of abstract argumentation. In: Simari, G., Rahwan, I. (eds.) Argumentation in Artificial Intelligence, pp. 85–104. Springer, US (2009)
12. Emele, C.D., Norman, T.J., Parsons, S.: Argumentation strategies for plan resourcing. In: Proceedings of the Tenth International Conference on Autonomous Agents and Multiagent Systems (2011)
13. Erriquez, E., van der Hoek, W., Wooldridge, M.: An abstract framework for reasoning about trust. In: Proceedings of AAMAS 2011 (to appear, 2011)
14. Gómez Lucero, M.J., Chesñevar, C.I., Simari, G.R.: Modelling Argument Accrual in Possibilistic Defeasible Logic Programming. In: Sossai, C., Chemello, G. (eds.) ECSQARU 2009. LNCS, vol. 5590, pp. 131–143. Springer, Heidelberg (2009)
15. Janssen, J., Cock, M.D., Vermeir, D.: Fuzzy argumentation frameworks. In: Proceedings of IMPU 2008 (12th International Conference on Information Processing and Management of Uncertainty in Knowledge-Based Systems), pp. 513–520 (2008)
16. Kohlas, J., Haenni, R.: Assumption-based reasoning and probabilistic argumentation systems. Tech. Rep. 96–07, Institute of Informatics, University of Fribourg, Switzerland (1996)
17. Krause, P., Ambler, S., Elvang-Goransson, M., Fox, J.: A logic of argumentation for reasoning under uncertainty. Computational Intelligence 11(1), 113–131 (1995)
18. Lewicki, P., Hill, T.: Statistics: Methods and Applications. StatSoft Inc. (2005)
19. Mitchell, T.M.: Machine Learning. McGraw-Hill Higher Education (1997)
20. Oren, N., Norman, T.J.: Semantics for evidence-based argumentation. In: Computational Models of Argument: Proceedings of COMMA 2008, Toulouse, France, May 28-30, pp. 276–284 (2008)
21. Oren, N., Norman, T.J.: Arguing Using Opponent Models. In: McBurney, P., Rahwan, I., Parsons, S., Maudet, N. (eds.) ArgMAS 2009. LNCS, vol. 6057, pp. 160–174. Springer, Heidelberg (2010)
22. Oren, N., Norman, T.J., Preece, A.: Arguing with confidential information. In: Proceedings of the 18th European Conference on Artificial Intelligence, Riva del Garda, Italy, pp. 280–284 (August 2006)
23. Parsons, S.: Normative Argumentation and Qualitative Probability. In: Gabbay, D.M., Kruse, R., Nonnengart, A., Ohlbach, H.J. (eds.) FAPR 1997 and ECSQARU 1997. LNCS, vol. 1244, pp. 466–480. Springer, Heidelberg (1997)

24. Patel, J., Teacy, W.T.L., Jennings, N.R., Luck, M., Chalmers, S., Oren, N., Norman, T.J., Preece, A., Gray, P.M.D., Shercliff, G., Stockreisser, P.J., Shao, J., Gray, W.A., Fiddian, N.J., Thompson, S.: Agent-based virtual organisations for the grid. Multiagent and Grid Systems 1(4), 237–249 (2006)
25. Pearl, J.: Probabilistic reasoning in intelligent systems: networks of plausible inference. Morgan Kaufmann Publishers Inc., San Francisco (1988)
26. Pechoucek, M., Marík, V., Bárta, J.: A knowledge-based approach to coalition formation. IEEE Intelligent Systems 17, 17–25 (2002)
27. Pollock, J.L.: Cognitive Carpentry. Bradford/MIT Press (1995)
28. Rahwan, T.: Algorithms for Coalition Formation in Multi-Agent Systems. Ph.D. thesis, University of Southampton (2007)
29. Riveret, R., Prakken, H., Rotolo, A., Sartor, G.: Heuristics in argumentation: A game theory investigation. In: Computational Models of Argument: Proceedings of COMMA 2008, Toulouse, France, May 28-30, pp. 324–335 (2008)
30. Rotstein, N., Oren, N., Norman, T.J.: Resource bounded argumentation frameworks. Tech. rep., University of Aberdeen (2011)
31. Sutton, R.S., Barto, A.G.: Reinforcement Learning: An Introduction (Adaptive Computation and Machine Learning). The MIT Press (March 1998)
32. Teacy, W.T.L., Patel, J., Jennings, N.R., Luck, M.: Travos: Trust and reputation in the context of inaccurate information sources. Autonomous Agents and Multi-Agent Systems 12(2), 183–198 (2006)

Splitting Argumentation Frameworks: An Empirical Evaluation

Ringo Baumann, Gerhard Brewka, and Renata Wong

Universität Leipzig, Johannisgasse 26, 04103 Leipzig, Germany
baumann@informatik.uni-leipzig.de

Abstract. In a recent paper Baumann [1] has shown that splitting results, similar to those known for logic programs under answer set semantics and default logic, can also be obtained for Dung argumentation frameworks (AFs). Under certain conditions a given AF A can be split into subparts A_1 and A_2 such that extensions of A can be computed by (1) computing an extension E_1 of A_1, (2) modifying A_2 based on E_1, and (3) combining E_1 and an extension E_2 of the modified variant of A_2. In this paper we perform a systematic empirical evaluation of the effects of splitting on the computation of extensions. Our study shows that the performance of algorithms may drastically improve when splitting is applied.

1 Introduction

Dung's abstract argumentation frameworks (AFs) [3] are widely used in formal approaches to argumentation. They provide several standard semantics, each capturing different intuitions about how to handle conflicts among (abstract) arguments. This makes them a highly useful tool in argumentation (see for instance Prakken's ASPIC [6] for a typical way of using AFs) and algorithms for computing extensions have received considerable interest.

In a recent paper, Baumann [1] has shown that splitting results, similar to those known for logic programs under answer set semantics [4] and default logic [8], can also be obtained for Dung argumentation frameworks. It turns out that under certain conditions a given AF A can be split into subparts A_1 and A_2 such that the computation of extensions of A can be divided into smaller subproblems: to compute an extension of A one has to (1) compute an extension E_1 of A_1, (2) modify A_2 based on E_1, and (3) combine E_1 and an extension E_2 of the modified variant of A_2.

Given these results, the obvious question is: does splitting an AF really pay off in practice? In this paper we aim to give an empirical answer to this question. We do this by systematically comparing the behavior of an algorithm for computing extensions with and without splitting. Our study shows that the performance of the algorithm indeed may drastically improve when splitting is applied.

Our evaluation is based on an implementation of Caminada's labelling algorithm [5], arguably the standard genuine algorithm for computing extensions.

S. Modgil, N. Oren, and F. Toni (Eds.): TAFA 2011, LNAI 7132, pp. 17–31, 2012.

We focus on preferred and stable semantics. We also include results for grounded semantics, but as this semantics is known to be polynomial an improvement of performance here was never expected, and our results confirm this.

The paper is organized as follows: we start in Sect. 2 with the necessary background on AFs, labellings, splitting and strongly connected components. We then describe in Sect. 3 the algorithms used in our evaluation. Sect. 4 contains the empirical evaluation and thus the main results of the paper. Sect. 5 concludes.

2   Background

2.1   Argumentation Frameworks

An *argumentation framework* $\mathcal{A}$ is a pair (A, R), where A is a non-empty finite set whose elements are called *arguments* and $R \subseteq A \times A$ a binary relation, called the *attack relation*.

In the following we consider a fixed countable set $\mathcal{U}$ of arguments, called the *universe*. Quantified formulae refer to this universe and all denoted sets are finite subsets of $\mathcal{U}$ or $\mathcal{U} \times \mathcal{U}$ respectively. Furthermore we will use the following abbreviations. Let $\mathcal{A} = (A, R)$ be an AF, B and B' subsets of A and $a \in A$. Then

1. $(B, B') \, \bar{\in} \, R \Leftrightarrow_{def} \exists b \exists b' : b \in B \wedge b' \in B' \wedge (b, b') \in R$,
2. a is defended by B in $\mathcal{A} \Leftrightarrow_{def} \forall a' : a' \in A \wedge (a', a) \in R \rightarrow (B, \{a'\}) \, \bar{\in} \, R$,
3. B is conflict-free in $\mathcal{A} \Leftrightarrow_{def} (B, B) \, \bar{\not\in} \, R$,
4. $cf(\mathcal{A}) = \{C \, | \, C \subseteq A, C \text{ conflict-free in } \mathcal{A}\}$.

The set of all extensions of $\mathcal{A}$ under semantics $\mathcal{S}$ is denoted by $\mathcal{E}_{\mathcal{S}}(\mathcal{A})$. We consider the classical semantics introduced by Dung, namely stable, preferred, complete and grounded (compare [3]).

Definition 1. *Let $\mathcal{A} = (A, R)$ be an AF and $E \subseteq A$. E is a*

1. *admissible extension[1] ($E \in \mathcal{E}_{ad}(\mathcal{A})$) iff*
   $E \in cf(\mathcal{A})$ *and each $a \in E$ is defended by E in $\mathcal{A}$,*
2. *complete extension ($E \in \mathcal{E}_{co}(\mathcal{A})$) iff*
   $E \in \mathcal{E}_{ad}(\mathcal{A})$ *and for each $a \in A$ defended by E in $\mathcal{A}$, $a \in E$ holds,*
3. *stable extension ($E \in \mathcal{E}_{st}(\mathcal{A})$) iff*
   $E \in \mathcal{E}_{co}(\mathcal{A})$ *and for every $a \in A \backslash E$, $(E, \{a\}) \, \bar{\in} \, R$ holds,*
4. *preferred extension (i.e. $E \in \mathcal{E}_{pr}(\mathcal{A})$) iff*
   $E \in \mathcal{E}_{co}(\mathcal{A})$ *and for each $E' \in \mathcal{E}_{co}(\mathcal{A})$, $E \not\subset E'$ holds,*
5. *grounded extension ($E \in \mathcal{E}_{gr}(\mathcal{A})$) iff*
   $E \in \mathcal{E}_{co}(\mathcal{A})$ *and for each $E' \in \mathcal{E}_{co}(\mathcal{A})$, $E' \not\subset E$ holds.*

---

[1] Note that it is more common to speak about admissible sets instead of the admissible semantics. For reasons of unified notation we used the less common version.

2.2 Labelling-Based Semantics

The labelling approach [2,5] provides an alternative possibility to describe extensions. Given an AF $\mathcal{A} = (A, R)$, a labelling is a total function $L : A \rightarrow \{in, out, undec\}$. We use $x(L)$ for $L^{-1}(\{x\})$, i.e. $x(L) = \{a \in A \mid L(a) = x\}$. This allows to rewrite a labelling L as a triple $(in(L), out(L), undec(L))$ which is frequently used. Analogously to $\mathcal{E}_S(\mathcal{A})$ we write $\mathcal{L}_S(\mathcal{A})$ for the set of all labellings prescribed by semantics S for an AF $\mathcal{A}$.

Definition 2. *Given an AF $\mathcal{A} = (A, R)$ and a labelling L of $\mathcal{A}$. L is called a complete labelling ($L \in \mathcal{L}_{co}(\mathcal{A})$) iff for any $a \in A$ the following holds:*

1. *If $a \in in(L)$, then for each $b \in A$ s.t. $(b, a) \in R$, $b \in out(L)$,*
2. *If $a \in out(L)$, then there is a $b \in A$ s.t. $(b, a) \in R$ and $b \in in(L)$,*
3. *If $a \in undec(L)$, then there is a $b \in A$ s.t. $(b, a) \in R$ and $b \in undec(L)$ and there is no $b \in A$ s.t. $(b, a) \in R$ and $b \in in(L)$.*

Now we are ready to define the remaining counterparts of the extension-based semantics in terms of complete labellings.

Definition 3. *Given an AF $\mathcal{A} = (A, R)$ and a labelling $L \in \mathcal{L}_{co}(\mathcal{A})$. L is a*

1. *stable labelling ($L \in \mathcal{L}_{st}(\mathcal{A})$) iff $undec(L) = \emptyset$,*
2. *preferred labelling ($L \in \mathcal{L}_{pr}(\mathcal{A})$) iff for each $L' \in \mathcal{L}_{co}(\mathcal{A})$, $in(L) \not\subset in(L')$,*
3. *grounded labelling ($L \in \mathcal{L}_{gr}(\mathcal{A})$) iff for each $L' \in \mathcal{L}_{co}(\mathcal{A})$, $in(L') \not\subset in(L)$.*

Theorem 1. *[5] Given an AF $\mathcal{A}$. For each $\sigma \in \{co, st, pr, gr\}$,*

1. *$E \in \mathcal{E}_\sigma(\mathcal{A})$ iff $\exists L \in \mathcal{L}_\sigma(\mathcal{A}) : in(L) = E$ and*
2. *$|\mathcal{E}_\sigma(\mathcal{A})| = |\mathcal{L}_\sigma(\mathcal{A})|$ holds.*

This theorem will be used to make the splitting results applicable for our algorithm.

2.3 Splitting Results

Baumann [1][2] showed that, under certain conditions, the computation of the extensions of an AF $\mathcal{A}$ can be considerably simplified: one splits the AF $\mathcal{A}$ into two subframeworks $\mathcal{A}_1$ and $\mathcal{A}_2$, computes an extension E_1 of $\mathcal{A}_1$, uses E_1 to reduce and modify $\mathcal{A}_2$, computes an extension E_2 of the modified and reduced version of $\mathcal{A}_2$ and then simply combines E_1 and E_2. We briefly recall the relevant definitions as they are crucial for the algorithms to be discussed later.

Definition 4. *Let $\mathcal{A}_1 = (A_1, R_1)$ and $\mathcal{A}_2 = (A_2, R_2)$ be AFs such that $A_1 \cap A_2 = \emptyset$. Let $R_3 \subseteq A_1 \times A_2$. We call the tuple $(\mathcal{A}_1, \mathcal{A}_2, R_3)$ a splitting of the argumentation framework $\mathcal{A} = (A_1 \cup A_2, R_1 \cup R_2 \cup R_3)$.*

---

[2] The full version is available at http://www.informatik.uni-leipzig.de/~baumann/

Definition 5. *Let $\mathcal{A} = (A, R)$ be an AF, A' a set disjoint from A, $S \subseteq A'$ and $L \subseteq A' \times A$. The (S, L)-reduct of $\mathcal{A}$, denoted $\mathcal{A}^{S,L}$ is the AF*

$$\mathcal{A}^{S,L} = (A^{S,L}, R^{S,L})$$

where

$$A^{S,L} = \{a \in A \mid (S, \{a\}) \not\subseteq L\}$$

and

$$R^{S,L} = \{(a, b) \in R \mid a, b \in A^{S,L}\}.$$

Definition 6. *Let $\mathcal{A} = (A, R)$ be an AF, E an extension of $\mathcal{A}$. The set of arguments undefined with respect to E is*

$$U_E = \{a \in A \mid a \notin E, (E, \{a\}) \not\subseteq R\}.$$

It can be checked that in case of $\sigma \in \{co, st, pr, gr\}$, U_E equals $undec(L)$, where L is the unique σ - labelling s.t. $in(L) = E$ holds (compare Theorem 1).

Definition 7. *Let $\mathcal{A} = (A, R)$ be an AF, A' a set disjoint from A, $S \subseteq A'$ and $L \subseteq A' \times A$. The (S, L)-modification of $\mathcal{A}$, denoted $mod_{S,L}(\mathcal{A})$, is the AF*

$$mod_{S,L}(\mathcal{A}) = (A, R \cup \{(b, b) \mid a \in S, (a, b) \in L\}).$$

We now present the splitting theorem in both extension-based and labelling-based semantics style. The labelling-based notation can be easily obtained by using the original extension-based splitting results (1.(a), 2.(a)), Theorem 1 and the observation below Def. 6.

Theorem 2. *($\sigma \in \{st, pr, co, gr\}$) Let $\mathcal{A} = (A, R)$ be an AF which possesses a splitting $(\mathcal{A}_1, \mathcal{A}_2, R_3)$ with $\mathcal{A}_1 = (A_1, R_1)$ and $\mathcal{A}_2 = (A_2, R_2)$.*

1. *(a)* $E_1 \in \mathcal{E}_\sigma(\mathcal{A}_1) \wedge E_2 \in \mathcal{E}_\sigma(mod_{U_{E_1}, R_3}(\mathcal{A}_2^{E_1, R_3})) \Rightarrow E_1 \cup E_2 \in \mathcal{E}_\sigma(\mathcal{A})$
   (b) $L_1 \in \mathcal{L}_\sigma(\mathcal{A}_1) \wedge L_2 \in \mathcal{L}_\sigma(mod_{undec(L_1), R_3}(\mathcal{A}_2^{in(L_1), R_3}) \Rightarrow$
   $\exists! \ L \in \mathcal{L}_\sigma(\mathcal{A}) : in(L) = in(L_1) \cup in(L_2)$
2. *(a)* $E \in \mathcal{E}_\sigma(\mathcal{A}) \Rightarrow E \cap A_1 \in \mathcal{E}_\sigma(\mathcal{A}_1) \wedge E \cap A_2 \in \mathcal{E}_\sigma(mod_{U_{E \cap A_1}, R_3}(\mathcal{A}_2^{E \cap A_1, R_3}))$
   (b) $L \in \mathcal{L}_\sigma(\mathcal{A}) \Rightarrow \exists! \ L_1 \in \mathcal{L}_\sigma(\mathcal{A}_1) : in(L_1) = in(L) \cap A_1 \wedge$
   $\exists! \ L_2 \in \mathcal{L}_\sigma(mod_{undec(L) \cap A_1, R_3}(\mathcal{A}_2^{in(L) \cap A_1, R_3})) : in(L_2) = in(L) \cap A_2$

2.4   Splittings and Strongly Connected Components (SCC)

To generate a splitting we use the related graph-theoretic concept of *strongly connected components*. A directed graph is strongly connected if there is a path from each vertex to every other vertex. The SCCs of a graph $\mathcal{A}$ ($SCC(\mathcal{A})$ for short) are its maximal strongly connected subgraphs. Contracting every SCC to a single vertex leads to an acyclic graph. It is well-known that an acyclic graph induces a partial order on the set of vertices. Based on this order every SCC-decomposition can be easily transformed into a splitting. The most obvious possibility is to take the union of the initial nodes of the decomposition ($= \mathcal{A}_1$) and the union of the remaining subgraph ($= \mathcal{A}_2$).

The following figure exemplifies the idea. We sketch three different splittings, namely S_1, S_2 and S_3. Note that these are not all possible splittings.

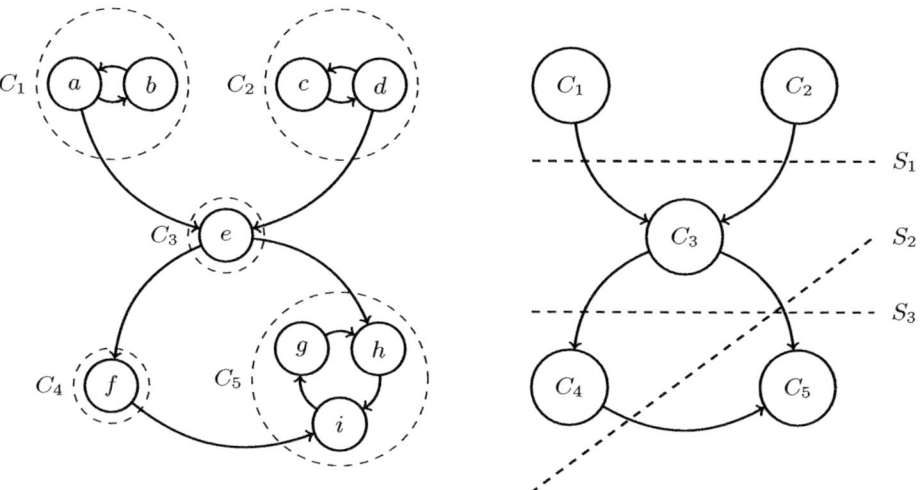

Fig. 1. SCCs and Splittings

3   Algorithms

Our implementation (available at wwwstud.rz.uni-leipzig.de/~bss01gsc/) is based on the labelling algorithms for grounded, preferred and stable semantics in [5] and on the standard Tarjan algorithm for computing strongly connected components from [7]. For the latter we refer the reader to the original paper. We briefly describe the former to make the paper more self-contained.

3.1   Labelling Algorithms

The grounded labelling (L_{gr}) is generated as follows: all arguments which are not attacked are assigned the label *IN*. The next step is to assign the label *OUT* to all those arguments that are attacked by at least one of the arguments just labeled *IN*. We continue assigning the label *IN* to any argument having all of its attackers labeled *OUT*. The iteration stops when no further assignment can be made. The set *undec*(L_{gr}) is the set of arguments from A which were not labeled during the iteration.

In order to present the algorithms for preferred and stable labellings, further terminological explanations are in place.

Definition 8. *Given an AF $\mathcal{A} = (A, R)$ and a labelling $L \in \mathcal{L}_\sigma(\mathcal{A})$, an argument $a \in A$ is*

1. *legally IN iff x is labeled IN and* $\forall b : (b, a) \in R$, *b is labeled OUT,*
2. *legally OUT iff x is labeled OUT and* $\exists b : (b, a) \in R$ *and b is labeled IN,*
3. *illegally IN iff it is not legally IN,*
4. *illegally OUT iff it is not legally OUT,*
5. *super-illegally IN iff it is illegally IN and* $\exists b : (b, a) \in R$ *and b is legally IN or UNDEC.*

The algorithm for computing all preferred labellings (Algorithm 1) starts by assigning to all arguments the label *IN* (labelling L_{IN}), and initializing an empty set in which candidate labellings are to be stored. Then, by way of the main procedure *find_labellings* arguments that are *illegally IN* in L_{IN} are identified. To each of these arguments a procedure called *transition_step* is applied, by which the label of the given argument is changed from *IN* to *OUT*. If such an argument whose label has been changed from *IN* to *OUT* or if any argument(s) it attacks is *illegally OUT*, it will be relabeled as *UNDEC*. Thus we have obtained a new labelling which contains one less *IN*-argument. Then the entire process repeats again by passing any new labelling onto the main procedure, and the process continues until an acceptance or rejection condition is met. A labelling which does not have any argument which is *illegally IN* will be added to the candidate labellings, unless at any previous stage in the recursion it is detected that a better labelling has been found, i.e. a labelling with a larger *in*-set is already contained in *candidate_labellings*. If such a labelling with a larger cardinality of the *in*-set exists, the current labelling will not be processed.

In order to avoid the situation in which incomplete labellings are being generated by any incorrect assignment of labels, the algorithm is designed to always extract first those arguments that are *super-illegally IN*, i.e. arguments having at least one attacker *legally IN* or *UNDEC*, whenever we try to extract arguments that are *illegally IN*.

The algorithm for computing all stable labellings is obtained by rewriting line 1.5 of the algorithm for preferred labellings to read "**if** $undec(L) \neq \emptyset$ **then return**". If the set of arguments labeled *UNDEC* in a labelling is not empty, i.e. it violates the requirement for a stable labelling, the labelling will not be further processed.

3.2   Computation of Splitting

Our splitting algorithm consists of two parts: The first part (Algorithm 2) is executed prior to the first call of a labelling algorithm for a semantics and computes $\mathcal{A}_1$, $\mathcal{A}_2$ and the set R_3. The second part (Algorithm 3) is executed after receiving an extension from the labelling algorithm. The tuple $\mathcal{A}_2$ is then modified in accordance with the extension.

The first task is set to look for all the initial arguments ($\mathcal{A}_1$) of our framework ($\mathcal{A}$). We use the set of *strongly connected components* returned by the Tarjan algorithm. The algorithm starts by introducing a Boolean variable *scc_attacked* which will be initialized to *false* for every SCC in $SCC(\mathcal{A})$. Given an SCC, once an argument in this SCC is attacked by some argument in another SCC, the variable will be set to *true* and the execution of the algorithm for this SCC

Algorithm 1. Computation of Preferred Labellings

---

      input : $L_{IN} = (in(L_{IN}) = A, out(L_{IN}) = \emptyset, undec(L_{IN}) = \emptyset)$

1.1 $candidate\_labellings := \emptyset$
1.2 $find\_labellings(L_{IN})$

1.3 **PROCEDURE** $find\_labellings(L)$
1.4 **begin**
1.5   **if** $\exists L' \in candidate\_labellings : in(L) \subset in(L')$ **then return**;
1.6   **if** L does not contain an argument illegally IN **then**
1.7   **foreach** $L' \in candidate\_labellings$ **do**
1.8   **if** $in(L') \subset in(L)$ **then**
1.9 $candidate\_labellings := candidate\_labellings - \{L'\}$
1.10 $candidate\_labellings := candidate\_labellings \cup \{L\}$
1.11   **return**;
1.12   **else**
1.13   **if** L has an argument that is super-illegally IN **then**
1.14 $x :=$ some argument that is super-illegally IN in L
1.15 $find\_labellings(transition\_step(L, x))$
1.16   **else**
1.17   **foreach** x that is illegally IN in L **do**
1.18 $find\_labellings(transition\_step(L, x))$

---

stops. Then the algorithm starts processing the next SCC. Only if *scc_attacked* remains *false*, which means that the corresponding SCC is not attacked, will all the arguments of this SCC be added to A_1. This way of splitting corresponds to S_1 in Fig. 1.

The splitting operation described above may result in subframeworks which differ a lot in size. We also provide a possibility to equalize the cardinalities along the partial ordering dictated by $SCC(\mathcal{A})$. The algorithm, called *optimize*, accepts the already computed arguments of A_1 and adds new ones under certain conditions. The first criterion used is the cardinality of A_1. Since the addition of new arguments relies on the partial order, it may not always be possible. Therefore, choosing 45% as a starting condition for equalization was an attempt to optimally equalize the numbers of arguments on the one hand, and on the other not to slow down the splitting process unnecessarily. Another condition limits the number of arguments added to A_1 by imposing a relative restriction on the added SCC's cardinality, i.e. if $|SCC| + |A_1| > |A| * 60\%$, the SCC will not be accepted. The algorithm runs recursively until no further arguments can be added (i.e. when $|optimal\_set| = |A_1|$). This way of splitting corresponds to S_3 in Fig. 1.

On the basis of the set A_1 we can then compute the sets R_1, A_2, R_2 as well as the set of attacks along which the framework is split (R_3). The pseudo code for these operations is not included here due to their obvious simplicity.

The processing of the tuple $\mathcal{A}_1$ by a labelling algorithm may return an extension as part of a labelling, if it exists. This extension (E_1) will in turn be used for modifying the AF $\mathcal{A}_2$ in the second part of the splitting algorithm. We start with the set A_2' which is A_2 minus all the arguments in A_2 that are attacked by E_1, and we call it the modified set of A_2.

Next we apply the second algorithm on E_1, starting with an empty set, in order to compute a reduced set of undefined arguments (U_{E_1}). Note that we are not concerned with all the undefined arguments as stated in Definition 6, but only with those that are sources of an attack in R_3. Whenever an argument is a source of an attack in R_3, if it neither is an element of the extension E_1 nor is attacked by E_1, it will be added to the set U_{E_1}.

We then proceed to the final step in the modification of the AF $\mathcal{A}_2$. Given U_{E_1}, for every argument of A_2' which is attacked by U_{E_1}, a loop is added. By this addition, we have modified the set R_2. We call this modified set R_2', and now we can define $\mathcal{A}_2'$ as the tuple (A_2', R_2'). With the given definition, $\mathcal{A}_2'$ is to be processed by a labelling algorithm.

4   Experimental Results

Our evaluation of the runtime for grounded, preferred and stable semantics is based on the sampling of 100 random frameworks[3]. The tests were performed on a Samsung P510 notebook with a Pentium Dual Core Processor, CPU speed: 2.0 GHz, CPU Caches: 32 KB (L1) and 1024 KB (L2), RAM: 2 GB. We focused in our experiments on frameworks where the number of attacks (n) exceeds the number of arguments (m) by a factor between 1.5 and 3.

The reasons for this restriction are as follows. First of all, even leaving execution times aside[4], by further increasing the number of attacks the probability of generating frameworks consisting of a single SCC grows, thus rendering the experiment inconclusive as splitting has no effect on AFs with a single SCC. For example, initial tests showed that if 500 or more attacks (n) are given for 100 arguments (m), then almost all of the randomly generated frameworks will consist of only a single SCC and no effect of splitting is to be expected.

On the other hand, choosing an n smaller than m would not lead to significant differences in execution time between AFs with and without splitting as execution times tend to be fast under such conditions anyway.

With the above limitations in mind, a total of 100 examples were collected, with 20 examples extracted from each of the following m/n combinations: 10/30, 50/100, 100/175, 200/375 and 500/750. A brief description of the results obtained will be presented below together with a tabular summary of statistical

---

[3] The attacks were created by randomly selecting the source and the target argument for frameworks with given number of arguments and attacks.

[4] For example, our preliminary testing showed that for AFs with 100 arguments, if 200 attacks are specified, the percentage of frameworks with runtime over 3 min for preferred semantics without splitting was about 70%.

Algorithm 2. Computation of Splitting, part 1

---

    input : set of strongly connected components $(SCC(\mathcal{A}))$

    output: A_1, R_1, A_2, R_2, R_3

2.1 **PROCEDURE** $compute\_A_1(SCC(\mathcal{A}))$

2.2 **begin**

2.3   **foreach** SCC $\in SCC(\mathcal{A})$ **do**

2.4 $scc\_attacked := false$

2.5   loop:

2.6   **foreach** $a \in$ SCC **do**

2.7   **foreach** b $s.t.$ $(b,a) \in R$ **do**

2.8   **if** $b \notin$ SCC **then**

2.9 $scc\_attacked = true$

2.10   **break** loop;

2.11   **if** $scc\_attacked = false$ **then** add SCC to A_1

2.12   **return** A_1

2.13 **PROCEDURE** $optimize(SCC(\mathcal{A}), A_1)$

2.14 **begin**

2.15 $optimal\_set := A_1$

2.16 $illegal\_attacks := false$

2.17   **foreach** SCC $\in SCC(\mathcal{A})$ **do**

2.18   **if** $|A_1| < |A| * 0.45$ **then**

2.19   pick an $a \in$ SCC

2.20   **if** $a \notin A_1$ and $|A_1| + |SCC| < |A| * 0.6$ **then**

2.21 $illegal\_attacks = false$

2.22   loop:

2.23   **foreach** $a \in$ SCC **do**

2.24   **foreach** $(b,a) \in R$ **do**

2.25   **if** $b \notin A_1$ and $b \notin$ SCC **then**

2.26 $illegal\_attacks = true$

2.27   **break** loop;

2.28   **if** $illegal\_attacks = false$ **then** add SCC to A_1

2.29   **if** $|A_1| < |A| * 0.45$ and $|optimal\_set| \neq |A_1|$ **then**

2.30 $optimize(SCC(\mathcal{A}), A_1)$

2.31   **return** A_1

---

data for each combination. Each table contains average-runtime results (in milliseconds) and gain-in-time results (in %)[5] for the grounded, preferred and stable semantics. Under "average runtime", the first column contains results from executing without splitting, the second from executing with non-optimized splitting

---

[5] For convenience, in the presented data we use "0 ms" to mean "close to 0 ms" and "100%" to mean "close to 100%".

Algorithm 3. Computation of Splitting, part 2

---

    input : an extension of $\mathcal{A}_1$ (E_1), A_2, R_1, R_2, R_3
    output: $\mathcal{A}_2' = (A_2', R_2')$

3.1 $compute\_modified\_A_2(E_1, A_2, R_3)$
3.2 $compute\_U_{E_1}(E_1, R_1, R_3)$
3.3 $compute\_modified\_R_2(U_{E_1}, R_2, R_3)$

3.4 **PROCEDURE** $compute\_modified\_A_2(E_1, A_2, R_3)$
3.5 **begin**
3.6 $A_2' := A_2$
3.7   **foreach** $a \in E_1$ **do**
3.8   **foreach** $(a, b) \in R_3$ **do**
3.9   **if** $b \in A_2'$ **then** remove b from A_2'

3.10   **return** A_2'

3.11 **PROCEDURE** $compute\_U_{E_1}(E_1, R_1, R_3)$
3.12 **begin**
3.13 $U_{E_1} := \emptyset$
3.14   **foreach** $(a, b) \in R_3$ **do**
3.15   **if** $a \notin E_1$ and $(E_1, \{a\}) \bar{\notin} R_1$ **then** add a to U_{E_1}

3.16   **return** U_{E_1}

3.17 **PROCEDURE** $compute\_modified\_R_2(U_{E_1}, R_2, R_3)$
3.18 **begin**
3.19 $R_2' := R_2 - \{(x, y) | (x, y) \in R_2 \text{ and } (x \notin A_2' \text{ or } y \notin A_2')\}$
3.20   **foreach** $a \in U_{E_1}$ **do**
3.21   **foreach** $(a, b) \in R_3$ **do** add (b, b) to R_2'

3.22   **return** R_2'

---

and the third from executing with optimized splitting. Under "gain in time", minimal, maximal and average gain results, each in relation to non-optimized and optimized splitting, are distinguished.

The 10/30 combination was the only case in which we experienced no runtime that was over 3 min.[6] Thanks to the low number of arguments we were given a possibility of structural analysis. Although 20 examples is a small sample size, we were able to distinguish 4 characteristics based on the structure of the framework and the corresponding difference in runtime between executions without and with splitting. The analysis below applies to the preferred and stable semantics as the execution of the grounded semantics did not show any difference.

First, in 3 cases out of 20 a single SCC was generated. As splitting has no effect on AFs consisting of a single SCC, there was no runtime improvement for

---

[6] It comes as no surprise since the computation of preferred labellings for an AF with 10 arguments and 100 attacks takes around 260,000 ms.

all 3 semantics. However, no noticeable runtime delay in relation to the splitting process was recorded either.

Second, 3 further examples had the form of a single argument SCC attacking a large SCC. Here we recorded no improvement or only a slight improvement in the runtime when splitting was applied: 0-20%.

Third, yet 3 further cases consisted of a single argument SCC with a self-loop attacking a large SCC. The only difference regarding the single argument between this form and the previous one was that we now had a loop attack. However in terms of runtime the gap was significant. In the second case it was between 68-71% for preferred semantics and between 99-100% for stable semantics.

And last, 11 of the random AFs had the form of a larger SCC attacking a single argument SCC, a single argument SCC with a self-loop or two SCCs; or the form of two SCCs, with at least one attack each, attacking a single SCC. The difference in execution without and with splitting ranged here between 80-99% for preferred semantics and between 59-100% for stable semantics.

The limited data suggest that splitting can render computation significantly faster for frameworks with certain characteristics. It seems that the most relevant are those AFs having one or more SCCs, each with at least one attack (i.e. a single argument SCC with a loop or an SCC with at least 2 arguments), attacking one or more SCCs whose structure in itself is not relevant.[7]

In general we obtained an average acceleration of 60% for both types of splitting in comparison to an execution without splitting. It is partly due to the fact that for the 10/30 combination both non-optimized splitting and optimized splitting usually overlap, which in turn is a result of the existence of large SCCs that limits the possibility of having different splittings. In no case was the execution with splitting slower than the one without.

Table 1. Evaluation results for 10 arguments and 30 attacks

m = 10	average runtime (in ms)			gain in time (in %)					
n = 30	w/o spl.	w/ spl.	opt. spl.	min	min/op	max	max/op	avg.	avg./op
grounded	1	1	1	0	0	0	0	0	0
pref.	3871	886	890	0	0	99	99	60	61
stable	1040	267	262	0	0	100	100	59	60

---

[7] An additional test on an AF of 10 arguments, of which 9 constituted an SCC with 81 attacks and all 9 attacked the 10th argument, recorded a 90% runtime difference for both preferred and stable semantics. This additional result lies nicely within the ranges of the previously obtained 80-99% and 59-100% respectively. A further test of a single argument with a self-loop attacking each argument of an SCC with 9 arguments and 81 attacks showed a 90% runtime difference for preferred semantics and 100% for stable semantics. The performance was evidently better than the previously obtained result for preferred semantics (68-71%). Having removed the loop attack we obtained a runtime of 1 *ms* for preferred and stable semantics, both with and without splitting. Again, these results are also in compliance with the ones obtained in the sample test using 20 examples.

Table 2. Evaluation results for 50 arguments and 100 attacks

m = 50	average runtime (in ms)			gain in time (in %)					
n = 100	w/o spl.	w/ spl.	opt. spl.	min	min/op	max	max/op	avg.	avg./op
grounded	2	2	2	0	0	0	0	0	0
pref.	35860	23237	23352	0	0	99	99	29	26
stable	663	487	480	0	0	99	99	29	29

The runtimes for the 50/100 combination were very diversified: from 1 *ms* (for stable) and 2 *ms* (for preferred semantics) to 381,512 *ms* (preferred)[8] and 4,456 *ms* (stable). The grounded labelling was computed at the speed of 1-3 *ms* in each case, no improvement nor delay was recorded for executions with splitting in comparison to those without.

In 9 out of the 20 cases, the computation time for preferred and stable labellings without and with splitting was very short (below 20 *ms*). No significant difference was observed. The time gain for these cases was given as 0%, which had a negative effect on the average gain in time as shown in Table 2: it dropped to only 26-29%. Note that the maximal gain in time for both semantics was at 99%.

For the stable semantics we observed dramatic improvements in cases where no labellings existed. Through splitting of the framework, the time needed to find the first argument of the *undec* set, hence breaking the execution of the labelling algorithm, was at times very short. In 8 out of 15 cases where no labelling existed, the execution times lay below 20 *ms* which as mentioned above had 0% gain. Among the remaining 7 cases, 2 recorded an improvement of 99%, the rest between 17-75%. In none of the 20 examples was the execution without splitting faster than the one with splitting. Neither significant improvement nor delay was found for optimized splitting as compared to regular splitting.

Some 40% of the frameworks generated with 100 arguments and 175 attacks had a computation time of at least 3 *min* for the preferred semantics without splitting. They were not taken into consideration for the reason stated at the beginning of this section. In the collected examples, the runtimes varied from around 20 *ms* to slightly below 40,000 *ms*. No stable labelling existed in 19 out of the 20 examples. In 9 out of these 19 examples, we obtained an improvement of 90-100% for the stable semantics and 0-50% for the remaining 10. No slow down due to the process of splitting was noticeable.

Here, for the first time, we recorded a significant improvement in runtime when the optimized version of splitting was applied. It was 13% for the preferred semantics and 5% for the stable semantics, both of which were better than the non-optimized variant. On average, an execution with splitting was better than one without splitting by 56-69% for the preferred semantics and by 60-65% for the stable semantics.

---

[8] This example had already been included in the data before the imposition of the 3-minute limit, and so this is the only example with a runtime above 3 *mins*.

Table 3. Evaluation results for 100 arguments and 175 attacks

m = 100	average runtime (in ms)			gain in time (in %)					
n = 175	w/o spl.	w/ spl.	opt. spl.	min	min/op	max	max/op	avg.	avg./op
grounded	2	2	2	0	0	0	0	0	0
pref.	8335	3701	2502	0	0	93	99	56	69
stable	499	297	262	0	0	100	99	60	65

Table 4. Evaluation results for 200 arguments and 375 attacks

m = 200	average runtime (in ms)			gain in time (in %)					
n = 375	w/o spl.	w/ spl.	opt. spl.	min	min/op	max	max/op	avg.	avg./op
grounded	3	3	3	0	0	0	0	0	0
pref.	9333	6531	6296	14	16	99	98	47	45
stable	352	236	222	26	26	96	93	56	56

The computation time for preferred and stable labellings without splitting in frameworks of 200 arguments and 375 attacks was in general above 15 *ms*, thus making a more precise comparison possible. All the generated AFs showed a runtime improvement of at least 14% (pref.) and 26% (stable) when the execution with splitting is compared to the execution without splitting. Here too the gain in time reached in some cases 99% for the preferred labellings and 96% for the stable labellings.

With an average runtime of 3 *ms* for the grounded semantics, no difference between execution without and with splitting was found. The computation of stable labellings with applied splitting took on average 56% less time than that without. For the preferred semantics, the gain was somewhat less, it was 45% with optimized splitting and 47% with non-optimized splitting.

It was relatively comfortable testing the 500/750 combination since only about 20% of the randomly generated frameworks had a runtime above 3 *min* for preferred labellings without splitting. The execution time was quite steady. The lowest runtime for preferred semantics without splitting was 53 *ms* and 58 *ms* for stable semantics without splitting. The absence of drastic highs and lows was mirrored in all the average runtimes for preferred semantics, which were much lower than the average runtimes measured for 200/375. Here we observed also a steady improvement after splitting was applied. The lowest of which was 35% for preferred semantics and 33% for stable. The upper range was also less drastic with up to 86% for preferred and 97% for stable. The average differences were quite high with 57-61% for preferred labellings and 62-66% for stable. There was a drop in efficiency for the optimized type of splitting as compared to the non-optimized type (by 4% for both preferred and stable labellings). However, in AFs with a runtime above 700 *ms*, the optimized type ran faster than the one without optimization. In no case though was an execution with splitting slower than the one without splitting.

While in frameworks with 200 arguments and lower the grounded semantics did not perform worse after splitting, here we observed a visible slowdown. There was an average loss of 2% in the case of the non-optimized variant and an average loss of 36% in the case of the optimized variant.

Table 5. Evaluation results for 500 arguments and 750 attacks

m = 500	average runtime (in ms)			gain in time (in %)					
n = 750	w/o spl.	w/ spl.	opt. spl.	min	min/op	max	max/op	avg.	avg./op
grounded	10	10	13	-12	-60	15	-15	-2	-36
pref.	2785	1697	1168	36	35	86	78	61	57
stable	232	120	99	33	47	97	89	66	62

5   Conclusions

Based on our evaluations of 100 randomly generated AFs, we have made the following observations:

1. Among the 100 AFs, we observed an average improvement by 50-51% and by 54% for preferred and stable semantics respectively. The data contained some inconclusive examples which had "marred" the results to some extent.
2. No instance, neither for preferred semantics nor for stable, was found in which the execution with splitting lasted longer than the one without. This shows that the additional overhead introduced by splitting is negligible.
3. The optimized type of splitting did better than the non-optimized type in cases when the AF without splitting had a relatively long runtime. When the runtime was relatively short, the type without optimization usually performed better.
4. Splitting may significantly improve runtime for stable semantics in frameworks where no stable labellings exist. By splitting the framework, we were able to complete the execution of the algorithm a lot faster because it took less time to find a labelling with the *undec* set that was not empty.
5. It seems that there exist certain regularities between the structure of frameworks and the corresponding runtime. Having an SCC with at least one attack (or several SCCs with at least one attack each) attacking the rest of the framework can improve runtime significantly. We especially hope that this will greatly affect computation of large frameworks with large SCCs, which so far we were unable to test due to the required long computation time.

In future work we plan not only to extend our evaluation to larger AFs, we would also like to see whether there is an impact of moving from randomly generated to "natural" argumentation frameworks arising in realistic argumentation scenarios. Moreover, our results together with the theoretical considerations from the beginning of Sect. 4 suggest an advanced version of the algorithm where splitting is (1) performed iteratively on the identified subparts and (2) conditioned on the number of arguments and ratio between arguments and attacks, that is, only if the number of arguments is above a threshold and this ratio is in the "interesting" range splitting is performed.

References

1. Baumann, R.: Splitting an Argumentation Framework. In: Delgrande, J.P., Faber, W. (eds.) LPNMR 2011. LNCS, vol. 6645, pp. 40–53. Springer, Heidelberg (2011)
2. Caminada, M.: On the Issue of Reinstatement in Argumentation. In: Fisher, M., van der Hoek, W., Konev, B., Lisitsa, A. (eds.) JELIA 2006. LNCS (LNAI), vol. 4160, pp. 111–123. Springer, Heidelberg (2006)
3. Dung, P.M.: On the acceptability of arguments and its fundamental role in non-monotonic reasoning, logic programming and n-person games. Artif. Intell. 77(2), 321–358 (1995)
4. Lifschitz, V., Turner, H.: Splitting a logic program. In: ICLP, pp. 23–37 (1994)
5. Modgil, S., Caminada, M.: Proof theories and algorithms for abstract argumentation frameworks. In: Rahwan, I., Simari, G.R. (eds.) Argumentation in Artificial Intelligence, pp. 105–132. Springer, Heidelberg (2009)
6. Prakken, H.: An abstract framework for argumentation with structured arguments. Argument and Computation 1, 93–124 (2010)
7. Tarjan, R.E.: Depth-first search and linear graph algorithms. SIAM J. Comput. 1(2), 146–160 (1972)
8. Turner, H.: Splitting a default theory. In: Proc. AAAI 1996, pp. 645–651 (1996)

On the Complexity of Computing the Justification Status of an Argument[*]

Wolfgang Dvořák

Technische Universität Wien, Institute for Information Systems 184/2
Favoritenstrasse 9-11, 1040 Vienna, Austria
dvorak@dbai.tuwien.ac.at

Abstract. We address the problem of determining the acceptance status of an argument w.r.t. labeling-based semantics. Wu and Caminada recently proposed a labeling-based justification status of arguments to distinguish different levels of acceptability for arguments. We generalize their approach, which was originally restricted to complete semantics, to arbitrary argumentation semantics and provide a comprehensive study of the computational properties.

1 Introduction

We study the problem of computing the acceptance status of an argument in abstract argumentation frameworks (AFs) [12], following the approach of labeling-based justification statuses by Wu and Caminada [23].

Dung [12] introduced abstract argumentation frameworks together with semantics which specify subsets of arguments, so called extensions, distinguishing the arguments which are accepted from the arguments which are not. Towards a more fine-grainted distinction of arguments several kind of argumentation labelings have been proposed either for algorithmic or logical purposes (see, e.g. [8,21,22]). In this work we follow the approach of three-valued labelings as proposed by Caminada [8]. Roughly speaking such labelings partition the arguments of an framework into three sets. Similar as in the concept of extensions, there are the acceptable arguments (which are labeled by *in*) but further labelings distinguish two kinds of arguments which are not accepted: those which are attacked by an accepted argument (and labeled *out*) and those which are neither accepted nor attacked (and labeled *undec*).

Traditional extension-based approaches for deciding the acceptance status of arguments distinguish between skeptically accepted arguments, i.e. arguments contained in each extension, credulously accepted arguments, i.e. arguments contained in at least one extension and arguments which are in no extension at all. However such a characterization completely ignores the additional information provided by labelings. To take this information into account, Wu and Caminada [23] proposed their labeling-based justification status of an argument, which allows to distinguish different levels of acceptance (and rejection) for arguments based on the labelings of an argumentation framework. The main idea of this justification status for an argument a is to consider all labels l such that at least one complete labeling of the AF assigns l to a.

---

[*] This work was supported by the Vienna Science and Technology Fund (WWTF) under grant ICT08-028.

S. Modgil, N. Oren, and F. Toni (Eds.): TAFA 2011, LNAI 7132, pp. 32–49, 2012.

As mentioned, Wu and Caminada only consider the justification status concerning complete semantics. In this paper we first generalize the concept of justification status to arbitrary argumentation semantics and then consider instantiations for several important semantics, namely the semantics defined in [12], semi-stable [4], stage [21] and resolution-based grounded [1] semantics. We provide a detailed complexity analysis for the concept of justification status w.r.t. the afore mentioned semantics (which has not been done in [23]). Further we show general properties of these justification statuses as well as relations between justification statuses for different semantics.

The structure of the remainder of the paper is as follows: In Section 2 we introduce abstract argumentation frameworks, the semantics we consider in the paper and the concept of labelings. We also highlight known complexity results for these semantics. Section 3 gives the definition of the justification status of an argument, as well as basic results about the properties of such justification statuses for different semantics. In Section 4 we provide a comprehensive complexity analysis for the decision problems associated to the justification status of an argument. Finally, in Section 5 we conclude the paper with a summary and discussion of the presented results.

2   Preliminaries

In this section we introduce abstract argumentation frameworks [12] and the concept of labelings [8]. Further we recall some of the most important semantics for abstract argumentation (see [3]). Finally, we highlight complexity results for typical decision problems associated to such frameworks.

Definition 1. *An* argumentation framework *(AF) is a pair $F = (A, R)$ where A is a finite set of arguments and $R \subseteq A \times A$ is the attack relation. For a given AF $F = (A, R)$ we use A_F to denote the set A of its arguments and R_F to denote its attack relation R. We sometimes use $a \rightarrowtail^R b$ instead of $(a, b) \in R$. For $S \subseteq A$ and $a \in A$, we also write $S \rightarrowtail^R a$ (resp. $a \rightarrowtail^R S$) in case there exists $b \in S$, such that $b \rightarrowtail^R a$ (resp. $a \rightarrowtail^R b$). In case no ambiguity arises, we use $\rightarrowtail$ instead of $\rightarrowtail^R$.*

Using the extension-based approach, one assigns a set $\sigma(F) \subseteq 2^A$ of extensions to each AF $F = (A, R)$. For σ consider the functions *stb*, *adm*, *prf*, *com*, *grd*, *stg*, and *sem* which stand for stable, admissible, preferred, complete, ground stage, and semi-stable semantics. Before actually defining the semantics, we have to introduce a few more formal concepts.

Definition 2. *Given an AF $F = (A, R)$, an argument $a \in A$ is defended (in F) $S \subseteq A$ if for each $b \in A$, such that $b \rightarrowtail a$, also $S \rightarrowtail b$ holds. The characteristic function $\mathcal{F}_F : 2^A \to 2^A$, is defined as $\mathcal{F}_F(S) = \{x \in A_F \mid x \text{ is defended by } S\}$. Moreover, for a set $S \subseteq A$, we denote by S_R^+ the set $S \cup \{a \mid S \rightarrowtail^R a\}$.*

We are now ready to define the semantics.

Definition 3. *Let $F = (A, R)$ be an AF. A set $S \subseteq A$ is conflict-free (in F) (denoted as $S \in cf(F)$), iff there are no $a, b \in S$, such that $(a, b) \in R$. For $S \in cf(F)$, we define:*

- *$S \in stb(F)$, if $S_R^+ = A$;*
- *$S \in adm(F)$, if $S \subseteq \mathcal{F}_F(S)$;*

- $S \in com(F)$, *if $S = \mathcal{F}_F(S)$;*
- $S \in prf(F)$, *if $S \in adm(F)$ and there is no $T \in adm(F)$ with $T \supset S$;*
- $S \in grd(F)$, *if $S \in com(F)$ and there is no $T \in com(F)$ with $T \subset S$;*
- $S \in sem(F)$, *if $S \in adm(F)$ and there is no $T \in adm(F)$ with $T_R^+ \supset S_R^+$.*
- $S \in stg(F)$, *if there is no $T \in cf(F)$ in F, such that $T_R^+ \supset S_R^+$;*

We recall that for each AF F, $stb(F) \subseteq sem(F) \subseteq prf(F) \subseteq com(F) \subseteq adm(F)$, and that for each of the considered semantics σ (except stable) $\sigma(F) \neq \emptyset$ holds. Moreover we have that for each AF F there is an unique grounded extension, which is the least fixed-point of $\mathcal{F}_F$. Further in case that an AF has at least one stable extension then its stable, semi-stable, and stage extensions coincide.

Example 1. Consider the AF $F = (A, R)$, with $A = \{a, b, c, d, e\}$ and $R = \{(a, b), (c, b), (c, d), (d, c), (d, e), (e, e)\}$. The graph representation of F is given as follows.

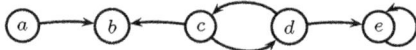

We have $stb(F) = stg(F) = sem(F) = \{\{a, d\}\}$. Further we have the admissible sets $\{\}, \{a\}, \{c\}, \{d\}, \{a, c\}, \{a, d\}$, and thus $prf(F) = \{\{a, c\}, \{a, d\}\}$. Finally the complete extensions of F are $\{a\}$, $\{a, c\}$ and $\{a, d\}$, with $\{a\}$ being the grounded extension. $\Diamond$

On the base of these semantics one can define the family of resolution-based semantics [1], with the resolution-based grounded semantics being the most popular instance.

Definition 4. *Given AF $F = (A, R)$. A resolution $\beta \subset R$ is a set of attacks such that $(a, b) \in \beta$ iff $(b, a) \notin \beta$ and $\{(a, b), (b, a)\} \subseteq R$. A set $E \subseteq A$ is a resolution-based grounded extension, denoted as $E \in grd^*(F)$, if (i) there is a resolution β such that $grd((A, R \setminus \beta)) = E$ and (ii) there is no resolution β' such that $grd((A, R \setminus \beta')) \subset E$.*

For the AF F in example 1 we have that $grd^*(F) = \{\{a, c\}, \{a, d\}\}$.

An extension separates the accepted arguments from the non-accepted arguments, but there may be a substantial difference whether an argument is rejected because of a conflict with the arguments in the extension or because the extension does not defend it. The more fine-grained concept of labelings, as introduced by Caminada [5], generalizes extensions and captures the above observation.

Definition 5. *Let $F = (A, R)$ be an AF. A labeling for F is a function $\mathcal{L} : A \to \{in, out, undec\}$. We will denote labelings $\mathcal{L}$ also by triples $(\mathcal{L}_{in}, \mathcal{L}_{out}, \mathcal{L}_{undec})$, where $\mathcal{L}_l = \{a \in A \mid \mathcal{L}(a) = l\}$.*

The intuition behind these labels is the following. An argument is labeled with: *in* if it is accepted; *out* if there are strong reasons to reject it, i.e. its attacked by an accepted argument; and by *undec* if the argument is undecided, i.e. neither accepted nor attacked by accepted arguments. In [7,8], the authors define labeling-based semantics for AFs independent of the extension-based semantics and then showed a strong correspondence to extension based semantics. For simplicity we use this correspondence to directly define labeling-based semantics via extensions. To this end, for each AF we define a function mapping sets of arguments to labelings.

Definition 6. *Let $F = (A, R)$ be an AF. We define the function $Ext2Lab_F : 2^A \to$ $\{in, out, undec\}^A$ such that $Ext2Lab_F(E) = (E, E_R^+ \setminus E, A \setminus E_R^+)$ for $E \subseteq A$.*

In [8], $Ext2Lab$ is restricted to conflict-free sets. Here it is well-defined for arbitrary sets of arguments, while it is equivalent to original definition when restricted to conflict-free sets. Now one can interpret an extension-based semantics as labeling-based semantics, using the function $Ext2Lab$ to map each extension to a labeling. In particular $Ext2Lab$ is a one-to-one mapping [8], i.e. different extensions yield different labelings.

Definition 7. *Let $F = (A, R)$ be an AF and σ an extension-based semantics. The corresponding labeling-based semantics σ_L is defined as follows $\sigma_L(F) = \{Ext2Lab(E) \mid E \in \sigma(F)\}$. If no ambiguity arises we will use $\sigma(F)$ instead of $\sigma_L(F)$.*

We mention that our definition of adm_L doesn't match the definition of admissible labelings by Caminada and Gabbay [8], but what they call JV-labelings. However both semantics propose the same justification statuses.

Next we turn to the complexity of reasoning in AFs. We assume the reader has knowledge about standard complexity classes, i.e. P, NP and LOGSPACE (L), but we briefly recapitulate the concept of oracle machines and some related complexity classes. Let $\mathcal{C}$ be some complexity class. By a $\mathcal{C}$-oracle machine we mean a Turing machine which can access an oracle that decides a given (sub)-problem in $\mathcal{C}$ within one step. We denote the class of problems decidable in polynomial time when using such a $\mathcal{C}$-oracle machine, as $P^{\mathcal{C}}$ if the underlying Turing machine is deterministic and $NP^{\mathcal{C}}$ if the underlying Turing machine is nondeterministic. Moreover we consider deterministic oracle machines where the number of allowed oracle calls is bounded by a constant k, and denote the corresponding complexity classes as $P^{\mathcal{C}[k]}$. We now turn to concrete complexity classes. The class $\Sigma_2^P = NP^{NP}$, denotes the problems which can be decided by a nondeterministic polynomial time algorithm that has access to an NP-oracle. The class $\Pi_2^P = coNP^{NP}$ is defined as the complementary class of Σ_2^P, i.e. $\Pi_2^P = co\Sigma_2^P$. Finally we define the classes D^P and D_2^P. A decision problem L is in the class D^P iff L can be characterized as $L_1 \cap L_2$ for decision problems $L_1 \in NP$ and $L_2 \in coNP$. Similarly $L \in D_2^P$ iff L can be characterized as $L_1 \cap L_2$ for $L_1 \in \Sigma_2^P$ and $L_2 \in \Pi_2^P$. Next we give an overview of relations between the complexity classes used in this paper.

$$L \subseteq P \subseteq \begin{matrix} NP \\ coNP \end{matrix} \subseteq D^P \subseteq \begin{matrix} \Sigma_2^P \\ \Pi_2^P \end{matrix} \subseteq P^{\Sigma_2^P[1]} \subseteq D_2^P$$

The typical decision problems for a semantics σ are:

- $Cred_\sigma$: Given AF F and $a \in A_F$. Is a contained in some $S \in \sigma(F)$?
- $Skept_\sigma$: Given AF F and $a \in A_F$. Is a contained in each $S \in \sigma(F)$?
- $Skept'_\sigma$: Given AF F and $a \in A_F$. Is a contained in each $S \in \sigma(F)$ and $\sigma(F) \neq \emptyset$?
- Ver_σ: Given AF F and $S \subseteq A_F$. Is $S \in \sigma(F)$?
- $Exists_\sigma$: Given AF F. Is $\sigma(F) \neq \emptyset$?
- $Exists_\sigma^{\neg\emptyset}$: Given AF F. Does there exist a set $S \neq \emptyset$ such that $S \in \sigma(F)$?

From the literature [1,10,11,12,15,16,17,19,20], we obtain the complexity-landscape of abstract argumentation as given in Table 1. We mention that most of the semantics σ (except stable) always provide at least one extension. For these σ the problem $Exists_\sigma$ can be trivially answered with yes and further the problems $Skept'_\sigma$ and $Skept_\sigma$ coincide.

Table 1. Complexity of abstract argumentation ($\mathcal{C}$-c denotes completeness for class $\mathcal{C}$)

σ	Cred_σ	Skept_σ	Skept'_σ	Ver_σ	Exists_σ	$\text{Exists}_\sigma^{\neg\emptyset}$
grd	P-c	P-c	P-c	P-c	trivial	in L
stb	NP-c	coNP-c	D^P-c	in L	NP-c	NP-c
adm	NP-c	trivial	trivial	in L	trivial	NP-c
com	NP-c	P-c	P-c	in L	trivial	NP-c
prf	NP-c	Π_2^P-c	Π_2^P-c	coNP-c	trivial	NP-c
sem	Σ_2^P-c	Π_2^P-c	Π_2^P-c	coNP-c	trivial	NP-c
stg	Σ_2^P-c	Π_2^P-c	Π_2^P-c	coNP-c	trivial	in L
grd^*	NP-c	coNP-c	coNP-c	P-c	trivial	in P

3   Justification Status of Arguments

Here we consider the task of reasoning in AFs with labeling-based semantics. When using extensions the usual reasoning modes are skeptical acceptance, accepting an argument if it is in all extensions, and credulous acceptance, i.e. accepting an argument if it is an least one extension. Applying these acceptance criterions to labelings would not take the labels *out*, *undec* in to account. To overcome this, Wu and Caminada [23] recently proposed a more elaborate concept, the so called *justification status* of an argument, to reason with complete labelings. The main idea is to define the acceptance status of an argument a, by the set of labels l such that there exists a labeling mapping a to l. In the following definition we generalize this concept to arbitrary semantics.

Definition 8. *Let $F = (A, R)$ be an AF and σ a semantic. The* justification status *of an argument $a \in A$ in F is defined as $\mathcal{JS}_\sigma(F, a) = \{\mathcal{L}(a) \mid \mathcal{L} \in \sigma(F)\}$.*

Definition 8 proposes eight different justification statuses for arguments. Following [23], we call $\{in\}$ strong accept; $\{in, undec\}$ weak accept; $\{out\}$ strong reject; $\{out, undec\}$ weak reject; and consider the remaining possibilities as borderline cases.

The justification status of an argument generalizes the idea of skeptical and credulous reasoning in the sense that one can encode skeptical and credulous acceptance as a property of the justification status of the argument. An argument a, is skeptically accepted iff $\mathcal{JS}_\sigma(F, a) = \{in\}$, whereas it is credulously accepted iff $in \in \mathcal{JS}_\sigma(F, a)$.

While in general there are eight possible justification statuses for each argument, none of the semantics under our considerations is able to generate all of them. In the following we give a compact analysis of the possible justification statuses for each semantics. We start with a general observation for unique status semantics:

Lemma 1. *Let σ be an unique status semantics, $F = (A, R)$ and $a \in A$. Then $\mathcal{JS}_\sigma(F, a) \in \{\{in\}, \{out\}, \{undec\}\}$.*

Proof. Immediate by the fact that there is exactly one labeling $\mathcal{L} \in \sigma(F)$. ☐

Lemma 1 can be interpreted such that for unique status semantics the justification status approach does not provide an additional value. This applies to grounded semantics as well as to ideal [13], eager [6] and in general parametrized ideal [18] semantics.

The following proposition fully characterizes the possible justification statuses for the semantics under our considerations.

Proposition 1. *Let $F = (A, R)$ be an AF and $a \in A$. Then we have that:*

1. $\mathcal{JS}_{grd}(F, a) \in \{\{in\}, \{out\}, \{undec\}\}$
2. $\mathcal{JS}_{adm}(F, a) \in \{\{undec\}, \{in, undec\}, \{out, undec\}, \{in, out, undec\}\}$
3. $\mathcal{JS}_{com}(F, a) \in 2^{\{in, out, undec\}} \setminus \{\{in, out\}, \emptyset\}$
4. $\mathcal{JS}_{stb}(F, a) \in \{\{in\}, \{out\}, \{in, out\}, \emptyset\}$
5. $\mathcal{JS}_{prf}(F, a) \in 2^{\{in, out, undec\}} \setminus \{\emptyset\}$
6. $\mathcal{JS}_{sem}(F, a) \in 2^{\{in, out, undec\}} \setminus \{\emptyset\}$
7. $\mathcal{JS}_{stg}(F, a) \in 2^{\{in, out, undec\}} \setminus \{\emptyset\}$
8. $\mathcal{JS}_{grd^*}(F, a) \in 2^{\{in, out, undec\}} \setminus \{\emptyset\}$

Proof. We have that (1) holds because *grd* is a unique status semantics. For (2) we use that $(\emptyset, \emptyset, A)$ is always an admissible labeling and thus $undec \in \mathcal{JS}_{adm}(F, a)$. The case of complete semantics has already been studied by Wu and Caminada [23]. The restricted domain steams from the fact that the grounded labeling $\mathcal{L}_{grd}$ is the unique $\subseteq$-minimal w.r.t. both $\mathcal{L}_{in}$ and $\mathcal{L}_{out}$ complete labeling. So if $\mathcal{L}_{grd}(a) = in$ or $\mathcal{L}_{grd}(a) = out$ we have that also $\mathcal{JS}_{com}(F, a) = \{in\}$ or resp. $\mathcal{JS}_{com}(F, a) = \{out\}$. Otherwise we have that $\mathcal{L}_{grd}(a) = undec$ and therefore $undec \in \mathcal{JS}_{com}(F, a)$. To show (4) we use that $\mathcal{L}_{undec} = \emptyset$ for every stable labeling and thus $undec \notin \mathcal{JS}_{stb}(F, a)$. The remaining follows from the fact that for each semantic $\sigma \in \{prf, sem, stg, grd^*\}$ it is the case that $\sigma(F) \neq \emptyset$ and thus also that $\mathcal{JS}_{\sigma}(F, a) \neq \{\}$. □

Proposition 1 shows that several justification statuses are not possible w.r.t. some semantics. It remains to show that the remaining ones are possible.

Example 2. Here we use the AFs illustrated in Figure 1. First we consider the justification statuses $\{in\}$, and $\{out\}$. For semantics $\sigma \in \{grd, com, stb, prf, sem, stg, grd^*\}$ we have that $\mathcal{JS}_{\sigma}(F_1, a) = \{in\}$ and $\mathcal{JS}_{\sigma}(F_1, b) = \{out\}$. Next let us consider $\{undec\}$ and $\sigma \in \{grd, adm, com, prf, sem, stg, grd^*\}$, we have that $\mathcal{JS}_{\sigma}(F_2, a) = \{undec\}$. The following examples complete the picture for admissible semantics: $\mathcal{JS}_{adm}(F_1, a) = \{in, undec\}$, $\mathcal{JS}_{adm}(F_1, b) = \{out, undec\}$, $\mathcal{JS}_{stb}(F_2, a) = \{\}$ and $\mathcal{JS}_{\sigma}(F_3, a) = \{in, out, undec\}$ for $\sigma \in \{adm, com\}$.

Let us now consider the remaining justification statuses for complete semantics. To this end let us consider AF_4 where we have that $\mathcal{JS}_{\sigma}(F_4, b) = \{out, undec\}$ and $\mathcal{JS}_{\sigma}(F_4, c) = \{in, undec\}$ for $\sigma \in \{com, prf, sem\}$.

Next we study the remaining justification statuses for stable, preferred, semi-stable and stage semantics. First to get the justification status $\{in, out\}$ we use the AF F_3 and obtain that $\mathcal{JS}_{\sigma}(F_3, a) = \{in, out\}$ for $\sigma \in \{stb, prf, sem, stg\}$. The picture is completed by the following observations on F_5 and F_6: $\mathcal{JS}_{stg}(F_5, a) = \{in, undec\}$, $\mathcal{JS}_{stg}(F_5, e) = \{out, undec\}$, $\mathcal{JS}_{stg}(F_5, c) = \{in, out, undec\}$ and $\mathcal{JS}_{prf}(F_6, e) = \mathcal{JS}_{sem}(F_6, e) = \{in, out, undec\}$.

Finally for resolution based grounded semantics we have that $\mathcal{JS}_{grd^*}(F_7, a) = \{in, out\}$, $\mathcal{JS}_{grd^*}(F_7, c) = \{out, undec\}$, $\mathcal{JS}_{grd^*}(F_7, d) = \{in, undec\}$ and $\mathcal{JS}_{grd^*}(F_7, e) = \{in, out, undec\}$. ◇

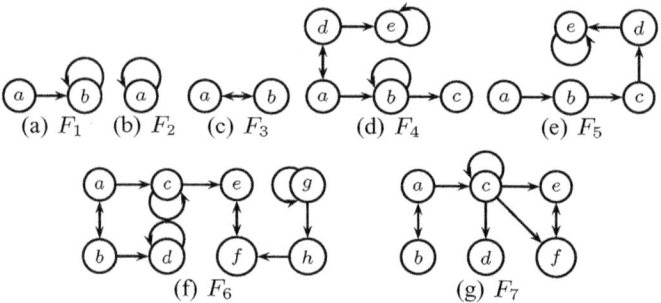

Fig. 1. Example argumentation frameworks F_1-F_7

Wu and Caminada [23] mention that the complete labelings of an AF can be seen as *subjective but reasonable positions* given the conflicting information of the AF. We are now interested how the justification statuses are affected if we change our opinion about what are reasonable positions, i.e. we change semantics.

Proposition 2. *Let* $F = (A, R)$ *be an AF and* $a \in A$ *then:*

$$\mathcal{JS}_{stb}(F, a) \subseteq \mathcal{JS}_{sem}(F, a) \subseteq \mathcal{JS}_{prf}(F, a) \subseteq \mathcal{JS}_{com}(F, a) \subseteq \mathcal{JS}_{adm}(F, a)$$
$$\mathcal{JS}_{grd}(F, a) \subseteq \mathcal{JS}_{com}(F, a) \subseteq \mathcal{JS}_{adm}(F, a)$$

Proof. The above $\subseteq$ follows from well-known relations between the semantics [4,12], namely that: each stable extensions is also a semi-stable extension; each semi-stable extensions is also a preferred extension; each preferred extension is a complete extension; each complete extension is an admissible set; and the grounded extension is a complete extension. $\square$

Proposition 2 indicates that there are two kinds of skepticism. First there are different levels of skepticism, about what is an reasonable positions in an argument, that is captured by the semantics. Second depending on ones skepticism one tolerates different kinds of dispute about an argument. In an extension based setting this is captured be the reasoning modes, i.e. skeptical and credulous acceptance. In a labeling based setting this is captured by the justification statuses one is willing to accept. It has to be mentioned that this kinds of skepticism differ conceptual from those introduced in [2], which are about relations between the extensions of different semantics whereas we are interested in the status of an argument.

We can show that the justification statuses w.r.t. admissible, complete and preferred semantics only differ on the *undec* labels.

Proposition 3. *Let* $F = (A, R)$ *be an AF and* $a \in A$.

1. $\mathcal{JS}_{adm}(F, a) = \mathcal{JS}_{com}(F, a) \cup \{undec\}$
2. $\mathcal{JS}_{adm}(F, a) = \mathcal{JS}_{prf}(F, a) \cup \{undec\}$
3. $\mathcal{JS}_{com}(F, a) = \begin{cases} \mathcal{JS}_{grd}(F, a) & \text{if } a \in grd(F)^+ \\ \mathcal{JS}_{adm}(F, a) & \text{otherwise} \end{cases}$

$$4.\ \mathcal{JS}_{com}(F,a) = \begin{cases} \mathcal{JS}_{grd}(F,a) & \text{if } a \in grd(F)^+ \\ \mathcal{JS}_{prf}(F,a) \cup \{undec\} & \text{otherwise} \end{cases}$$

Proof. For (1) & (2) we first use the fact that $(\emptyset, \emptyset, A) \in adm(F)$ and hence adding *undec* is always valid. Further by the definitions of the semantics we have that for each $\mathcal{L} \in adm(F)$ there exists $\mathcal{L}' \in com(F)$ (resp. $\mathcal{L}' \in prf(F)$) such that $\mathcal{L}_{in} \subseteq \mathcal{L}'_{in}$ and $\mathcal{L}_{out} \subseteq \mathcal{L}'_{out}$. Hence $\mathcal{JS}_{adm}(F,a) \subseteq \mathcal{JS}_{com}(F,a) \cup \{undec\}$. Further as each complete (resp. preferred) labeling is also an admissible labeling we also get $\mathcal{JS}_{adm}(F,a) \supseteq \mathcal{JS}_{com}(F,a) \cup \{undec\}$. (3) In [23] it is observed that if $a \in grd(F)^+$ then $\mathcal{JS}_{com}(F,a) = \mathcal{JS}_{grd}(F,a)$. Further they observe that if $a \notin grd(F)^+$ then $undec \in \mathcal{JS}_{com}(F,a)$ and by (1) then $\mathcal{JS}_{adm}(F,a) = \mathcal{JS}_{com}(F,a)$. Finally we obtain (4) by combining (2) and (3). □

4 Complexity Analysis

In this section we provide a formal analysis of the computational properties of the justification status. To this end we define the corresponding decision problems. The canonical decision problem is deciding if a given argument has a given justification status.

Definition 9. *The* justification status decision problem JS_σ: *Given an AF* $F = (A, R)$, $L \subseteq \{in, out, undec\}$ *and an argument* $a \in A$. *Is* $\mathcal{JS}_\sigma(F,a) = L$.

We can express skeptical acceptance as instance of JS_σ by choosing $L = \{in\}$, but we can't neither express credulous acceptance nor if an argument is at least weakly accepted. Thus we generalize the above definition to capture these cases:

Definition 10. *The* generalized justification status decision problem GJS_σ: *Given an AF* $F = (A, R)$, $L, M \subseteq \{in, out, undec\}$ *and an argument* $a \in A$. *Is* $L \subseteq \mathcal{JS}_\sigma(F,a)$ *as well as* $\mathcal{JS}_\sigma(F,a) \cap M = \emptyset$.

4.1 General Complexity

One can encode an instance of JS_σ as GJS_σ instance, using $M = \{in, out, undec\}\setminus L$. Thus we have that GJS_σ is at least as computationally hard as JS_σ and as we will show, for the semantics σ under our considerations the problems JS_σ as GJS_σ have the same complexity. In the following we give hardness proofs for JS_σ, and membership proofs, i.e .algorithms, for GJS_σ which then implies the completeness of both problems.

We start with some general observations about the problems JS_σ and GJS_σ and then present results for specific $\sigma \in \{grd, adm, com, stb, prf, sem, stg, grd^*\}$. Our first result allows us to propagate hardness results from skeptical acceptance to JS_σ.

Proposition 4. *If one of the problems* coExists$_\sigma$, Skept$'_\sigma$ *is C-hard then* JS_σ *is C-hard.*

Proof. Immediate by the fact that both coExists$_\sigma$ and Skept$'_\sigma$ can be easily encoded as instances of JS_σ. For the first, simply set $M = \{\}$ and for the latter set $M = \{in\}$. □

Next we provide a (generic) algorithm to decide GJS_σ:

Theorem 1. *If for semantics σ the problem* Ver$_\sigma$ *is in a complexity class C then the problem GJS_σ is in the complexity class* $\text{NP}^C \wedge \text{coNP}^C$.

Proof. Let us consider an arbitrary instance of GJS_σ, with AF F, an argument $a \in A_F$ and sets $L, M \subseteq \{in, out, undec\}$. To prove $GJS_\sigma \in NP^C \land coNP^C$ we first give an NP^C algorithm to decide $L \subseteq \mathcal{JS}_\sigma(F, a)$:

1. For each $l \in L$ guess a labeling $\mathcal{L}_l$ with $\mathcal{L}_l(a) = l$
2. Test whether $\mathcal{L}_l \in \sigma(F)$ or not, using the C-oracle.
3. Accept if for each $l \in L$, $\mathcal{L}_l \in \sigma(F)$

As there are at most three labels $l \in L$ the guess is polynomial in the size of the input AF and thus we have an NP^C algorithm.

Next we show that $\mathcal{JS}_\sigma(F, a) \cap M = \emptyset$ can be decided in $coNP^C$. To this end we give an NP^C algorithm that decides the complementary problem $\mathcal{JS}_\sigma(F, a) \cap M \neq \emptyset$.

1. For each $l \in M$ guess a labeling $\mathcal{L}_l$ with $\mathcal{L}_l(a) = l$
2. Test whether $\mathcal{L}_l \in \sigma(F)$ or not
3. Accept if there exists an $l \in M$ such that $\mathcal{L}_l \in \sigma(F)$

Now as $GJS_\sigma(F, a)$ iff $L \subseteq \mathcal{JS}_\sigma(F, a)$ and $\mathcal{JS}_\sigma(F, a) \cap M = \emptyset$ we finally observe that $GJS_\sigma \in NP^C \land coNP^C$. □

The algorithm in the above proof is (worst case) optimal for semantics $\sigma \in \{adm, com, stb, sem, stg, grd^*\}$, but, as we will show, not for $\sigma \in \{grd, prf\}$. We now turn to study the complexity for specific semantics and start with grounded semantics:

Proposition 5. *The problems JS_{grd}, GJS_{grd} are P-complete.*

Proof. The membership in the class P follows by the fact that the unique grounded labeling can be computed in polynomial time (see Table 2). The hardness follows from the fact that $Skept'_{grd}$ is P-hard and Proposition 4. □

Next we consider admissible and complete semantics. By Proposition 3 the justification statuses of admissible and complete semantics are closely related, thus it is not surprising that we can use the same construction to show hardness for both semantics.

Proposition 6. *The problems JS_{com}, GJS_{com}, JS_{adm}, GJS_{adm} are D^P-complete.*

Proof. The membership is an immediate consequence of Theorem 1 and the fact that Ver_{com}, Ver_{adm} are in L.

We prove hardness by reducing the (D^P-hard) SAT-UNSAT problem to JS_{com} (resp. JS_{adm}). We assume that the two CNF formulas are given as a set of clauses, where each clause is a set over atoms and negated atoms (denoted by $\bar{x}$). For such CNFs $\varphi(X) = \bigwedge_{c \in C_\varphi} c$, $\psi(Y) = \bigwedge_{c \in C_\psi} c$ over variables X resp. Y with $X \cap Y = \emptyset$, define the AF $F_{\varphi,\psi} = (A, R)$ [1] with $A = X \cup \bar{X} \cup Y \cup \bar{Y} \cup C_\varphi \cup C_\psi \cup \{t, t'\}$ and

$$R = \{(z, \bar{z}), (\bar{z}, z) \mid z \in X \cup Y\} \cup \{(l, c) \mid l \in c, c \in C_\varphi \cup C_\psi\} \cup$$
$$\{(c, t), (c, c) \mid c \in C_\varphi\} \cup \{(c, t'), (c, c) \mid c \in C_\psi\} \cup \{(t, t'), (t', t)\}$$

---

[1] This reduction builds slightly modified standard translations (as defined in [17]) of both formulas and adds a mutual attack between them.

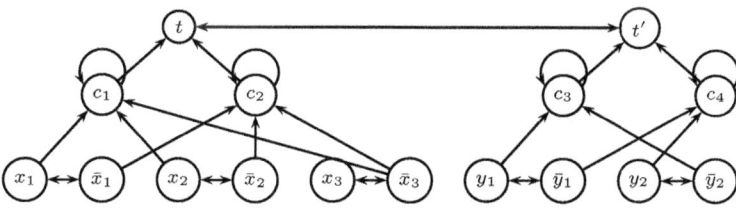

Fig. 2. AF $F_{\varphi,\psi}$ with $c_{\varphi,1} = \{x_1, x_2, \bar{x}_3\}$, $c_{\varphi,2} = \{\bar{x}_1, \bar{x}_2, \bar{x}_3\}$, $c_{\psi,3} = \{y_1, \bar{y}_2\}$, $c_{\psi,4} = \{\bar{y}_1, y_2\}$

where $\bar{X} = \{\bar{x} \mid x \in X\}$, $\bar{Y} = \{\bar{y} \mid y \in y\}$ and t, t' are fresh arguments. See Figure 2 for an illustrating example. Let us first mention that $grd(F) = \emptyset$ and thus for each $a \in A$ it holds that $JS_{com}(F, a) = JS_{adm}(F, a)$. Hence it suffices to show that $(\varphi, \psi) \in$ SAT-UNSAT iff $JS_{com}(F, t) = \{in, undec\}$.

As $grd(F) = \emptyset$ we have that $undec \in JS_{com}(F, t)$, independently of the instance (φ, ψ). It is easy to see that[2] $in \in JS_{com}(F, t)$ iff φ is satisfiable and that $in \in JS_{com}(F, t')$ iff ψ is satisfiable. Further t' is the only not self-conflicting argument attacking t. Thus we have that $out \in JS_{com}(F, t)$ iff $in \in JS_{com}(F, t')$ iff ψ is satisfiable. To sum up, $JS_{com}(F, t) = \{in, undec\}$ iff φ is satisfiable and ψ is unsatisfiable, i.e. $(\varphi, \psi) \in$ SAT-UNSAT. $\qquad\square$

Next we consider stable and resolution-based grounded semantics.

Proposition 7. *The problems JS_{stb}, GJS_{stb} are D^P-complete.*

Proof. The membership follows from Theorem 1 and the fact that Ver_{stb} is in L. Further as $Skept'_{stb}$ is D^P-hard [17] we can use Proposition 4 to obtain D^P-hardness for JS_{stb} and thus D^P-completeness. $\qquad\square$

Proposition 8. *The problems JS_{grd*}, GJS_{grd*} are D^P-complete.*

Proof. The membership follows by the fact that that Ver_{grd*} is in P and Theorem 1. Again we show hardness by reducing the (D^P-hard) SAT-UNSAT problem to JS_{grd*}. Consider two CNFs $\varphi(X) = \bigwedge_{c \in C_\varphi} c$, $\psi(Y) = \bigwedge_{c \in C_\psi} c$ over variables X resp. Y with $X \cap Y = \emptyset$ and additional assume that φ and ψ have at least one counter model. Define the AF $F_{\varphi,\psi} = (A, R)$ with $A = X \cup \bar{X} \cup Y \cup \bar{Y} \cup C_\varphi \cup C_\psi \cup \{t, t'\}$ and

$$R = \{(z, \bar{z}), (\bar{z}, z) \mid z \in X \cup Y\} \cup \{(l, c) \mid l \in c, c \in C_\varphi \cup C_\psi\} \cup$$
$$\{(c, t), (c, c) \mid c \in C_\varphi\} \cup \{(c, t') \mid c \in C_\psi\} \cup \{(t', t)\}$$

where $\bar{X} = \{\bar{x} \mid x \in X\}$, $\bar{Y} = \{\bar{y} \mid y \in y\}$ and t, t' are fresh arguments. See Figure 3 for an illustrating example. We show that $(\varphi, \psi) \in$ SAT-UNSAT iff $JS_{grd*}(F, t) = \{in, undec\}$.

In analogy to [1] we observe that resolving a symmetric attack between z and $\bar{z}$ with $z \in X \cup Y$ selects either z or $\bar{z}$ for being in the grounded extension. Thus each resolution yields a distinct extension and $\subseteq$-minimality does not came in play for $F_{\varphi,\psi}$. Now we

---

[2] Following the argument for the standard translation in [17].

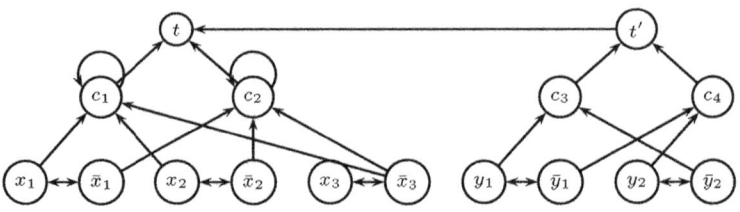

Fig. 3. AF $F_{\varphi,\psi}$ with $c_{\varphi,1}=\{x_1,x_2,\bar{x}_3\}$, $c_{\varphi,2}=\{\bar{x}_1,\bar{x}_2,\bar{x}_3\}$, $c_{\psi,3}=\{y_1,\bar{y}_2\}$, $c_{\psi,4}=\{\bar{y}_1,y_2\}$

have a one to one correspondence between pairs of true-assignments (M_X, M_Y) for (φ, ψ) and resolutions for $F_{\varphi,\psi}$. One can check that t is labeled in iff M_X satisfies φ and M_Y falsifies ψ, and that t is labeled out iff M_Y satisfies ψ.

Now given that $(\varphi, \psi) \in$ SAT-UNSAT we have that $in \in JS_{grd*}(F,t)$ and $out \notin JS_{grd*}(F,t)$. Further by the assumption that φ is not valid there is a pair (M_X, M_Y) that falsifies both φ, and ψ, hence $undec \in JS_{grd*}(F,t)$.

Given that $JS_{grd*}(F,t) = \{in, undec\}$. By the fact that $in \in JS_{grd*}(F,t)$ we conclude that φ has a model and by $out \notin JS_{grd*}(F,t)$ we have that ψ has no model, hence $\varphi \in$ SAT and $\psi \in$ UNSAT. $\qquad\square$

In the case of preferred semantics Proposition 1 does not yield the optimal membership, i.e. it yields D_2^P membership instead of $P^{\Sigma_2^P[1]}$-membership. Hence we have to provide an algorithm for preferred semantics to show membership.

Proposition 9. *The problems JS_{prf}, GJS_{prf} are $P^{\Sigma_2^P[1]}$-complete.*

Proof. To show *membership* we give a $P^{\Sigma_2^P[1]}$ algorithm for GJS_{prf}. To this end, let us consider an arbitrary instance with AF $F = (A, R)$, an argument $a \in A$ and sets $L, M \subseteq \{in, out, undec\}$. First we observe that testing whether $in \in \mathcal{JS}_\sigma(F, a)$ (resp. $out \in \mathcal{JS}_\sigma(F, a)$) can be done with an NP-algorithm. Such an algorithm simple guesses a labeling $\mathcal{L} = (\mathcal{L}_{in}, \mathcal{L}_{out}, \mathcal{L}_{undec})$ and then tests if $\mathcal{L} \in \mathcal{L}_{adm}(F)$ and $a \in \mathcal{L}_{in}$ (resp. $a \in \mathcal{L}_{out}$). But testing for $undec \in \mathcal{JS}_\sigma(F, a)$ is not that easy, as it doesn't suffice to consider admissible labelings. Our algorithm proceeds as follows: If $undec \in L \cup M$ (w.l.o.g. we assume $L \cap M = \emptyset$) we start the following Σ_2^P algorithm.

1. Guess labeling $\mathcal{L}$ with $\mathcal{L}(a) = undec$.
2. Test whether $\mathcal{L}$ is preferred or not, using a coNP-oracle.
3. Test whether $L \setminus \{undec\} \subseteq \mathcal{JS}_\sigma(F, a)$, using a NP-oracle.
4. Test whether, $(M \setminus \{undec\}) \cap \mathcal{JS}_\sigma(F, a) = \emptyset$, using a coNP-oracle.
5. If $undec \in L$ accept iff 2,3 and 4 hold;
   If $undec \in M$ accept iff 2 holds, or either 3 or 4 fails.

If $undec \in L$ we use the answer of the Σ_2^P algorithm as answer for the problem GJS_{prf}. Otherwise, i.e. $undec \in M$, we negate the answer of the Σ_2^P algorithm. In the case where $undec \notin L \cup M$ we omit steps 1 & 2 of the above algorithm, which leads to an D^P-algorithm deciding GJS_{prf}.

We show *hardness* by reducing the QBF_2-problem, i.e. the problem of deciding if a QBF with one quantifier alternation is valid, to JS_{prf}. Let Φ be an instance of QBF_2, then Φ is either of the form (i) $\Phi = \forall X \exists Y \varphi(X,Y)$ or (ii) $\Phi = \exists X \forall Y \varphi(X,Y)$. In case (i) we can use the reduction from the Π_2^P-hardness proof of $Skept'_{prf}$ [15] to encode ϕ as $Skept'$-problem, which itself can be encoded as instance of JS_{prf}. (see proof of Proposition 4). In case (ii) we can encode $\neg \Phi$ as $coSkept'$-problem in the same way as in (i), but it remains to show that the problem $coSkept'_{prf}$ can be reduced to a JS_{prf}-instance. To this end consider an arbitrary AF $F = (A, R)$ with argument $t \in A$. If $(t,t) \in R$ then $coSkept'$ is trivially true, thus for the remainder of the proof we assume $(t,t) \notin R$. We build the AF $F' = (A \cup \{g, u, v, w\}, R \cup \{(t,g), (g,g), (g,u), (v,w), (w,v)\} \cup \{(v,a) \mid a \in A \setminus \{t\}\})$ and claim that $JS_{prf}(F', u) = \{in, undec\}$ iff the argument t is not skeptical accepted in F.

As $\{v, t, u\}$ is conflict-free and attacks all the other arguments it is a preferred extension of F' and thus $in \in JS_{prf}(F', u)$. Further as g is the only attacker of u and $(g,g) \in R'$ we also have that $out \notin JS_{prf}(F', u)$. It remains to show that $undec \in JS_{prf}(F', u)$ iff t is not skeptical accepted in F.

To show the "if"-part, consider $E \in prf(F)$ with $t \notin E$. It holds that $E \cup \{w\} \in adm(F')$ and thus there exists $E' \in prf(F')$ such that $E \cup \{w\} \subseteq E'$. As $w \in E'$ we have that $v \notin E'$ and as w, u, g do not attack any attacker of t, we also have that $t \notin E'$. Hence also $u \notin E'$ and thus $undec \in JS_{prf}(F', u)$.

Next, to show the "only if"-part, assume $undec \in JS_{prf}(F', u)$. Then there exists an $E' \in prf(F')$ such that $u \notin E'$, thus by construction F' also $t \notin E'$ and $v \notin E'$. Moreover we have that $E = E' \cap A \in adm(F)$ and $E' = E \cup \{w\}$. Next we show that also $E \in prf(F)$. Let us assume, towards a contradiction, that $E \notin prf(F)$ then there exists $S \in prf(F)$ with $E \subset F$. But by construction $S' = S \cup \{w\} \in adm(F')$ and as $E \subset S$ also $E' \subset S'$, a contradiction to our assumption $E' \in prf(F')$. We have that E is a preferred extension of F with $t \notin E$, i.e. t is not skeptical accepted in F.   □

Finally we consider semi-stable and stage semantics. To this end we first recall some concepts and results from the literature.

Definition 11. *Given a $QBF_\forall^2$ formula $\Phi = \forall Y \exists Z \bigwedge_{c \in C} c$, with C a set of clauses, we define the AF $S_\Phi = (A, R)$, where*

$$A = \{t, \bar{t}, b\} \cup C \cup Y \cup \bar{Y} \cup Y' \cup \bar{Y}' \cup Z \cup \bar{Z}$$
$$R = \{(c,t) \mid c \in C\} \cup \{(t, \bar{t}), (\bar{t}, t), (t, b), (b, b)\} \cup$$
$$\{(x, \bar{x}), (\bar{x}, x) \mid x \in Y \cup Z\} \cup$$
$$\{(y, y'), (\bar{y}, \bar{y}'), (y', y'), (\bar{y}', \bar{y}') \mid y \in Y\} \cup$$
$$\{(l, c) \mid literal\ l\ occurs\ in\ c \in C\}.$$

Theorem 2 ([19]). *For a semantics $\sigma \in \{sem, stg\}$, a $QBF_\forall^2$ formula Φ is a valid iff t is skeptically accepted in F_Φ w.r.t. σ iff $\bar{t}$ is not credulously accepted in F_Φ w.r.t. σ.*

Using this we obtain the exact complexity of JS_{sem}, GJS_{sem}, JS_{stg} and GJS_{stg}.

Proposition 10. *The problems JS_{sem}, GJS_{sem} are D_2^P-complete.*

Proof. The membership follows from Theorem 1 and the fact that $Ver_{sem} \in coNP$. To show hardness we reducing the (D_2^P-hard) $QBF_\forall^2$-$coQBF_\forall^2$ problem to JS_{sem}. The

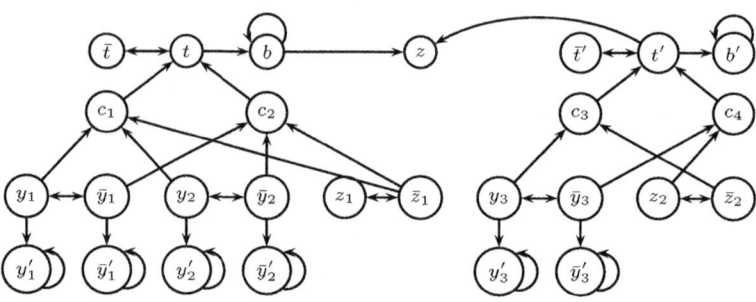

Fig. 4. AF $F_{\varphi,\psi}$ with $c_{\varphi,1}=\{y_1,y_2,\bar{z}_1\}$, $c_{\varphi,2}=\{\bar{y}_1,\bar{y}_2,\bar{z}_1\}$, $c_{\psi,3}=\{y_3,\bar{y}_3\}$, $c_{\psi,4}=\{\bar{z}_1,z_2\}$

$QBF_\forall^2$-co$QBF_\forall^2$ problem remains hard when the $QBFs$ Φ, Ψ are in CNF with satisfiable clause sets ϕ, ψ. In the following we will tacitly assume Φ, Ψ are of that form.

We define the AF $F_{\Phi,\Psi} = \mathcal{S}_\Phi \dot{\cup} \mathcal{S}_\Psi \cup (\{z\}, \{(b,z),(t',z)\})$. Here the symbol "$\dot{\cup}$" denotes the disjoint union, i.e. we rename each argument of $\mathcal{S}_\Psi$ that also occurs in $\mathcal{S}_\Phi$, and denote the renamed argument $t \in \mathcal{S}_\Psi$ by t'. One can see that $E_1 \in sem(\mathcal{S}_\Phi)$, $E_2 \in sem(\mathcal{S}_\Psi)$ iff either $E_1 \dot{\cup} E_2 \in sem(\mathcal{S}_\Psi)$ or $E_1 \dot{\cup} E_2 \cup \{z\} \in sem(F_{\Phi,\Psi})$. We have to show that $(\Phi, \Psi) \in QBF_\forall^2$-co$QBF_\forall^2$ iff $\mathcal{J}\mathcal{S}_{sem}(F_{\Phi,\Psi}, z) = \{in, out\}$.

For the "only if"-part, let us assume (Φ, Ψ) is a valid instance of $QBF_\forall^2$-co$QBF_\forall^2$. Then by Theorem 2 for each $E_1 \in sem(\mathcal{S}_\Phi)$ holds that $t \in E_1$ and thus also for each $E \in sem(F_{\Phi,\Psi})$ holds $t \in E$. As Ψ is not valid, by Theorem 2, there exists $E_2 \in sem(\mathcal{S}_\Psi)$ such that $t' \notin E_2$, hence $\bar{t}' \in E_2$. Thus there exist an $E \in sem(F_{\Phi,\Psi})$ such that $t \in E$ and $\bar{t}' \in E$. But then E defends z and by the maximality of semi-stable extension we have $z \in E$ and therefore $in \in \mathcal{J}\mathcal{S}_{sem}(F_{\Phi,\Psi}, z)$. Further as ψ is satisfiable there also exists an $E_2 \in sem(\mathcal{S}_\Psi)$ such that $t' \in E_2$, hence an $E \in sem(F_{\Phi,\Psi})$ with $t' \in E$ and thus $out \in \mathcal{J}\mathcal{S}_{sem}(F_{\Phi,\Psi}, z)$. As for each $E \in sem(F_{\Phi,\Psi})$ clearly either $t' \in E$ or $\bar{t}' \in E$ we have $undec \notin \mathcal{J}\mathcal{S}_{sem}(F_{\Phi,\Psi}, z)$.

For the "if"-part, let us assume $(\Phi, \Psi) \notin QBF_\forall^2$-co$QBF_\forall^2$. Then either (i) Ψ is valid or (ii) Φ and Ψ are invalid. We have to show that $\mathcal{J}\mathcal{S}_{sem}(F_{\Phi,\Psi}, z) \neq \{in, out\}$.

(i) If Ψ is valid then, by Theorem 2, for each $E_2 \in sem(\mathcal{S}_\Psi)$ it holds that $t' \in E_2$ and thus also for each $E \in sem(F_{\Phi,\Psi})$. Hence $\mathcal{J}\mathcal{S}_{sem}(F_{\Phi,\Psi}, z) = \{out\}$.
(ii) If Φ is not valid then, by Theorem 2, there exists $E_1 \in sem(\mathcal{S}_\Phi)$ such that $t \notin E_1$. If also Ψ is not valid then there exists an $E \in sem(F_{\Phi,\Psi})$ with $t \notin E$ and $t' \notin E$. Thus z is neither attacked by E nor defended E, hence $undec \in \mathcal{J}\mathcal{S}_{sem}(F_{\Phi,\Psi}, z)$.

□

Proposition 11. *The problems JS_{stg}, GJS_{stg} are* D_2^P*-complete.*

Proof. The membership is immediate by Theorem 1 and the fact that $Ver_{stg} \in coNP$.

To prove hardness we reducing the (D_2^P-hard) $QBF_\forall^2$-co$QBF_\forall^2$ problem to JS_{stg}. and as before we assume that the clause sets of the QBFs are satisfiable. For our reduction we define the AF $F_{\Phi,\Psi} = \mathcal{S}_\Phi \dot{\cup} \mathcal{S}_\Psi \cup (\{y,z,g\}, \{(t,y),(t,g),(t',z),(g,g),(y,g),$

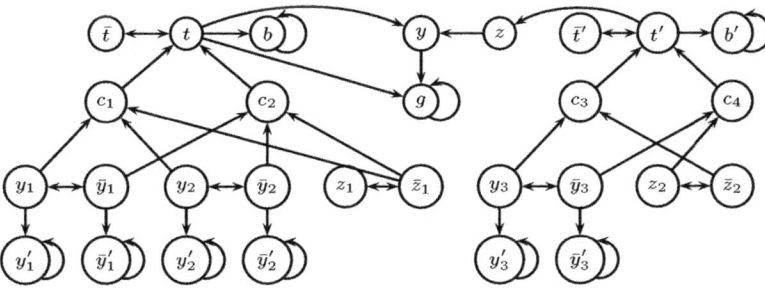

Fig. 5. AF $F_{\varphi,\psi}$ with $c_{\varphi,1} = \{y_1, y_2, \bar{z}_1\}$, $c_{\varphi,2} = \{\bar{y}_1, \bar{y}_2, \bar{z}_1\}$, $c_{\psi,3} = \{y_3, \bar{y}_3\}$, $c_{\psi,4} = \{\bar{z}_1, z_2\}$

$(z,y)\}$) (again $\dot{\cup}$ denotes the disjoint union). One can see that $E_1 \in stg(\mathcal{S}_\Phi)$, $E_2 \in stg(\mathcal{S}_\Psi)$ iff $E_1 \dot{\cup} E_2 \cup T \in stg(F_{\Phi,\Psi})$ with $T \subset \{y,z\}$. We have to show that (Φ, Ψ) is a valid instance of $QBF_\forall^2$-co$QBF_\forall^2$ iff $\mathcal{JS}_{stg}(F_{\Phi,\Psi}, z) = \{in, out\}$.

For the "only if"-part, let us assume $(\Phi, \Psi) \in QBF_\forall^2$-co$QBF_\forall^2$. Then by Theorem 2 for each $E_1 \in stg(\mathcal{S}_\Phi)$ holds that $t \in E_1$ and also for each $E \in stg(F_{\Phi,\Psi})$ holds $t \in E$. As Ψ is not valid, by Theorem 2, there exists $E_2 \in stg(\mathcal{S}_\Psi)$ such that $t' \notin E_2$, hence $\bar{t}' \in E_2$. Thus there exist an extension $E \in stg(F_{\Phi,\Psi})$ such that $t \in E$ and $\bar{t}' \in E$. But then E attacks y and $E \cup \{z\}$ is conflict-free. By the maximality of stage extension we have $z \in E$ and thus $in \in \mathcal{JS}_{stg}(F_{\Phi,\Psi}, z)$. Further as ψ is satisfiable there also exists an $E_2 \in stg(\mathcal{S}_\Psi)$ such that $t' \in E_2$, hence an $E \in stg(F_{\Phi,\Psi})$ with $t' \in E$ and thus $out \in \mathcal{JS}_{stg}(F_{\Phi,\Psi}, z)$. As for each $E \in stg(F_{\Phi,\Psi})$ clearly either $t' \in E$ or $\bar{t}' \in E$ we have $undec \notin \mathcal{JS}_{stg}(F_{\Phi,\Psi}, z)$.

For the "if"-part, let us assume that $(\Phi, \Psi) \notin QBF_\forall^2$-co$QBF_\forall^2$. Then either (i) Ψ is valid or (ii) Φ and Ψ are invalid. We have to show that $\mathcal{JS}_{stg}(F_{\Phi,\Psi}, z) \neq \{in, out\}$.

(i) If Ψ is valid then, by Theorem 2, for each $E_2 \in stg(\mathcal{S}_\Psi)$ it holds that $t' \in E_2$ and thus also for each $E \in stg(F_{\Phi,\Psi})$. Hence $\mathcal{JS}_{sem}(F, z) = \{out\}$.

(ii) If Φ is not valid then, by Theorem 2, there exists $E_1 \in sem(\mathcal{S}_\Phi)$ such that $t \notin E_1$. If also Ψ is not valid then there exists an $E_2 \in sem(\mathcal{S}_\Psi)$ such that $t' \notin E_2$. We have that $E = E_1 \cup E_2 \cup \{y\} \in stg(F_{\Phi,\Psi})$, as it is the only conflict free superset of $E_1 \cup E_2$ such that $g \in E \cup E^+$. As neither $z \in E$ nor $E \rightarrowtail z$ we have that $undec \in \mathcal{JS}_{sem}(F, z)$.

$\square$

We showed that for all semantics σ under our considerations, except grounded semantics, the problems JS_σ, GJS_σ are even harder than NP. Thus one might be interested in tractable fragments, i.e. classes of problem instances that can be solved in polynomial time. First, there are AFs having a special graph structure [14]. It is easy to see that the known tractable fragments for $Cred_\sigma$ and $Skept_\sigma$ (i.e. acyclic AFs, symmetric AFs, bipartite AFs, AFs of bounded tree-width / clique-width) are also tractable fragments for JS_σ, GJS_σ. Second, one can consider instances that test for a fixed justification status. For instance consider $JS_{com}(F, a) = \{in\}$ which can be decided in P instead of using a D^P-algorithm. While a full analysis of such fragments is beyond the scope of this paper, we study the exact complexity of weak acceptance in detail.

4.2   The Complexity of Weak Acceptance

Here we consider the problem of deciding whether an argument is at least weakly accepted. This reasoning mode is in particular of interest, as its more skeptical as credulous acceptance but not as skeptical as skeptical acceptance.

Definition 12. *Given an AF F and an argument $a \in A_F$. We say that a is weakly accepted (in F) if $\mathcal{JS}_\sigma(F, a) \in \{\{in\}, \{in, undec\}\}$. The corresponding weak acceptance problem Weak_σ is, given an AF $F = (A, R)$ and an argument $a \in A$, deciding whether a is weakly accepted.*

Clearly the weak acceptance problem Weak_σ is a version of GJS_σ, where $L = \{in\}$ and $M = \{out\}$. This gives us upper bounds for the complexity of Weak_σ.

Proposition 12. *We have that:*

1. Weak_{grd} *is P-complete,*
2. $\mathsf{Weak}_{adm} = \mathsf{Weak}_{com} = \mathsf{Weak}_{prf}$ *is D^P-complete,*
3. Weak_{stb} *is D^P-complete,*
4. Weak_{sem}, Weak_{stg} *are Π_2^P-complete, and*
5. Weak_{grd^*} *is D^P-complete.*

Proof. (1) By Proposition 1 we have that $\{in, undec\} \notin \mathcal{JS}_{grd}(F, a)$ and therefore $\mathsf{Weak}_{grd} = \mathsf{Skept}_{grd}$. Hence Weak_{grd} is P-complete as well.

(2) By Proposition 3 we have that the justification statuses of an argument w.r.t. admissible, complete, and preferred semantics only differ on the *undec* labels. As weak acceptance of an argument only depends on the *in* and *out* labels we have that $\mathsf{Weak}_{adm} = \mathsf{Weak}_{com} = \mathsf{Weak}_{prf}$. Hence it suffices to consider the Weak_{com} here.

The membership in D^P follows immediately from the complexity of GJS_{com}. To obtain D^P-hardness, let us recall the proof of Proposition 6: There we reduced a pair (φ, ψ) of propositional formulas to an AF F and proved that $(\varphi, \psi) \in$ SAT-UNSAT iff $\mathcal{JS}_{com}(F, t) = \{in, undec\}$. Further we observed that independently of (φ, ψ) it always holds that $undec \in \mathcal{JS}_{com}(F, t)$. That is we have that t is weakly accepted iff $\mathcal{JS}_{com}(F, t) = \{in, undec\}$ iff $(\varphi, \psi) \in$ SAT-UNSAT. Hence Weak_{com} is D^P-hard.

(3) The D^P-completeness for Weak_{stb} follows from the fact that, by Proposition 1, we have that $\{in, undec\} \notin \mathcal{JS}_{stb}(F, a)$ and thus $\mathsf{Weak}_{stb} = \mathsf{Skept}'_{stb}$.

(4) To prove the membership in the class Π_2^P we provide an Σ_2^P-algorithm for the complementary problem, i.e. disproving that an argument a is weakly accepted. To this end we use an alternative characterization of weak acceptance, i.e. an argument a is weakly accepted iff $\{in, out\} \cap \mathcal{JS}_\sigma(F, a) \neq \emptyset$ and $out \notin \mathcal{JS}_\sigma(F, a)$. This may seems a bit weird but can be directly checked by the following Σ_2^P-algorithm:

1. Use the NP-oracle to check whether $\{in, out\} \cap \mathcal{JS}_\sigma(F, a) = \emptyset$ and if so accept
2. Guess a labeling $\mathcal{L}$ such that $\mathcal{L}(a) = out$
3. Use the NP-oracle to verify that $\mathcal{L}$ is a semi-stable (resp. stage) labeling.
4. If $\mathcal{L}$ is a semi-stable (resp. stage) accept otherwise reject.

We first treat the NP membership of step (1). We have that $\{in, out\} \cap \mathcal{JS}_\sigma(F, a) \neq \emptyset$ iff there exists an admissible (resp. conflict-free) set S such that $a \in S^+$. Hence we can test $\{in, out\} \cap \mathcal{JS}_\sigma(F, a) \neq \emptyset$ by guessing a set S, checking whether it is admissible (conflict-free) and $a \in S^+$. This clearly can be done in NP.

For the hardness part, recall the reduction from Definition 11. It is easy to see (and was shown in [19]) that each semi-stable (stage) extension of S_Φ either contains t or $\bar{t}$. So there is no extension that labels t with $undec$ and thus the argument t is weakly accepted in S_Φ iff it is skeptically accepted in S_Φ. By Theorem 2 we obtain Π_2^P-hardness.

(5) The membership in the class D^P follows immediately from the complexity of the corresponding problem GJS_{grd*}. To obtain D^P-hardness, let us recall the proof of Proposition 8: There we reduced a pair (φ, ψ) of formulas to an AF F and proved that $(\varphi, \psi) \in$ SAT-UNSAT iff $\mathcal{JS}_{grd*}(F, t) = \{in, undec\}$ iff t is weakly accepted (as $undec \in \mathcal{JS}_{grd*}(F, t)$ anyway). Hence D^P-hardness carries over to Weak$_{grd*}$. □

Table 2. Overview of Complexity results ($\mathcal{C}$-c denotes completeness for class $\mathcal{C}$)

σ	grd	adm	com	stb	prf	sem	stg	grd*
Cred$_\sigma$	P-c	NP-c	NP-c	NP-c	NP-c	Σ_2^P-c	Σ_2^P-c	NP-c
Skept$'_\sigma$	P-c	trivial	P-c	D^P-c	Π_2^P-c	Π_2^P-c	Π_2^P-c	coNP-c
Weak$_\sigma$	P-c	D^P-c	D^P-c	D^P-c	D^P-c	Π_2^P-c	Π_2^P-c	D^P-c
JS_σ	P-c	D^P-c	D^P-c	D^P-c	$P^{\Sigma_2^P[1]}$-c	D_2^P-c	D_2^P-c	D^P-c
GJS_σ	P-c	D^P-c	D^P-c	D^P-c	$P^{\Sigma_2^P[1]}$-c	D_2^P-c	D_2^P-c	D^P-c

5 Conclusion

In this paper we generalized the labeling-based justification-status, introduced by Wu and Caminada [23], to arbitrary argumentation semantics. In particular we considered the justification-status w.r.t. grounded, admissible, complete, preferred, semi-stable and stage semantics and provided a comparison between different semantics.

We have studied the computational complexity of decision problems associated to the justification status of an argument w.r.t. different semantics (see Table 2). The overall picture is that complexity has slightly increased compared to the complexity of credulous and skeptical acceptance. The main reason for this is that, in contrast to credulous / skeptical acceptance, we have to do both: to determine for labels to be in the justification status, and to determine for other labels not to be in the justification status. In general this causes two orthogonal sources of complexity, with two notable exceptions. First, when considering grounded semantics, we have just one labeling which can be computed in polynomial time and thus reasoning problems remains simple. Second, for preferred semantics we have that one can decide the labels in, out using admissible sets which is much easier than using preferred extensions. However for deciding the label $undec$ admissible sets are not sufficient and we have to use (proof procedures for) preferred extensions. Thus the complexity in case of preferred semantics is dominated by the complexity of deciding *undec*-label. This is also mirrored by the fact that weak acceptance w.r.t. preferred semantics is only D^P-complete.

Our complexity results show that for several semantics weak acceptance is significantly easier than validating the justification status of an argument in general. For preferred semantics weak acceptance is even easier than skeptical acceptance. A similar observation can be made for credulous and skeptical acceptance, which can also be stated in terms of justification statuses, the overall picture of computational complexity is the following: Introducing justification statuses increases the complexity in general, but if we fix the justification statuses we are interested in, we might get back the lower complexity. For practical issues this means that it might be a bad idea (at least from the complexity point of view) to design a general purpose algorithm handling all the cases, but it probably make sense to use procedures that apply simpler algorithms if an instance falls in one of the easier problem class.

An related topic is the combination of weak acceptance with ideal reasoning [18]. The basic idea behind ideal reasoning is to find the maximal admissible set that is part of the skeptical accepted arguments w.r.t. some base-semantics. In [9], the authors introduce the credulous outcome aggregation operator, which basically does ideal reasoning build on weak acceptance instead of skeptical acceptance. So it would be interesting, how switching from skeptical to weak acceptance influences the ideal acceptance of arguments as well as the computational properties of ideal reasoning. Restating the results in [9,18], we have that for standard ideal [13] and eager [6] semantics (and in general for parameterized ideal semantics with a *prf*-preserving base-semantics [18]) switching to weak acceptance does not change anything. On the other hand for complete semantics we have that ideal reasoning with skeptical acceptance leads to grounded semantics, while ideal reasoning with weak acceptance leads to standard ideal semantics. The situation for stage and resolution-based grounded semantics is not that clear, but it is easy to construct examples where ideal reasoning with skeptical acceptance and ideal reasoning with weak acceptance lead different extensions. Thus an interesting direction for future research would be the interaction of acceptance criterions based on a choice of justification statuses, e.g. weak acceptance, with ideal reasoning.

Finally let us mention that the complexity results in Table 2 strongly correlate with expressibility results presented in [20]. There the authors study (faithful) translations between different argumentation semantics and the results there indicate an expressibility hierarchy of admissibility-based semantics. That is we have four levels of expressibility: grounded semantics are on the first level; admissible, complete and stable semantics on the second level; preferred semantics on the third level; and semi-stable semantics are on the fourth level. Ordering the semantics w.r.t. complexity of JS_σ (resp. GJS_σ) leads exactly to the same hierarchy of semantics whereas this neither holds for credulous nor for skeptical acceptance.

Acknowledgments. The author is grateful to Martin Caminada, Stefan Woltran and the anonymous referees for fruitful discussions and valuable comments which helped to improve the paper.

References

1. Baroni, P., Dunne, P.E., Giacomin, M.: On the resolution-based family of abstract argumentation semantics and its grounded instance. Artif. Intell. 175(3-4), 791–813 (2011)
2. Baroni, P., Giacomin, M.: Comparing Argumentation Semantics with Respect to Skepticism. In: Mellouli, K. (ed.) ECSQARU 2007. LNCS (LNAI), vol. 4724, pp. 210–221. Springer, Heidelberg (2007)
3. Baroni, P., Giacomin, M.: Semantics of abstract argument systems. In: Argumentation in Artificial Intelligence, pp. 25–44. Springer, Heidelberg (2009)
4. Caminada, M.: Semi-stable semantics. In: Proc. COMMA 2006, pp. 121–130 (2006)
5. Caminada, M.: On the Issue of Reinstatement in Argumentation. In: Fisher, M., van der Hoek, W., Konev, B., Lisitsa, A. (eds.) JELIA 2006. LNCS (LNAI), vol. 4160, pp. 111–123. Springer, Heidelberg (2006)
6. Caminada, M.: Comparing two unique extension semantics for formal argumentation: ideal and eager. In: Proc. BNAIC 2007, pp. 81–87 (2007)
7. Caminada, M.: A labelling approach for ideal and stage semantics. Argument & Computation 2, 1–21 (2011)
8. Caminada, M., Gabbay, D.M.: A logical account of formal argumentation. Studia Logica 93(2), 109–145 (2009)
9. Caminada, M., Pigozzi, G.: On judgment aggregation in abstract argumentation. Autonomous Agents and Multi-Agent Systems 22(1), 64–102 (2011)
10. Coste-Marquis, S., Devred, C., Marquis, P.: Symmetric Argumentation Frameworks. In: Godo, L. (ed.) ECSQARU 2005. LNCS (LNAI), vol. 3571, pp. 317–328. Springer, Heidelberg (2005)
11. Dimopoulos, Y., Torres, A.: Graph theoretical structures in logic programs and default theories. Theor. Comput. Sci. 170(1-2), 209–244 (1996)
12. Dung, P.M.: On the acceptability of arguments and its fundamental role in nonmonotonic reasoning, logic programming and n-person games. Artif. Intell. 77(2), 321–358 (1995)
13. Dung, P.M., Mancarella, P., Toni, F.: Computing ideal sceptical argumentation. Artif. Intell. 171(10-15), 642–674 (2007)
14. Dunne, P.E.: Computational properties of argument systems satisfying graph-theoretic constraints. Artif. Intell. 171(10-15), 701–729 (2007)
15. Dunne, P.E., Bench-Capon, T.J.M.: Coherence in finite argument systems. Artif. Intell. 141(1/2), 187–203 (2002)
16. Dunne, P.E., Caminada, M.: Computational Complexity of Semi-Stable Semantics in Abstract Argumentation Frameworks. In: Hölldobler, S., Lutz, C., Wansing, H. (eds.) JELIA 2008. LNCS (LNAI), vol. 5293, pp. 153–165. Springer, Heidelberg (2008)
17. Dunne, P.E., Wooldridge, M.: Complexity of abstract argumentation. In: Simari, G., Rahwan, I. (eds.) Argumentation in Artificial Intelligence, pp. 85–104. Springer, US (2009)
18. Dvořák, W., Dunne, P.E., Woltran, S.: Parametric Properties of Ideal Semantics. In: Proc. IJCAI 2011, pp. 851–856 (2011)
19. Dvořák, W., Woltran, S.: Complexity of semi-stable and stage semantics in argumentation frameworks. Inf. Process. Lett. 110(11), 425–430 (2010)
20. Dvořák, W., Woltran, S.: On the intertranslatability of argumentation semantics. J. Artif. Intell. Res. 41, 445–475 (2011)
21. Verheij, B.: Two approaches to dialectical argumentation: admissible sets and argumentation stages. In: Proc. NAIC 1996, pp. 357–368 (1996)
22. Verheij, B.: A Labeling Approach to the Computation of Credulous Acceptance in Argumentation. In: Proc. IJCAI 2007, pp. 623–628 (2007)
23. Wu, Y., Caminada, M.: A labelling-based justification status of arguments. Studies in Logic 3(4), 12–29 (2010)

Arguments over Co-operative Plans

Rolando Medellin-Gasque[1], Katie Atkinson[1],
Peter McBurney[2], and Trevor Bench-Capon[1]

[1] University of Liverpool, Department of Computer Science, Liverpool UK
{medellin,katie,tbc}@liverpool.ac.uk
[2] King's College London, Department of Informatics, London UK
peter.mcburney@kcl.ac.uk

Abstract. Autonomous planning agents that share a common goal should be able to propose, justify and share information about plans. To reach an agreement on the best plan, strategies for persuasion and negotiation could be used by agents in order to share their beliefs about the world and resolve conflicts between the agents. We present an argumentation scheme and associated critical questions to create and justify plan proposals where plans are combinations of actions requiring several agents for their execution. An analysis of different ways in which actions can combine is presented and then associated with the argumentation scheme and the critical questions. We believe these elements are necessary to enable agents to engage in rational debate over co-operative plan proposals.

Keywords: plan proposal, argumentation schemes, critical questions, co-operation.

1 Introduction

Planning in Artificial Intelligence is concerned with the automatic synthesis of action strategies from a description of actions, sensors and goals [11]. The planning literature has been focusing in recent years on overcoming strong assumptions about plan generation. The complexity of distributed systems restricts the application of single-agent planning strategies to distributed problems usually because a local agent view is not sufficient. A common assumption in AI planning is that the planner has accurate and complete knowledge of the world and the capabilities of other agents.

Our goal is to provide autonomous agents (with different views of the world) with a strategy to propose and justify plans in terms of acceptable arguments and enable them to critique and defend plans in order to choose the best option among these. An argumentation based dialogue then is suitable to support some planning tasks such as choosing the best plan, plan modification and even establishing coordination strategies for the execution of a plan.

In this paper we present an argumentation scheme to propose and justify plans based on the argumentation scheme for action proposals of Atkinson *et al.* in [3]. We extend the concept of action used in [3] with action-elements taken from the PDDL 2.1 Planning Specification[1] presented in [10] such as time constraints and invariant conditions. Thus, this work extends the action proposal model of [3] to more complex types

---

[1] Planning Domain Description Language (PDDL) is an attempt to standardize planning domain and problem description languages developed for the International Planning Competitions.

S. Modgil, N. Oren, and F. Toni (Eds.): TAFA 2011, LNAI 7132, pp. 50–66, 2012.
© Springer-Verlag Berlin Heidelberg 2012

of action-proposals involving several durative actions performed by several agents. An analysis over different ways to combine actions to form plans is also presented in order to create more specific critical questions. The analysis is based on interval algebra proposed by Allen in [1]. Allen's interval relations define the basic relations between time intervals. We relate time intervals with the action duration in order to define ways in which actions may be combined in a co-operative plan. Furthermore, we present critical questions grouped in 6 categories that address specific elements of the plan-proposal.

As a basis to formalize our argumentation scheme we will use a formal model developed in [18], an Action-based Alternating Transition System (AATS). This transition system defines actions that may be performed by agents through the states from which these could be performed and the states that will result, with a particular focus on the simultaneous action of a group of agents. This makes AATSs especially suitable for situations where co-operation is important.

The paper is structured as follows: Section 2 presents the action representation and proposal including action combinations. Sub-section 2.3 introduces the AATS notation to formalize the action proposal in sub-section 2.4. In section 3 we present the plan proposal as an argumentation scheme of AATS models together with critical questions associated to the extended action and the ways in which actions can combine. In section 4 we develop an example using the elements presented in this paper. Section 5 comments on related work and finally, in section 6, we conclude the paper and discuss future research work.

2   Action Representation and Proposal

2.1   Action Representation

Actions usually are represented as operations an agent is able to perform from a state where some preconditions hold. We want to extend this action definition and incorporate elements useful when representing and reasoning about temporal plans. We use actions as presented in the PDDL 2.1 specification [10] which have elements to express temporal domain descriptions over plans. In the PDDL 2.1 specification, instead of having an action with preconditions and effects, actions are represented as durative actions with elements to express more precisely temporal conditions and effects. The durative action representation is as follows: initially, the action can start at a point in time when a set of preconditions hold, at the commencement of the action, " start effects" become true. Action has a duration and "invariant conditions" (distinct from preconditions) and are accessible throughout the duration of the action. Actions are not black-boxes and access to start effects is available during the performance of the action. The end of the action is given by "termination conditions" where upon, end effects become true[2]. The planning community is still developing ways to create planners that handle temporally extended actions. Our intention in presenting this durative action representation is to consider all the elements needed by agents to engage in argumentative dialogues

---

[2] This model still represents a simplified model of time; durative actions could be extended to allow effects to be asserted at arbitrary points during the interval of execution, or to be a function of duration ("until" actions).

over co-operative plans. In the next sub-section we will define different ways in which actions may combine to form a plan.

2.2   Action Combinations

We now define the way in which actions can be combined to form plans. By action combinations we mean the different ways in which atomic actions could be combined in a plan in terms of concurrency, repetition and temporal aspects. Even if there is just one action there could be variants such as its periodicity or whether the action execution is optional. Two or more actions could be defined in a plan as a sequence (as in classical AI planning) or as a set of actions with no particular order (partial-order planning) that could, but need not, overlap. We want to cover both cases and others focusing on aspects such as the order of the actions and their periodicity.

The analysis is based on the interval algebra proposed by Allen in [1]. Interval algebra is based on the 13 possible primitive relationships (6 of which are inverses) between two time intervals (Figure 1). We apply a similar model to combinations of actions. Most of the interest in Allen's representation for time intervals comes from a mechanism by which the time relationships between the pairs of intervals can be propagated through the collection of all intervals. The notion of disjunction of interval relationships can be used to declare multiple paths and interactions. This idea gives us reason to think this analysis could be extended for larger plans. We add to Allen's list cases focusing on specific properties such as the periodicity, optionality and interleaving of actions. The 14 cases are presented in the following list, for arbitrary actions α and β:

AC1.- Action α occurs exactly k times, where k is a non-negative integer ($\alpha(k)$).
AC2.- Optionally, action α occurs exactly k times ($\alpha(k, o)$).
AC3.- Action α occurs from 1 to k times, where $k > 1$ ($\alpha((1 - k)$).
AC4.- Optionally, action α occurs from 1 to k times ($\alpha(1 - k, o)$).
AC5.- Action α precedes β ($precedes(\alpha, \beta)$) (Figure 1a).
AC6.- Action α meets β ($meets(\alpha, \beta)$) (Figure 1b).
AC7.- Action α overlaps β ($overlaps(\alpha, \beta)$) (Figure 1c).
AC8.- Action α starts β. ($starts(\alpha, \beta)$) (Figure 1d).
AC9.- Action α is entirely in action β ($entirely(\alpha, \beta)$ (Figure 1e).
AC10.- Action α finishes β. ($finishes(\alpha, \beta)$ (Figure 1f).
AC11.- Action α equals β. ($equals(\alpha, \beta)$ (Figure 1g).
AC12.- Action α or action β but not both ($\alpha|\beta)$)
AC13.- Both actions interleaving concurrently (overlapping) over periods of time until completion of both ($iC(\alpha, \beta)$.
AC14.- Both actions executed not concurrently over periods of time until completion of both. $i(\alpha, \beta)$.

The purpose of this analysis is to cover all of the ways in which actions may be combined in a plan with questions that match the specific action combinations. This analysis covers cases where plans are formed of one or two atomic actions. Perhaps, plans comprising one or two actions seem too simple for the purposes of creating real-world plans,

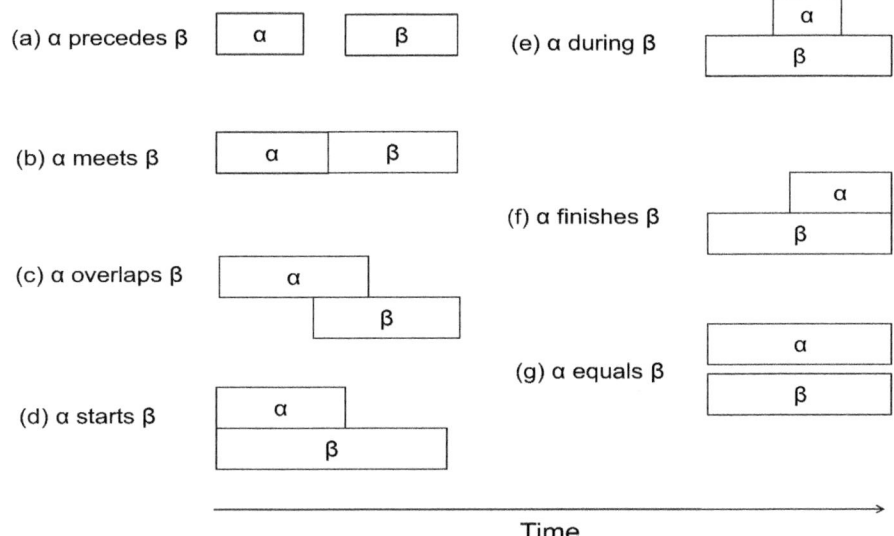

Fig. 1. Allen's possible primitive time relationships between two intervals labelled α and β. Time is represented by the horizontal axis. (a) to (f) have inverses.

but nevertheless, we want to identify basic cases in which actions may be combined before extending this to larger plans.

2.3   Action-Based Alternating Transition Systems

We use Action-based Alternating Transition Systems (AATS) as introduced in [18] as a basis for our formalism to represent action and plan proposals. AATS models define joint-actions that may be performed by agents in a state and the effects of these actions. In particular, an AATS model defines semantic structures useful to represent joint-actions for multiple agents, their preconditions and the states that will result from the transition. An *AATS* is an (n+7)-tuple of the form:

$$S = \langle Q, q_0, Ag, Ac_1, ..., Ac_n, \rho, \tau, \Phi, \pi \rangle$$

where:

- Q is a finite non-empty set of states;
- $q_0 \in Q$ is the initial state;
- $Ag = \{1, ..., n\}$ is a finite non-empty set of agents;
- Ac_i is a finite, non-empty set of actions, for each $i \in Ag$, where $Ac_i \cap Ac_j = \emptyset$ for all $i \neq j \in Ag$; Now we can say that a joint action j_{Ag} for the set of agents Ag is a tuple $(\alpha_i, .., \alpha_n)$ where for each $\alpha_j (j \leq n)$ there is some $i \in Ag$ such that $\alpha_j \in Ac_i$. We denote the set of all joint-actions J_{AG}. Given an element j of J_{AG} and an agent $i \in Ag$, $i's$ action in j is denoted by j_i .

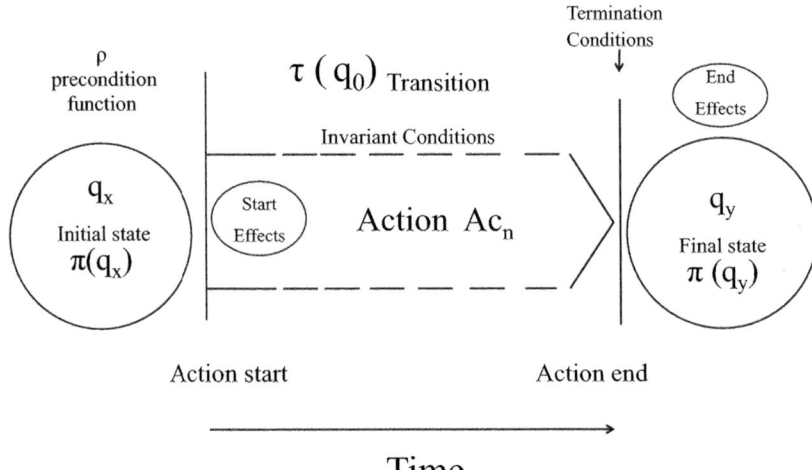

Fig. 2. Action Proposal Representation

- $\rho: Ac_{Ag} \to 2^Q$ is an action precondition function, which for each action $\alpha \in Ac_{Ag}$ defines the set of states $\rho(\alpha)$ from which α may be executed;
- $\tau: Q \times J_{Ag} \to Q$ is a partial system transition function, which defines the state $\tau(q, j)$ that would result by the performance of j from state q, note that, as this function is partial, not all joint actions are possible in all states (cf. the pre-condition function above);
- Φ is a finite, non-empty set of atomic propositions; and
- $\pi: Q \to 2^\Phi$ is an interpretation function, which gives the set of primitive propositions satisfied in each state: if $p \in \pi(q)$, then this means that the propositional variable p is satisfied (equivalently, true) in state q.

In [2] Atkinson and Bench-Capon extended this transition system to enable representation of a theory of practical reasoning related to arguments about action through which values[3] were added to the system. The extensions are:

- Av_i, is a finite, non-empty set of values $Av_i \subseteq V$, for each $i \in Ag$.
- $\delta: Q \times Q \times Av_{Ag} \to \{+, -, =\}$ is a valuation function which defines the status (promoted(+), demoted(-) or neutral (=)) of a value $v_u \in Av_{Ag}$ ascribed by the agent to the transition between two states: $\delta(q_x, q_y, v_u)$ labels the transition between q_x and q_y with one of $\{+, -, =\}$ with respect to the value $v_u \in Av_{Ag}$.

2.4 Proposals for Action

Argumentation schemes are stereotypical patterns of defeasible reasoning used in everyday conversational argumentation. In an argumentation scheme, arguments are presented as general inference rules where under a given set of premises a conclusion can

---

[3] Our use of the term values follows [4] where values are qualitative social interests of agents.

be presumptively drawn [20]. Artificial Intelligence has become increasingly interested in argumentation schemes due to their potential for making significant improvements in the reasoning capabilities of artificial agents [7] and for automation of agent interactions. In [21], Walton explains: *"...arguments need to be examined within the context of an ongoing investigation in dialogue in which questions are being asked and answered"*. Critical questions are a way to examine the acceptability of arguments instantiating schemes. Depending on the nature of the critical question, they can be used to critique several aspects of the argument. Usually, critical questions provide pointers which would make the argumentation scheme inapplicable or could lead to a valid way to attack the argument, either defeating the argument on one of its premises or on its presumptive conclusion.

The action proposal presented in [3] is as follows: In the current circumstances R, we should perform action A to achieve new circumstances S which will realize some goal G which will promote some value v. Furthermore, in [2] the authors re-stated the argumentation scheme in terms of the extended AATS. Figure 2 presents an action as in the PDDL 2.1 specification (presented in section 2.1). So, we can extend the action proposal from [3] with elements from the PDDL 2.1 specification. The extended action proposal and AATS representation are presented in Table 1.

Table 1. Argumentation scheme for actions

Action Proposal	as an AATS model $AS2$
In the current circumstances R	In the initial state $q_0 = q_x \in Q$
we should perform action A at time t with duration d	agent $i \in Ag$ should participate in joint action $j_n \in J_{Ag}$
to achieve start effects from point t	where $j_{ni} = \alpha_i$
given invariant conditions	such that $\tau(q_x, j_n)$ is q_y
action finishing by termination conditions	such that $p_a \in \pi(q_y)$ and $p_a \notin \pi(q_x)$
to achieve new circumstances S	or $p_a \notin \pi(q_y)$ and $p_a \in \pi(q_x)$
which will realize some goal G	such that for some $v_u \in Av_i$, $\delta(q_x, q_y, v_u)$ is +.
which will promote some value v	

For the purpose of this paper, time is discrete and actions take a single time step, thus we will not represent durative actions elements from section 2.1 in the plan proposal in the next section. Nevertheless, in the critical questions' section, time elements are considered. Future work will be focused on representing durative actions within the action and plan proposal and the representation of action elements such as the propositions satisfied during the transition, which do not arise in [3].

3   Plan Proposal and Critical Questions

We now present our argumentation scheme in terms of the action elements presented above. Our plan proposal ASP is as follows:

Given a social context X, in the current circumstances q_0 holding preconditions $\pi(q_0)$, the plan PL should be performed to achieve new circumstances q_x , that will hold postconditions $\pi(q_x)$ which will realize the plan-goal G which will promote value(s) V_G.

The valid instantiation of the scheme pre-supposes the existence of a regulatory environment or a social context X in which the proponent has some rights to engage in a dialogue with the co-operating agent. The "social context" was an extension to the argumentation scheme presented in [5] where agents use a social structure to issue valid commands between them. Current circumstances are represented by the initial state q_0. An agent could instantiate the scheme to propose plan PL as a finite set of linked action-combinations. The plan leads to a state in which post-conditions $\pi(q_x)$ hold and the plan-goal G is achieved (where G is an assignment of truth values to a set of propositions $p \subseteq \Phi$) and a non-empty set of values is promoted/demoted.

Our objective specifying a set of values V_G rather than a single value, comes from the idea that a plan (and the set of actions of which is conformed) might include different preferences for different actions. In other words, a value may be promoted by the first action of a plan and a different value promoted in the second action. So, the set of values promoted by the plan is just the set of values promoted by all the actions that comprise the plan. Indeed, this feature could be extended to allow a more complex value representation for the set of actions, this representation is out of the scope of this paper.

Table 2 presents the plan proposal and the AATS model representation.

Table 2. Plan Proposal ASP

Plan Proposal	**as an AATS model**
Given a social context X, in the current circumstances q_x holding preconditions $\pi(q_x)$ plan PL should be performed to achieve new circumstances q_y that will hold postconditions $\pi(q_y)$ which will realize the plan-goal G which will promote value(s) V_G.	Given social context Δ, In the initial state $q_0 = q_x \in Q$, where $\pi(q_0)$, agents $i, j \in Ag$ should execute plan PL, where PL is a finite set of joint-actions j_n such that $PL = \{j_0, .., j_n\}$ and $\{j_0, .., j_n\} \in J_{Ag}$ and $j_n = \{\alpha_i, .., \alpha_j\}$ with transition given by $\tau(q_x, PL)$ is q_y, where $\tau(q_0, \{j_1, .., j_n\}) = \tau(\tau(q_0, j_1), (j_2, ..j_n))$ and $\tau(q_x, \{\}) = q_x$ such that $p_a \in \pi(q_x)$ and $p_a \notin \pi(q_y)$ where $G = p$ and $(V_G \subseteq V$ such that $v_1 \in V_G$ iff $\delta(q_x, q_y, v_1)$ is $+)$ and $V_G \neq \emptyset$

3.1 Critical Questions for Plan Proposals

A benefit of having critical questions associated with an argument scheme is that the questions enable dialogue participants to identify points of challenge in a debate or locate premises in an instantiation of the argument scheme that can be recognized as questionable. Most of the critical questions are created from argumentation scheme elements and represent a valid way to challenge proposals that could identify sources of disagreement about a particular element of the argumentation scheme. A question can be seen as a weak form of attack on a particular element of the argument scheme given different beliefs about the world of the agent posing the question. Critical questions then could be used to create Dialogue Games for agents where the participants put forward arguments instantiating the argumentation scheme and opponents to the argument challenge it through objections based on critical questions. Argumentation-based dialogues are used to formalize dialogues between autonomous agents based on theories of argument exchange. In [19] a classification is given based on the role the question plays in the context of the argumentation scheme. A question could be used to: criticize a scheme premise, point to exceptional situations in which the scheme should not be used, set conditions for the proper use of the scheme, or point to other arguments that might be used to attack the scheme. Furthermore, questions could argue for an incompatible conclusion like: Are there (better) reasons for not to do plan A? We classify our set of critical questions into 6 layers (also presented in Figure 3).

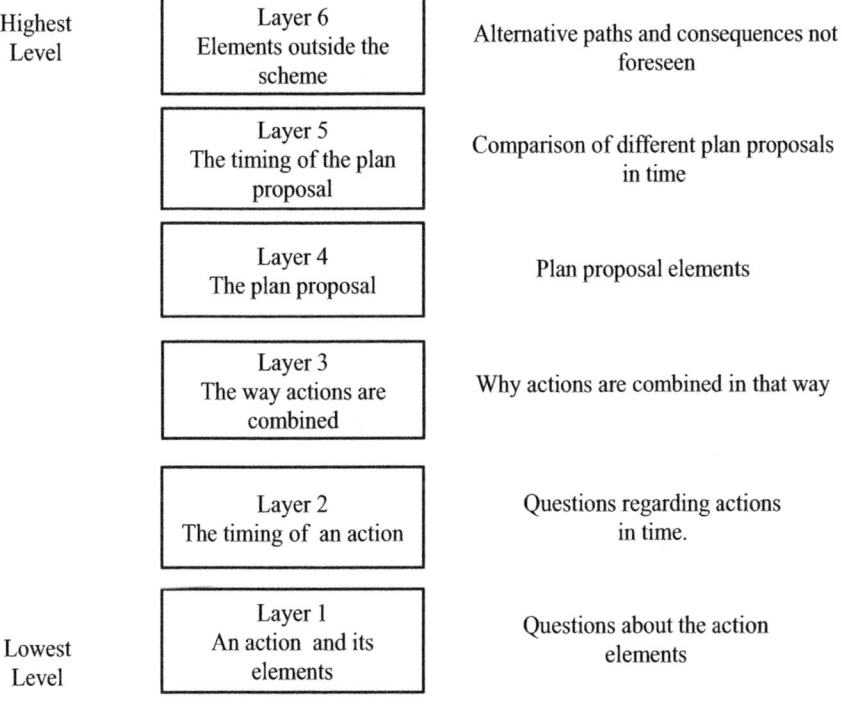

Fig. 3. Critical Question Layers

- Layer 1.- An action and its elements (Lowest level).
- Layer 2.- The timing of a particular action.
- Layer 3.- The way actions are combined.
- Layer 4.- The plan proposal overall.
- Layer 5.- The timing of the plan proposal.
- Layer 6.- Elements outside the scheme (alternative paths or consequences not foreseen) (Highest Level).

The layers are derived from the different categories of critical questions that relate to the different elements of the argumentation scheme. Each layer groups questions according to the level of detail on which they focus. At the plan proposal level, for example, the critical questions are all those that are independent of the way in which actions are composed inside the plan i.e. the way in which actions are combined. This classification allows us then, to consider questions at each layer separately. Furthermore, this classification gives us elements to create a strategy to select critical questions in a dialogue. Having the critical questions classified, an agent could pick a layer and narrow the scope of available questions. An agent then could focus on a specific level of the proposal *e.g.* either the plan proposal or specific sequences of actions. A strategy like this involves a dialogue-protocol where rules to issue such questions are specified. It could be that the answer to a critical question in one layer imposes constraints within another layer, so this may affect the optimal ordering in which the layers are addressed. Appropriate participant strategies, and their possible relationships with the dialogue protocol are the next step with the work.

Our set of critical questions is based on the set developed for action proposals in command dialogues presented in [5]. The complete list of 66 critical questions necessary to comprehensively question all relevant aspects of the plan proposals is presented in [15]. We believe this analysis enables plan proposals to be questioned in a comprehensive way in order to be fully and explicitly justified. We present here some example questions for each layer.

Layer 1. An action α and its elements (9 questions).
Questions aim to find inconsistencies for a particular action questioning or attacking the validity and possibility of its elements.
- CQA-01. Is the action α possible?
- CQA-02. Are the action preconditions as stated by proponent?
- CQA-04. Are the action invariants conditions as stated by proponent?
- CQA-07. Are the termination conditions as described possible?

Layer 2. The timing of action α (10 questions).
Questions also focus on possibility but for a particular time point for which the action has been specified.
- CQAT-02. Is the action possible with the specified duration?
- CQAT-06. What is the earliest time the action α can start?
- CQAT-08. Is the action α possible to finish at the specified time?
- CQAT-09. What is the earliest time the action α can end?

Layer 3. The way actions are combined (7 questions).
The analysis of time intervals from section 2.2 is used here and questions aim to reveal any inconsistencies given the way actions are combined in the plan. - CQAC-01. (For sequential actions) Could actions α and β be performed concurrently?
- CQAC-02. (For sequential actions) Can the order of the actions be changed?
- CQAC-03. (For concurrent actions) Is there a conflict in any of the invariant conditions of the actions?

- CQAC-06. (For concurrent actions) Is there a maximum duration for actions to perform concurrently?

Layer 4. The plan proposal (18 questions).
The questions in this layer aim to question the plan as a single entity with the elements that support it.
- CQPP-01. Is the plan possible?
- CQPP-04. Are the current plan circumstances R as stated by proponent?
- CQPP-12. Assuming believed conditions are true, will the plan bring about the desired state?
- CQPP-14. Can the desired goal G be realized?
- CQPP-16. Are the values V_G legitimate values?

Layer 5. The timing of the plan proposal (11 questions).
Questions focus in the plan possibility given the time specified.
- CQPPT-01. Is the starting point for the plan PL fixed? If not, what is the range allowed?
- CQPPT-05. Can the plan duration be longer?
- CQPPT-06. Is the plan PL possible with the specified duration?
- CQPPT-16. Is the plan PL possible at the specified start time?

Layer 6. Elements outside the scheme (11 questions).
Questions in this layer try to consider other alternatives and side effects not considered.
- CQOS-01. Does performing the plan PL have a side effect which demotes the value v_n?
- CQOS-03. Is there an alternative plan PL that promotes the same value v_n?
- CQOS-05. Is there an alternative plan PL to realize the same goal G?
- CQOS-06. Has the plan been already performed?
- CQOS-07. Does performing the plan promote some other value?
- CQOS-09. Is there another agent that could perform action α?

4   Example

To illustrate our approach we will use our argumentation scheme in the context of agents representing organizations in a conflict zone. The example was first introduced in [8] and also used in [17] to illustrate a similar problem regarding planning and dialogues for autonomous agents. The situation is the following: two agents, one representing a Non-Governmental Organization (NGO) and one representing a peace keeping force (KF), are working in a conflict zone.

The initial conditions are: Agent NGO is based at zone A and agent KF is based at zone C. The joint-goal is that agent NGO reaches zone J safely to help the villagers there. A initial sub-goal is to meet in zone F. The values involved are: v_1 representing humanitarian help and v_2 representing NGO security. The restrictions are: NGO can traverse the routes (A,B),(B,F),(F,H),(I,J) independently, but for all the other routes it

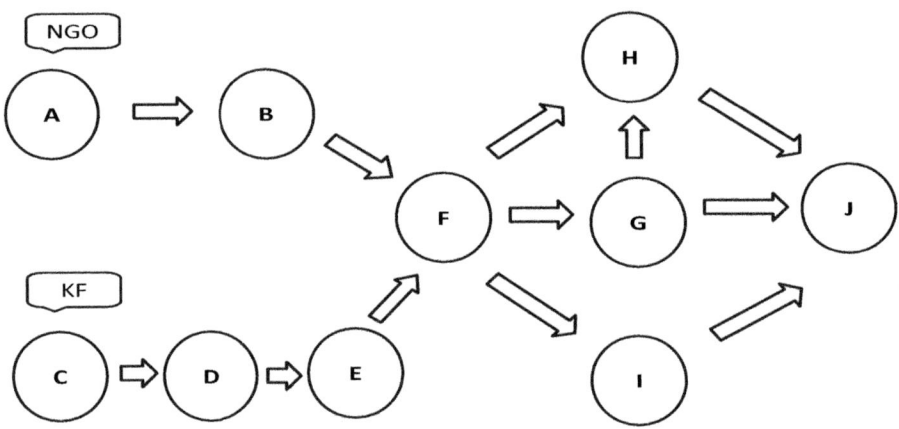

Fig. 4. Example NGO

needs to be accompanied by KF. KF can traverse any route. At any time, some disruption may flare up at zone G. If this happens, only the KF agent has the surveillance data to know this is happening, and must go to zone G to suppress the disturbance. Furthermore, NGO cannot traverse the routes where zone G is involved if there is a conflict. Finally agent NGO is able to see all the zones and routes only when in zone F. The routes between zones are shown as arcs in Figure 4. The list of possible actions and joint-actions is presented in Table 3.

Table 3. Actions and joint-actions

Actions	Joint-actions
$\alpha_0 = move_{NGO}(X, Y)$	$j_0 = (idle_{NGO}, idle_{KF})$
$\alpha_1 = idle_{NGO}(X)$	$j_1 = (idle_{NGO}, control_{KF})$
$\alpha_2 = move_{KF}(X)$	$j_2 = (idle_{NGO}, move_{KF})$
$\alpha_3 = move_{KF}(X, Y)$	$j_3 = (move_{NGO}, idle_{KF})$
$\alpha_4 = control_{KF}(X)$	$j_4 = (move_{NGO}, control_{KF})$
	$j_5 = (move_{NGO}, move_{KF})$

Our strategy to coordinate the agents is based on a persuasion dialogue where the NGO agent proposes a plan and engages in a dialogue where KF needs to accept all the actions in the plan to execute it. Another strategy could involve agents creating a plan from the top following a deliberation dialogue. As mentioned in section 3.1, a dialogue-protocol for engaging in such dialogues is left for future work.

We present now a possible dialogue between 2 agents based on the scenario presented above. Three proposals are presented (plans PL_1, PL_2, PL_3 are detailed in table 4) and evaluated with some questions. Note that the agreement on sub-goals is out of the scope of this paper. Also note the dialogue does not follow any particular protocol. Intuitively the agents engage in a persuasion dialogue where their proposals are questioned. NGO agent acts as the proponent in the beginning and then in step 8 the proponent role switches to agent KF.

Table 4. Plans

Time	Plan
	Plan PL_1 to reach zone F
1	$j_5 = equals(move_{NGO}(A, B), move_{KF}(C, D))$
2	$j_5 = equals(move_{NGO}(B, F), move_{KF}(D, E))$
3	$j_2 = equals(idle_{NGO}(F), move_{KF}(E, F))$
	Plan PL_2 to reach zone J
4	$j_5 = equals(move_{NGO}(F, G), move_{KF}(F, G))$
5	$j_5 = equals(move_{NGO}(G, J), move_{KF}(G, J))$
	Plan PL_3 to reach zone J
4	$j_5 = equals(move_{NGO}(F, H), move_{KF}(F, G))$
5	$j_4 = sodfe(move_{NGO}(H, J), control_{KF}(G))$
	Modified plan PL_3 to reach zone J
4	$j_5 = equals(move_{NGO}(F, H), move_{KF}(F, G))$
5	$j_1 = sodfe(idle_{NGO}(H), control_{KF}(G))$
6	$j_2 = sodfe(idle_{NGO}(H), move_{KF}(G, H))$
7	$j_5 = sodfe(move_{NGO}(H, J), move_{KF}(H, J))$

1. NGO: *I propose PL_1 to reach zone F to promote v_1 humanitarian help.*
2. KF: **(CQOS-01)** *Does performing the plan PL_1 have a side effect which demotes the value $v_2 = security$?*
3. NGO: *None of the effects demotes value v_2.*
4. KF: **(CQPP-05)** *The initial state is not possible, I am not in zone C.*
5. NGO: *Your position is considered as zone C .*
6. KF: *OK, I accept plan PL_1.*
   (From this point we assume the plan PL_1 is executed and agents have a new view of the world. Agents are now in zone F.)
7. NGO: *I propose plan PL_2 to reach zone J (plan-goal) and promote values v_1 and v_2, (different sets of values involved may lead to different plans.)*
8. KF: **(CQPP-17)** *Plan demotes value v_2, I reject the proposal.*
   Agent KF detects a conflict in zone G.
9. NGO *OK.*
   Acknowledge KF rejection.
10. KF: *I propose plan PL_3 to reach zone J.*
11. NGO:*I accept action j_5 of plan PL_3* The plan is then partially accepted by NGO.
12. NGO: **(CQA-04)** *I reject joint-action j_4 = $sodfe(move_{NGO}(H, J), control_{KF}(G))$. Agent cannot travel alone on route (H-J).*
    Invariant conditions are not as stated by proponent.

Action sequences are specified using the action combination analysis of section 2.2. $(sodfe)(j_0, j_1)$ means: joint-action j_0 Starts, Overlaps, During, Finishes or Equals joint-action j_1. From here agents exchange arguments in the action level, assuming the first action of PL_3 was accepted.

13. KF: *I propose a modification to plan PL_3.*
14. NGO: (**CQA-01**) *Is the action control() possible?*
15. KF: *The action control() is possible from the state specified.*
16. NGO:*Accept modification to plan PL_3.*

We believe the exchange of arguments using critical questions in a dialogue allows agents to choose the best possible plan. The detailed dialogue representation is presented in table 5. In the table we represent joint-states with a sub-index $(q_0, .., q_{18})$. Each joint-state represents the state of agent NGO, the state of agent KF and the conflict status. For example, the initial state q_0 is given by the function:
$$\pi(q_0) = \{In(A)_{NGO}, In(C)_{KF}, conflict(0)\}.$$

Table 5. Possible Dialogue for the NGO example

	Agent	Locution	Variables	Comments
1	NGO	ProposePlan(PL_1)	(q_0, PL_1) is q_{11}	Initial state NGO in zone A, KF in zone C.
2			v_1+	Final state NGO and KF in zone F.
3				Promotion of value "humanitarian aid".
4	KF	Question(**CQOS-01**)	$\delta(q_0, q_1, v_2)$ is $-$	Is there a side effect that demotes v_2?
5	NGO	Provide()	$\delta(q_0, q_1, v_2)$ is $=$	Value v_2 is not demoted.
6	KF	Question(**CQPP-05**)	$q_0 \notin Q$	Is the initial state possible?
7	NGO	Provide()	$q_0 \in Q$	q_0 is possible.
8	KF	AcceptProposal(PL_1)		Proposal accepted for plan PL_1
9	NGO	ProposePlan(PL_2)	(q_{11}, PL_2) is q_{16}	Initial state NGO and KF in zone F.
10			$G \in \pi(q_{16})$	Final state NGO and KF in zone J
11			v_1+ and v_2+	goal reached in q_{16}, values promoted
12	KF	RejectProposal()(**CQPP-17**)	PL_2	Plan demoted value v_2
13		Provide()	$\delta(q_{11}, q_{16}, v_2)$ is $-$	Demotion of v_2
14		Provide()	$conflict \in \pi(q_2)$	
15	NGO	AcceptRejection()		
16	KF	ProposePlan(PL_3)	(q_{11}, PL_3) is q_{18}	NGO and KF in zone F.
17			v_1+ and v_2+	NGO and KF in zone J.
18	NGO	AcceptAction()	j_5	
19		RejectAction() (**CQA-04**)	j_4	Invariant conditions are not as stated by proponent.
20		Provide()	$(q_7, j_4) \notin Q$	NGO cannot travel alone proposed route.
21	KF	ProposeActions()	j_1, j_2, j_5	Goal reached in q_{17}
22	NGO	Question(**CQA-01**)	$j_1 \notin J_{AG}$	Is the action $control()$ possible?
23	KF	Provide()	$j_1 \in J_{AG}$	Action $control()$ is possible.
24	NGO	AcceptProposal(PL_3)	j_1, j_2, j_5	

A more complex dialogue could arise given this simple problem. Agents could pose more critical questions and challenge arguments making more rich and complex the interaction. The example is one illustrating how agents could interact with the elements provided in this paper.

5   Related Work

Our approach is influenced by work on argumentation for practical reasoning [2] and dialogues about plans [6,16,17]. Regarding dialogues and plans, Tang, Norman and Parsons in [17] establish a model for individual and joint agents' actions suitable for describing the behaviour of a multi-agent team, including communication actions. Tang *et al's* work has been focused on setting a basis for implementing multi-agent planning dialogues based on argumentation that take into account the communication needs for the plan to be executed successfully. The model uses policies to generate plans and the communication needs are embedded in the policy algorithm generation. From the work of Tang *et al.* we are particularly interested in the techniques used to combine planning and dialogue models using policies. In our approach agents propose and justify previously created plans and then engage in a dialogue to justify the actions and possibly modify the plan. The approach in Tang *et al.* embeds the communication policy in the planning algorithm.

In [6] Belesiotis, Rovatsos and Rahwan develop an argumentation mechanism for reconciling conflicts between agents over plan proposals. The authors extend a proto-col where argument-moves enable discussion about planning steps in iterated dispute dialogues as presented in [9]. The authors then introduce a logic for arguments about plans based on the situation calculus [14]. From this approach we are interested in their protocol based on iterated disputes. We plan to modify the approach extending the way a plan proposal makes use of critical questions in the dispute tree.

Another related approach is presented in [16]. Onaindia *et al.* present the problem of solving cooperative distributed planning tasks through an argumentation-based model. The model allows agents to exchange partial solutions, express opinions on the ade-quacy of candidate solutions and adapt their own proposals for the benefit of the overall task. The argumentation-based model is designed in terms of argumentation schemes and critical questions whose interpretation is given through the semantic structure of a partial order planning paradigm. The approach assumes a lack of uncertainty and deter-ministic planning actions, thus, focuses only on questions concerned with the choice of actions. The argumentation scheme, based on the scheme for action proposal from [3] is of the form:

In the current circumstances and considering the current base plan Π_i, agent ag_i should perform the refinement step Π', which will result in a new partial plan Π_j, which will realise some sub-goals G, which will promote some values V.

Our work is very similar in approach to this work in the sense that plans should be entities treated at a detailed level when arguing about them. We go further and con-sider plan proposals in more detail referring to action elements and combinations of actions. Furthermore, our argumentation scheme is related to a more comprehensive set of critical questions, giving an agent more options to critique and enhance a proposal. We believe these elements allow an agent to question and/or attack the argument in a more targeted fashion, facilitating the modification of more types of plans and specific identification of differences between participants.

6 Conclusion

Our research aims at contributing to solving problems related to multi-agent planning, where agents need to agree on plans given different views of the world and of other agents' capabilities. We believe our main contribution in this paper is that we have articulated a novel list of critical questions related to an argumentation scheme for plan proposals as different combination of actions including temporal aspects. The critical questions address each element of a proposed plan and so they are comprehensive with respect to the representation we have chosen for plan proposals. We believe every component and every interaction of components in our representation of a proposal for plan is subject to a possible critical question.

The importance of this work is that it enables a proposal for plan execution to be considered rationally and automatically by software agents engaged in deliberation over the plan of action. The critical questions enable the proposed plan to be questioned/challenged in a comprehensive and organized manner, and to be clarified or defended in response, as appropriate. Indeed, it is possible to use the critical questions as the basis for an agent dialogue game protocol in which one participating agent may propose, and then clarify or defend a plan of action, while other agents question or challenge this proposal. For example, Atkinson and colleagues in [3] develop such a dialogue protocol for proposals for single actions. Whether the proposed plan of action survives such questions and attacks in the dialogue will depend upon the facts about the world underlying the proposal, and the ability of the proponent agent to defend his proposal from attack. Consequently, the acceptability or otherwise of the proposed plan will depend upon the outcome of the multi-agent dialogue based upon the critical questions, and vice-versa.

The multi-agent dialogue is a form of game-theoretic semantics for the statement of a plan of action in the same way as Hintikka's game-theoretic semantics for first-order logic [12] interprets well-formed formulae involving existential and universal quantifiers as equivalent to two-party games between a proponent and an opponent of some proposition. Our approach will interpret proposals for plans in terms of dialogue games between agents defending and attacking the proposal. Our work in this paper can therefore be seen as part of a larger effort to develop computational semantics for plans of actions between interacting software agents [13].

Future work includes analysis on how to represent formally action elements not yet accounted for in the formalization, such as the duration of actions and action invariants. One limitation in this work is that we only considered plans comprising two actions, effectively a plan for each agent; how to decompose these plans into a number of actions and the issues that arise from the interaction of their components is something we will consider in the next phase of our research. To support this theory we will also implement a prototype where agents use a protocol that allows them to engage in dialogues about plan proposals in a single solution.

Acknowledgements. Rolando Medellin-Gasque is grateful for financial assistance from CONACYT of Mexico. Katie Atkinson and Trevor Bench-Capon were partially supported by the FP7-ICT-2009-4 Programme, IMPACT Project, Grant Agreement Number 247228. The views expressed are those of the authors and are not necessarily representative of the project.

References

1. Allen, J.F.: Towards a general theory of action and time. Artificial Intelligence 23(2), 123–154 (1984)
2. Atkinson, K., Bench-Capon, T.: Practical reasoning as presumptive argumentation using action based alternating transition systems. Artificial Intelligence 171(10-15), 855–874 (2007)
3. Atkinson, K., Bench-Capon, T., McBurney, P.: A dialogue game protocol for multi-agent argument over proposals for action. Autonomous Agents and Multi-Agent Systems 11(2), 153–171 (2005)
4. Atkinson, K., Bench-Capon, T., McBurney, P.: Computational representation of practical argument. Synthese 152(2), 157–206 (2006)
5. Atkinson, K., Girle, R., McBurney, P., Parsons, S.: Command Dialogues. In: Rahwan, I., Moraitis, P. (eds.) ArgMAS 2008. LNCS, vol. 5384, pp. 93–106. Springer, Heidelberg (2009)
6. Belesiotis, A., Rovatsos, M., Rahwan, I.: A Generative Dialogue System for Arguing About Plans in Situation Calculus. In: McBurney, P., Rahwan, I., Parsons, S., Maudet, N. (eds.) ArgMAS 2009. LNCS, vol. 6057, pp. 23–41. Springer, Heidelberg (2010)
7. Bex, F., Prakken, H., Reed, C., Walton, D.: Towards a formal account of reasoning about evidence: argumentation schemes and generalisations. Artificial Intelligence Law 11(2-3), 125–165 (2003)
8. Burnett, C., Masato, D., Mccallum, M., Norman, T.J., Giampapa, J., Kollingbaum, M.J., Sycara, K.: Agent support for mission planning under policy constraints. In: Proceedings of the Second Annual Conference of the ITA. Imperial College (2008)
9. Dunne, P.E., Bench-Capon, T.: Two party immediate response disputes: properties and efficiency. Artificial Intelligence 149(2), 221–250 (2003)
10. Fox, M., Long, D.: PDDL2.1: An extension to PDDL for Expressing Temporal Planning Domains. Journal Artificial Intelligence Research (JAIR) 20, 61–124 (2003)
11. Geffner, H.: Perspectives on Artificial Intelligence Planning. In: Eighteenth National Conference on Artificial intelligence, pp. 1013–1023. American Association for Artificial Intelligence, Menlo Park (2002)
12. Hintikka, J.: The Game of Language: Studies in Game-Theoretical Semantics and Its Applications. Synthese Language Library, vol. 22. D. Reidel, Dordrecht (1983)
13. McBurney, P., Parsons, S.: Dialogue games for agent argumentation. In: Rahwan, I., Simari, G. (eds.) Argumentation in Artificial Intelligence, ch. 13, pp. 261–280. Springer, Berlin (2009)
14. Mccarthy, J., Hayes, P.J.: Some philosophical problems from the standpoint of artificial intelligence. In: Machine Intelligence, pp. 463–502. Edinburgh University Press (1969)
15. Medellin-Gasque, R., Atkinson, K., McBurney, P., Bench-Capon, T.: Critical Questions for Plan Proposals. Technical Report ULCS-11-003, Department of Computer Science, University of Liverpool, UK (March 2011)
16. Onaindía, E., Sapena, O., Torreño, A.: Cooperative Distributed Planning through Argumentation. International Journal of Artificial Intelligence 4, 118–136 (2010)
17. Tang, Y., Norman, T.J., Parsons, S.: A model for integrating dialogue and the execution of joint plans. In: AAMAS 2009: Proceedings of The 8th International Conference on Autonomous Agents and Multiagent Systems, pp. 883–890. International Foundation for Autonomous Agents and Multiagent Systems, Richland (2009)

18. van der Hoek, W., Roberts, M., Wooldridge, M.: Social laws in alternating time: Effectiveness, feasibility, and synthesis. Synthese 156, 1–19 (2007)
19. Verheij, B.: Dialectical argumentation with argumentation schemes: an approach to legal logic. Artificial Intelligence and Law 11(2-3), 167–195 (2003)
20. Walton, D.N.: Argumentation Schemes for Presumptive Reasoning. Lawrence Erlbaum Associates, Mahwah (1996)
21. Walton, D.N.: Justification of argumentation schemes. Australasian Journal of Logic 3 (2005)

An Implemented Dialogue System for Inquiry and Persuasion

Luke Riley[1], Katie Atkinson[1], Terry Payne[1], and Elizabeth Black[2]

[1] Department of Computer Science,
University of Liverpool,
L69 3BX, UK
{L.J.Riley,katie,trp}@liverpool.ac.uk
[2] Department of Information and Computer Sciences,
Utrecht University,
80.089, NL
lizblack@cs.uu.nl

Abstract. In this paper, we present an implemented system that enables autonomous agents to engage in dialogues that involve inquiries embedded within a process of practical reasoning. The implementation builds upon an existing formal model of value-based argumentation, which has itself been extended to permit a wider range of arguments to be expressed. We present extensions to the formal underlying theory used for the dialogue system, as well as the implementation itself. We demonstrate the use of the system through a particular case study. We discuss a number of interesting issues that have arisen from the implementation and the experimental avenues that this test-bed will enable us to pursue.

1 Introduction

Communication through argumentation is one of the key strands of work on computational argumentation. Work on agent-based dialogue systems has been greatly influenced by the dialogue typology of Walton and Krabbe [12]. A number of proposals have been set out for dialogue systems that encompass the main dialogue categories, for example see: [4] for inquiry dialogues; [10] for negotiation; [9] for persuasion; [8] for deliberation. However, very little work has been done on specifying and implementing systems that combine two or more dialogue types. In [3] a formal framework was set out for multi-agent dialogues over actions in which inquiry dialogues over beliefs are combined with persuasion dialogues over actions. The dialogue system allows agents with heterogeneous knowledge to each give input into a decision about how to act to achieve a shared goal. The underlying representation of an argument is in terms of a formal version of an argumentation scheme for practical reasoning, and critical questions that agents can employ to challenge assertions made by their peers. Although this dialogue system has been set out in a formal specification [3], it has not previously been validated through an implementation. In this paper we present the details of an implementation of this dialogue system for inquiry and persuasion over action. For a full implementation to be realised it was necessary to extend the formalism presented in [3] to enable a richer set of arguments to be put forward, which we describe. The implemented system

S. Modgil, N. Oren, and F. Toni (Eds.): TAFA 2011, LNAI 7132, pp. 67–84, 2012.

we present provides not only a proof-of-concept in terms of an application of a formal specification, but we also note a number of issues that have been identified through this the exercise. Furthermore, we consider this implementation to be a starting point for further investigations into agent argumentation dialogues, in particular, with respect to coalition formation.

The paper is structured as follows. In Section 2 we recapitulate from [3] the background material about the dialogue system we have implemented. In Section 3 we present new material that extends the formalism of [3] by providing the full list of critical questions associated with the argumentation scheme that is used in the dialogue system. In Section 4 we describe the implementation and demonstrate it with an example. In Section 5 we discuss issues that have arisen from the implementation, future avenues this work will allow us to explore and we conclude the paper.

2 Background

Our dialogue system allows agents to inquire about beliefs (to determine the state of the world) and collectively perform practical reasoning over an action to perform in a given situation. To do this, agents in our system may have epistemic knowledge (beliefs), represented by Garcia and Simari's Defeasible Logic Programming [6], as well as normative knowledge about the effects of actions.

The following definitions provide the formal framework for modeling beliefs.

Definition 1: *A* **defeasible rule** λ *is denoted* $\alpha_1 \wedge \ldots \wedge \alpha_n \rightarrow \alpha_0$ *where* α_i *is a literal for* $0 \leq i \leq n$. *A* **defeasible fact** *is denoted* α *where* α *is a literal. A* **belief** *is either a defeasible rule or a defeasible fact. We define the following functions* DefeasibleSection$(\lambda) = \{\alpha_1, \ldots, \alpha_n\}$; DefeasibleProp$(\lambda) = \alpha_0$.

The definition of a defeasible derivation is adapted from [6] to work with our assumption that all beliefs are defeasible.

Definition 2: *Let* Ψ *be a set of beliefs and* α *a literal. A* **defeasible derivation** *of* α *from* Ψ, *denoted* $\Psi \mid\sim \alpha$, *is a finite sequence* $\alpha_1, \alpha_2, \ldots, \alpha_n$ *of literals s.t.:* α_n *is* α; *and each literal* α_m $(1 \leq m \leq n)$ *is in the sequence because either* α_m *is a defeasible fact in* Ψ, *or there exists a defeasible rule* $\beta_1 \wedge \ldots \wedge \beta_j \rightarrow \alpha_m$ *in* Ψ *s.t. every literal* β_i $(1 \leq i \leq j)$ *is an element* α_k *preceding* α_m *in the sequence* $(k < m)$.

A b-argument is a minimally consistent set of beliefs from which a claim can be defeasibly derived.

Definition 3: *A* **b-argument** *is denoted* $B = \langle \Phi, \phi \rangle$ *where* ϕ *is a defeasible fact and* Φ *is a set of beliefs s.t.: 1)* $\Phi \mid\sim \phi$; *2)* $\forall \phi, \phi'$ *s.t.* $\Phi \mid\sim \phi$ *and* $\Phi \mid\sim \phi'$, *it is not the case that* $\phi \cup \phi' \vdash \perp$ *(where* $\vdash$ *represents classical implication); and there is no subset of* Φ *satisfying (1 and 2).* Φ *is called the* **support** *of the b-argument and* ϕ *is called the* **claim**.

Each agent has a unique id x taken from a set $\mathcal{I}$ of agent identifiers. Each agent's belief base could be inconsistent.

Definition 4: *A* **belief base** *of an agent* x *is a finite set of beliefs, denoted* Σ^x.

For handling reasoning about the effects of actions, the following argumentation scheme for practical reasoning is used, taken from [1]:

In the current circumstances R, we should perform action A, which will realise goal G, which will result in the new circumstances S, which will promote some value V.

This scheme uses 'values' to describe a social interest an agent has, which will be promoted by moving to a state in which goal G becomes true [2]. An agent may propose an action including its justification by instantiating this scheme. Other agents can then challenge instantiations by posing critical questions (CQ) associated with the scheme. Seventeen critical questions are associated with the above scheme [1] which raise potential issues with: the validity of the elements instantiated in the scheme; the connections between the elements of the scheme; the side effects of actions; and the possible alternatives.

In [3] an agent's knowledge about the effects of actions is represented as a Value-based Alternating Transition System (VATS), a modified version of an Action-Based Transition System (AATS) [13], which has been extended to enable the inclusion of values.

Definition 5: *The VATS formalism is as follows: A* **VATS** *for an agent* x*, denoted* S^x*, is a 9-tuple* $\langle Q^x, q_0^x, Ac^x, Av^x, \rho^x, \tau^x, \Phi^x, \pi^x, \delta^x \rangle$ *s.t.:*

- Q^x *is a finite set of* states;
- $q_0^x \in Q^x$ *is the designated* initial state;
- Ac^x *is a finite set of* actions;
- Av^x *is a finite set of* values;
- $\rho^x : Ac^x \mapsto 2^{Q^x}$ *is an* action precondition function, *which for each action* $a \in Ac^x$ *defines the set of states* $\rho(a)$ *from which* a *may be executed;*
- $\tau^x : Q^x \times Ac^x \mapsto Q^x$ *is a partial* system transition function, *which defines the state* $\tau^x(q, a)$ *that would result by the performance of* a *from state* q*. As this function is partial, not all actions are possible in all states;*
- Φ^x *is a finite set of* atomic propositions;
- $\pi^x : Q^x \mapsto 2^{\Phi^x}$ *is an* interpretation function, *which gives the set of primitive propositions satisfied in each state: if* $p \in \pi^x(q)$*, then this means that the propositional variable* p *is satisfied (equivalently, true) in state* q*; and*
- $\delta^x : Q^x \times Q^x \times Av^x \mapsto \{+, -, =\}$ *is a* valuation function *which defines the* status *(promoted* $(+)$*, demoted* $(-)$*, or neutral* $(=)$*) of a value* $v \in Av^x$ *ascribed by the agent to the transition between two states:* $\delta^x(q, q', v)$ *labels the transition between* q *and* q' *with respect to the value* $v \in Av^x$*.*

Note, $Q^x = \emptyset \leftrightarrow Ac^x = \emptyset \leftrightarrow Av^x = \emptyset \leftrightarrow \Phi^x = \emptyset.$

With its VATS an agent can construct a-arguments for and against actions. Together a-arguments and CQs are referred to as arguments over actions (AOAs).

Definition 6: *An* **a-argument** *constructed by an agent* x *from its VATS* S^x *is a 6-tuple* $A = \langle q_x, a, q_y, p, v, s \rangle$ *s.t.:* $q_x = q_0^x$*;* $a \in Ac^x$*;* $\tau^x(q_x, a) = q_y$*;* $p \in \pi^x(q_y)$*;* $v \in Av^x$*;* $\delta^x(q_x, q_y, v) = s$ *where* $s \in \{+, -, =\}$*.*
We define the following functions: Action$(A) = a$*;* Goal$(A) = p$*;* Value$(A) = v$*;* EndState$(A) = q_y$*;* Polarity$(A) = s$*.*
If Polarity(A) *has the value* $+(-resp.)$*, then we say* A *is an a-argument for (against resp.) action* a *to achieve goal* p*. If* Polarity(A) *has the value "*$=$*", then we say* A *is an a-argument that is neutral with regards to action* a*.*

Our framework assumes a closed cooperative multi-agent system. Agents collaborate to find the best action to achieve the dialogue initiator's goal by entering a *persuasion over action (pAct)* dialogue, which provides the agents with an opportunity to persuade the others by putting forward AOAs for the known possible actions. However, before the AOAs can be asserted, each agent x must inquire over its known propositions so that its initial state can be found. Once this has occurred, the correct AOAs for the current system state can be uttered. To find its initial state each agent x participating in the dialogue first opens an inquiry (inq) sub-dialogue (i.e. a dialogue that is embedded within a top-level dialogue)[1] with the other agents in the system. The result is a truth value for all of agents x's propositions that have not already been discussed in another inq sub-dialogue.

Our dialogue is defined as a dialogue game. Dialogue games typically consist of a set of communicative acts (called moves) and a set of rules that firstly state which moves are legal for any point of the dialogue (the protocol), secondly define the effect of making a move and a lastly determine when a dialogue terminates [7,11]. Within our framework, a dialogue denoted $\mathcal{D}_r^t$, is a sequence of moves $m_r, \ldots, m_t$ where $r, \ldots, t \in \mathbb{N}$ represents the time-point at which each move was made, with r being the starting point of the dialogue and t the end point. If $r = 1$, then this dialogue is considered a top level dialogue whose type is $pAct$, which is opened by the dialogue initiator. If the top level dialogue is closed then the dialogue game is over[2]. If $r \neq 1$ then this is a sub dialogue whose type is inq. The following functions operate over a dialogue[3]:

- Current($\mathcal{D}_1^t$) returns the most recently opened dialogue that has not been closed.
- Type($\mathcal{D}_r^t$) returns the type of the dialogue $\mathcal{D}_r^t$ (i.e. $pAct$ or inq).
- Initiator($\mathcal{D}_r^t$) returns the agent who opened dialogue D_r^t.
- Participants($\mathcal{D}_r^t$) returns the set of agents in the dialogue D_r^t.
- Topic($\mathcal{D}_r^t$) returns the goal the agents are trying to achieve **iff** Type(D_r^t) = $pAct$.
- Topic($\mathcal{D}_r^t$) returns the set of propositions which the agents are jointly trying to find the truth value of **iff** Type(D_r^t) = inq .
- Turn($\mathcal{D}_r^t$) returns the identifer of the agent whose turn it is.

The moves that the agents can perform are presented in Table 1. Agents take it in turns to perform one move at a time. All agents' assertions are stored in their commitment stores (CS) that grow monotonically over time, as follows:

***Definition 7:* Commitment store update.**
For a pAct dialogue with participants $\{x_1, \ldots, x_n\}$, $\forall x \in \{x_1, \ldots, x_n\}$ and a commitment store of agent x at time-point t denoted CS_x^t,

$$CS_x^t = \begin{cases} \emptyset & \textit{iff } t = 0, \\ CS_x^{t-1} \cup \Upsilon & \textit{iff } m_t = \langle x, assert, \Upsilon \rangle, \\ CS_x^{t-1} & \textit{otherwise.} \end{cases}$$

---

[1] Further details of the inquiry sub-dialogues are discussed in Section 3.4.
[2] For future work we will look into opening more than one $pAct$ dialogue before the game is over.
[3] Further dialogue details are given in [3].

Table 1. The format for moves used in this dialogue game, where x represents the agent making the move and either $\theta = pAct$ and γ is a proposition (representing the dialogue goal), or $\theta = inq$ and γ is a set of propositions (that the agent is inquiring over); Λ is a list of agents ($\Lambda = [x_1, \ldots, x_n]$, $\{x_1, \ldots, n_n\} \subseteq \mathcal{I}$); Υ is either a set of a-arguments and critical questions (if $\theta = pAct$) or Υ is a set of b-arguments and beliefs (if $\theta = inq$); and x is an agent ($x \in \mathcal{I}$)

Move	Format
open	$\langle x, open, dialogue(\theta, \gamma, \Lambda)\rangle$
assert	$\langle x, assert, \Upsilon\rangle$
close	$\langle x, close, dialogue(\theta, \gamma, \Lambda)\rangle$

Definition 8: *The union of all the commitment stores is defined as*

$$CSs = \bigcup_{\forall x_i \in \{x_1, \ldots, x_n\}} CS_{x_i}.$$

Dialogues commence when an event triggers one agent to open a *pAct* dialogue through its *pAct* strategy (see Section 3.5) to identify the best action to achieve a given proposition p, where $p = \text{Topic}(D_1^t)$. The other agents that have been included in the open dialogue move then initiate their individual *pAct* strategies, which are guaranteed to find all the arguments related to the dialogue topic, via the use of the *pAct* protocol (Section 3.3) and the *inq* protocol (Section 3.4), before terminating. Both dialogue types will not complete until all agents have made a close move one after another without a different move separating them, as this ensures that the dialogue does not terminate until none of the agents have anything more they want to say.

Once the *pAct* dialogue has terminated, the system evaluates the arguments to determine the maximally consistent acceptable set. This is achieved within our framework using a *Value-Based Argumentation Framework* (VAF) [2], A VAF is an extension of the argumentation framework (AF) of Dung [5]. In an AF an argument is admissible with respect to a set of arguments S if all of its attackers are attacked by some argument in S, and no argument in S attacks an argument in S. In a VAF, an argument succeeds in defeating an argument it attacks only if its value is ranked as high, or higher, than the value of the argument attacked; a particular ordering of the values is characterised as an *audience*. Arguments in a VAF are admissible with respect to an audience A and a set of arguments S if they are admissible with respect to S in the AF which results from removing all the attacks that are unsuccessful given the audience A. A maximal admissible set of a VAF is known as a *preferred extension*.

The output of evaluating a VAF is a recommended action (or non-action) that should be performed to achieve the agents' shared goal. An action can only be recommended by the system if it is present in an AOA that is present in the *preferred extension* and the AOA states that the action promotes a value. In the event that there is more than one acceptable action the choice is offered to the dialogue initiator.

3   Extending the Formalisation of Critical Questions

The dialogue system set out in [3] handled only three of the possible seventeen critical questions associated with the practical reasoning argumentation scheme. Here we extend the dialogue system by specifying all the necessary critical questions and show their use within the dialogue system. The CQs formalised in Section 3.2 that follow Definition 2 are a-arguments also. All CQs can be asserted by any agent to challenge an assertion of any other agent (including itself). If all agents follow the $pAct$ protocol then AOAs can only be asserted if they have not previously been asserted.

3.1   The State Comparison Definition

One particular issue that arose when implementing the dialogue system was the need for a mechanism to clarify how agents who may be using different propositions to represent the state of the world can accurately compare states (since agents' VATS reflect only an individual's representation of the world). As such, we define that two agents, m and n can compare their respective states $q_m \approx q_n$ iff $\pi(q_m) \cap \Phi^n = \pi(q_n) \cap \Phi^m$, otherwise $q_m \not\approx q_n$. The intersection is used to eliminate propositions that reside in only one of the agent's beliefs.

When the above approximation holds, the two states q_n and q_m cannot reasonably be said to be different, as both states will agree for each shared proposition. However, these two states may not be identical as the same conclusion can be reached whatever the truth assignments of the distinct propositions. If the comparison does not hold then the states are different due to both agents holding inconsistent truth assignments for their shared propositions.

This comparison requires either; each agent to have an internal model of the other agent's beliefs, or an instantiated state in an assertion should make explicit all the propositions that the agent holds to be true or false. Both will allow an agent to access the beliefs of another. This paper chooses the latter option due to the ease of implementation of such a representation. There are no privacy issues for this framework as it is designed for a closed and cooperative system. The following shows how the state comparison definition works, when:

$$\Phi^m = \{p, q, r, t\}, \Phi^n = \{p, r, v\}, q_m = [p, \neg q, \neg r, t], q_n = [p, \neg r, v]$$

The state comparison definition:

$$\pi(q_m) \cap \Phi^n = \pi(q_n) \cap \Phi^m$$

The substitution:

$$\{p, t\} \cap \{p, r, v\} = \{p, v\} \cap \{p, q, r, t\}$$
$$\{p\} = \{p\}$$

$Conclusion$: No evidence to suggest the states are different.

3.2   The Additional Critical Questions

We now define arguments that instantiate the remaining critical questions given in [1] that are applicable to our framework (those that are not applicable are discussed subsequent to the presentation of the definitions). Within the formal definitions given below we also give the natural language representation of the questions. Accompanying the definitions are figures that illustrate a situation where each CQ could be posed. All illustrations usually assume (unless otherwise stated) that both agents have in the initial state $\neg p$ (via the interpretation function), the pAct dialogue topic is to achieve p and agent 1 takes the first turn.

Definition 9: **A cq2-argument.** *Answers the question* 'Assuming the circumstances, does the action have the stated consequences?'. *It is constructed from a VATS S^x and denoted* $\langle q_x, a, q_y \rangle$ *s.t.* $q_x = q_0^x$; $a \in Ac^x$, $\tau^x(q_x, a) = q_y$. *It challenges an AOA* $\langle q'_x, a', q'_y, p', v', s' \rangle$ *or an AOA* $\langle q'_x, a', q'_y, v', s' \rangle$ *iff* $q_x \approx q'_x$, $a = a'$, $q_y \not\approx q'_y$.

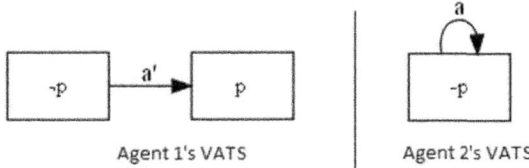

Agent 1's VATS   Agent 2's VATS

Fig. 1. Illustration of a cq2-argument (Definition 9) and a cq15-argument (Definition 21). Agent 2 will assert a cq-argument when agent 1 asserts an AOA to achieve p. The cq-argument that agent 2 chooses depends on which formal conditions are met. No value has been included for this example as values are not part of the definition of a cq2-argument or a cq15-arguments.

Definition 10: **A cq3-argument.** *Answers the question* 'Assuming the circumstances, and the action has the stated consequences, will the action bring about the desired goal?'. *It is constructed from a VATS S^x and denoted* $\langle q_x, a, q_y, \neg p \rangle$ *s.t.* $q_x = q_0^x$; $a \in Ac^x$; $\tau^x(q_x, a) = q_y$; $p \notin (q_y)$. *It challenges an AOA* $\langle q'_x, a', q'_y, p', v', s' \rangle$ *or an AOA* $\langle q'_x, a', q'_y, v', s' \rangle$ *iff* $q_x \approx q'_x$, $a = a'$, $q_y \approx q'_y$.

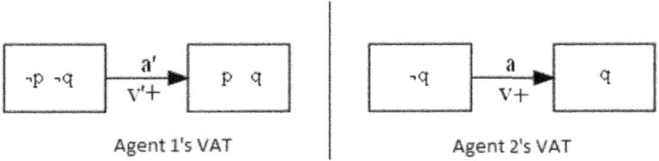

Agent 1's VAT   Agent 2's VAT

Fig. 2. Illustration of a cq3-argument (Definition 10). Agent 2 will assert a cq3-argument when agent 1 asserts an AOA to achieve p. The initial state of Agent 1 is $[\neg p, \neg q]$ and the initial state of Agent 2 is $[\neg q]$.

Definition 11: **A cq4-argument.** *Answers the question* 'Does the goal realise the value stated?'. *It is constructed from a VATS S^x and denoted* $\langle q_x, a, q_y, p, v, s \rangle$ *s.t.* $q_x = q_0^x$; $a \in Ac^x$; $\tau^x(q_x, a) = q_y$; $p \in \pi(q_y)$; $v \in Av^x$; $\delta^x(q_x, q_y, v) \neq +$; $s \in \{=, -\}$. *It challenges an AOA* $\langle q'_x, a', q'_y, p', v', s' \rangle$ *iff* $q_x \approx q'_x$, $p = p'$, $v = v'$, $s' = +$.

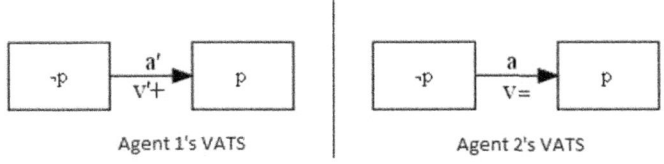

Agent 1's VATS   Agent 2's VATS

Fig. 3. Illustration of a cq4-argument (Definition 11). Agent 2 will assert a cq4-argument when agent 1 asserts an AOA to achieve p.

***Definition 12:* A cq5-argument.** *Answers the question* 'Are there alternative ways of realising the same consequences?'. *It is constructed from a VATS S^x and denoted* $\langle q_x, a, q_y \rangle$ *s.t.* $q_x = q_0^x$; $a \in Ac^x$; $\tau^x(q_x, a) = q_y$. *It challenges an AOA* $\langle q'_x, a', q'_y, v', s' \rangle$ *or an AOA* $\langle q'_x, a', q'_y, p', v', s' \rangle$ *iff* $q_x \approx q'_x$, $a \neq a'$, $q_y \approx q'_y$.

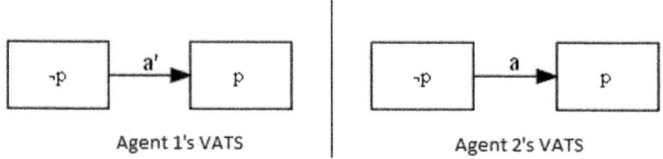

Agent 1's VATS   Agent 2's VATS

Fig. 4. Illustration of a cq5-argument (Definition 12). Agent 2 will assert a cq5-argument when agent 1 asserts an AOA to achieve p. No value has been included for this example as values are not part of the definition of a cq5-argument.

***Definition 13:* A cq6-argument.** *Answers the question* 'Are there alternative ways of realising the same goal?'. *It is constructed from a VATS S^x and denoted* $\langle q_x, a, q_y, p, v, + \rangle$ *s.t.* $q_x = q_0^x$; $a \in Ac^x$; $\tau^x(q_x, a) = q_y$; $p \in \pi(q_y)$; $v \in Av^x$; $\delta(q_x, q_y, v) = +$. *It challenges an AOA* $\langle q'_x, a', q'_y, p', v', s' \rangle$ *iff* $q_x \approx q'_x$, $a \neq a'$, $p = p'$, $s' = +$.

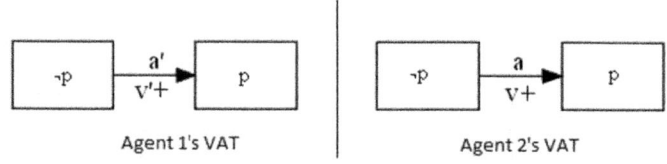

Agent 1's VAT   Agent 2's VAT

Fig. 5. Illustration of cq6-argument (Definition 13), a cq7 argument (Definition 14), a cq10 argument (Definition 17) and a cq-11 argument (Definition 18). Agent 2 will assert a cq-argument when agent 1 asserts an AOA to achieve p. The cq-argument that agent 2 chooses depends on which formal conditions are met. The main difference between CQ6 and CQ7 is whether the agent knows the new action achieves the goal (CQ6) or not (CQ7).

***Definition 14:* A cq7-argument.** *Answers the question* 'Are there alternative ways of promoting the same value?'. *It is constructed from a VATS S^x and denoted* $\langle q_x, a, q_y, v, + \rangle$ *s.t.* $q_x = q_0^x$; $a \in Ac^x$; $\tau^x(q_x, a) = q_y$; $v \in Av^x$; $\delta(q_x, q_y, v) = +$. *It challenges an AOA* $\langle q'_x, a', q'_y, p', v', s' \rangle$ *or an AOA* $\langle q'_x, a', q'_y, v', s' \rangle$ *iff* $q_x \approx q'_x$, $a \neq a'$, $v = v'$, $s' = +$.

See Fig. 5 for an illustration of example VATSs that could produce a cq7-argument.

***Definition 15:* A cq8-argument.** *Answers the question* 'Does doing the action have a side effect which demotes the value?'. *It is constructed from a VATS S^x and denoted* $\langle q_x, a, q_y, v, - \rangle$ *s.t.* $q_x = q_0^x$; $a \in Ac^x$; $\tau^x(q_x, a) = q_y$; $v \in Av^x$; $\delta^x(q_x, q_y, v) = -$. *It challenges an AOA* $\langle q'_x, a', q'_y, p', v', s' \rangle$ *or an AOA* $\langle q'_x, a', q'_y, v', s' \rangle$ *iff* $q_x \approx q'_x$, $a = a'$, $v = v'$, $s' = +$.

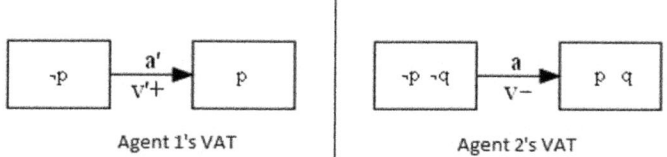

Fig. 6. Illustration of a cq8-argument (Definition 15) and a cq9-argument (Definition 16). Agent 2 will assert a cq-argument when agent 1 asserts an AOA to achieve p. The initial state of agent 2 is $[\neg p, \neg q]$ and the side effect is q. The cq-argument that agent 2 chooses depends on which formal conditions are met.

Definition 16: A **cq9-argument.** *Answers the question* 'Does doing the action have a side effect which demotes some other value?'. *It is constructed from a VATS S^x and denoted* $\langle q_x, a, q_y, v, - \rangle$ *s.t.:* $q_x = q_0^x$; $a \in Ac^x$; $\tau^x(q_x, a) = q_y$; $v \in Av^x$; $\delta^x(q_x, q_y, v) = -$. *It challenges an AOA* $\langle q'_x, a', q'_y, p', v', + \rangle$ *or an AOA* $\langle q'_x, a', q'_y, v', s' \rangle$ *iff* $q_x \approx q'_x$, $a = a'$, $v \neq v'$, $s' = +$.

See Fig. 6 for an illustration of example VATSs that could produce a cq9-argument.

Definition 17: A **cq10-argument** *Answers the question* 'Does doing the action have a side effect which promotes some other value?'. *It is constructed from a VATS S^x and denoted* $\langle q_x, a, q_y, v, + \rangle$ *s.t.:* $q_x = q_0^x$; $a \in Ac^x$; $\tau^x(q_x, a) = q_y$; $v \in Av^x$; $\delta^x(q_x, q_y, v) = +$. *It challenges an AOA* $\langle q'_x, a', q'_y, p', v', s' \rangle$ *or an AOA* $\langle q'_x, a', q'_y, v', s' \rangle$ *iff* $q_x \approx q'_x$, $a = a'$, $v \neq v'$, $s' = +$.

See Fig. 5 for an illustration of example VATSs that could produce a cq10-argument.

Definition 18: A **cq11-argument** *Answers the question* 'Does doing the action preclude some other action which would promote some other value?'. *It is constructed from a VATS S^x and denoted* $\langle q_x, a, q_y, v, + \rangle$ *s.t.* $q_x = q_0^x$; $a \in Ac^x$; $\tau^x(q_x, a) = q_y$; $v \in Av^x$; $\delta(q_x, q_y, v) = +$. *It challenges an AOA* $\langle q'_x, a', q'_y, p', v', s' \rangle$ *or an AOA* $\langle q'_x, a', q'_y, v', s' \rangle$ *iff* $q_x \approx q'_x$, $a \neq a'$, $v \neq v'$, $s' = +$.

See Fig. 5 for an illustration of example VATS that could produce a cq11-argument.

Definition 19: A **cq13-argument.** *Answers the question* 'Is the action possible?'. *It is constructed from a VATS S^x and denoted* $\langle a \rangle$ *s.t.* $a \notin Ac^x$. *It challenges any AOA that includes* q'_x, a', q'_y *in its definition, iff* $a = a'$.

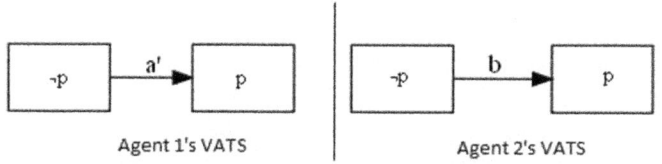

Fig. 7. Illustration of a cq13-argument (Definition 19). Agent 2 will assert a cq13-argument when agent 1 asserts an AOA to achieve p. No values are shown as no values occur in the definition of a cq13-argument.

Definition 20: **A cq14-argument.** *Answers the question* 'Are the consequences as described possible?'. *It is constructed from a VATS S^x and denoted $\langle q_x, a \rangle$ s.t. $q_x = q_0^x$; $a \in Ac^x$; $\tau^x(q_x, a) \notin Q^x$. It challenges any AOA that includes q_x', a', q_y' in its definition, iff $q_x \approx q_{x'}', a = a'$.*

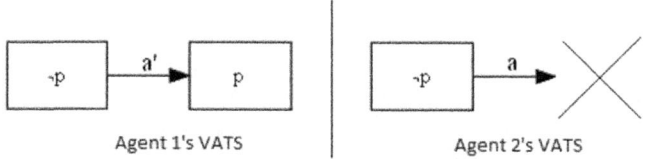

Agent 1's VATS   Agent 2's VATS

Fig. 8. Illustration of a cq14-argument (Definition 21). Agent 2 will assert a cq14-argument when agent 1 asserts an AOA to achieve p. No values are shown as no values occur in the definition of a cq14-argument.

Definition 21: **A cq15-argument.** *Answers the question* 'Can the desired goal be realised?'. *It is constructed from a VATS S^x and denoted $\langle \neg p \rangle$ s.t. $p \in \Phi^x$. It challenges an AOA $\langle q_x', a', q_y', p', v', s' \rangle$ iff $p = p'$ and $(\forall q \in Q^x)(p \notin \pi(q))$.*

See Fig. 1 for an illustration of example VATS that could produce a cq15-argument.

Definition 22: **A cq16-argument.** *Answers the question* 'Is the value indeed a legitimate value?'. *It is constructed from a VATS S^x and denoted $\langle v, - \rangle$ s.t. $v \notin Av^x$. It challenges an AOA $\langle q_x', a', q_y', p', v', s' \rangle$ or an AOA $\langle q_x', a', q_y', v', s' \rangle$ iff $v = v'$.*

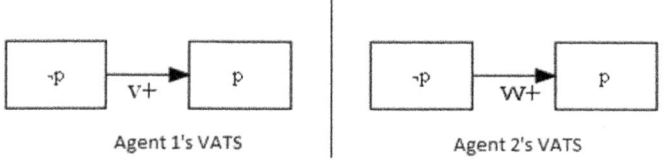

Agent 1's VATS   Agent 2's VATS

Fig. 9. Illustration of a cq16-Argument (Definition 22). Agent 2 will assert a cq16-argument when agent 1 asserts an AOA to achieve p. No actions are shown as an action does not occur in the definition of a cq16-argument.

Missing from the above list are CQ1 (are the believed circumstances true?), CQ12 (are the circumstances as described possible?) and CQ17 (is the other agent guaranteed to execute its part of the desired joint action?); their omission is explained here.

We assume that cooperative agents all accept the outcome of the inquiry dialogues and hence the representation issues concerning conflicting views of the initial state (as raised by CQ1 and CQ12) will be resolved. Note that under this assumption the outcome of the inquiry will be accepted by all agents even though it may be possible for an agent to construct a relevant counter argument.

The actual reasons for agents accepting one b-argument over another should be application dependant. For example in a safety critical system, the presence of one b-argument for a safety critical proposition maybe enough to convince the agents to accept that proposition over its negation. In other applications a simple majority vote could be sufficient. Lastly CQ17 is omitted as the system is not currently concerned with joint actions, though the use of an AATS lends itself to this, as we will explore in future work.

3.3   Extending the pAct Protocol

The protocol this implementation uses, named the *pAct* protocol extends the one presented in [3] by including the extra critical questions so that the agents can use them in a dialogue move. It returns the set of possible moves that are legal for each agent in the dialogue when the current dialogue is of the type *pAct*.

The *pAct* protocol takes the top level dialogue from the set of all dialogues $\mathcal{D}$, the identifier of the agent from the set of all identifiers $\mathcal{I}$ and returns the set of legal moves which is an element of the set of all subsets of the set of all moves $\mathcal{M}$. These definitions are the same for the inquiry protocol. Possible moves for an agent in the *pAct* dialogue are: an assertion of an a-argument to achieve the dialogue goal; an assertion of a cq-argument; a move to close the *pAct* dialogue or a move to open or close a nested *inq* dialogue.

3.4   Defining the Inquiry Protocol

An Inquiry Protocol needs to be formally defined as the details were left out of [3]. This protocol returns the set of possible moves that are legal for each agent in the dialogue when the current dialogue is of the type *Inq*.

This protocol will not allow any proposition to become a claim of a b-argument without supporting evidence. Supporting evidence takes the form of defeasible facts or a fully supported defeasible rule. A defeasible rule λ is fully supported when there is a defeasible derivation for the head of the rule that includes the rule and can be constructed from the union of all the commitment stores.

The protocol works by firstly allowing each agent to assert all its relevant beliefs that are not already present in the commitment store (Ξ_a). A belief can be either a defeasible rule or a defeasible fact. A defeasible fact is relevant if it is an element of the dialogue topic (Ξ_a (2)(ii,a)) or an element of a defeasible rule in the combined commitment store of all the agents (Ξ_a (2)(ii,b)). A defeasible rule is relevant if its consequent returns a defeasible fact that is an element of the dialogue topic (Ξ_a (3)(ii,a)) or an element of another defeasible rule in the CSs (Ξ_a(3)(ii,b)). Secondly in Ξ_b the agent checks to see if any of its asserted beliefs are now fully supported ($\Phi \subseteq CSs$). If they are, these beliefs get asserted as b-arguments in the form $B = \langle \Phi, \phi \rangle$ if they have not been already ($B \notin CSs$). Each agent only asserts a b-argument with claim ϕ if it asserted the belief that included ϕ ($\phi \in CS_x^t$). This is to eliminate multiple assertions of b-arguments. Lastly if the agents cannot assert anything new then the only move that will be returned is the 'close dialogue' move.

Definition 23: *The* **Inquiry protocol** *is a function* $\Xi : \mathcal{D} \times \mathcal{I} \mapsto \wp(\mathcal{M})$. *If* D_1^t *is a top-level dialogue s.t.* $\mathsf{Current}(D_1^t) = D_r^t$, $\mathsf{Turn}(D_r^t) = x$, $\mathsf{Participants}(D_r^t) = \Lambda = [x_1, \ldots, x_n]$, $CSs = \bigcup_{\forall x_i \in \{x_1, \ldots, x_n\}} CS_{x_i}^t$, $\mathsf{Type}(D_r^t) = Inq$, $\mathsf{Topic}(D_r^t) = \Phi^{\mathsf{Initiator}(D_r^t)}$ *and* $1 \leq t$, *then* $\Xi(D_1^t, x)$ *is*

$$\varXi_{\mathsf{a}}(D_1^t, x) \cup \varXi_{\mathsf{b}}(D_1^t, x) \cup \{\langle x, close, dialogue(Inq, \varPhi^{\mathsf{Initiator}(D_r^t)}, \varLambda)\rangle\}$$

where

$\varXi_{\mathsf{a}}(D_1^t, x) = \{\langle x, assert, \varPhi\rangle|$
    (1) $\varPhi \neq \emptyset$ where $\varPhi$ is a set of beliefs, and
    (2) $\forall \phi \in \varPhi$ where ϕ is a defeasible fact:
        (i) $\phi \notin CSs, \phi \in \varSigma^x$, and
        either (ii,a) $\phi \in \mathsf{Topic}(D_r^t)$,
        or (ii,b) $\exists \lambda \in CSs$ s.t. $\phi \in \mathsf{DefeasibleSection}(\lambda)$
    (3) $\forall \lambda \in \varPhi$ where λ is a defeasible rule:
        (i) $\lambda \notin CSs, \lambda \in \varSigma^x$, and
        either (ii,a) $\mathsf{DefeasibleProp}(\lambda) \in \mathsf{Topic}(D_r^t)$,
        or (ii,b) $\exists \lambda' \in CSs$ s.t. $\mathsf{DefeasibleProp}(\lambda) \in \mathsf{DefeasibleSection}(\lambda')$
$\varXi_{\mathsf{b}}(D_1^t, x) = \{\langle x, assert, \varUpsilon\rangle|$
    (1) $\varUpsilon \neq \emptyset, \varUpsilon$ is a set of b arguments , and
    (2) $\forall B \in \varUpsilon: B = \langle \varPhi, \phi\rangle$ is a b-argument, $\varPhi \subseteq CSs, \phi \in CS_x^t$ and $B \notin CSs$

3.5 pAct Strategy

Agents of this system use the *pAct* strategy. This strategy either opens a *pAct* dialogue if the agent is the dialogue initiator or selects one move out of the set of legal moves returned from the correct protocol (the *pAct* protocol if $\mathsf{Type}(\mathsf{Current}(\mathcal{D}_1^t)) == pAct$, else the *inquiry* protocol). The strategy is honest as agents assert only arguments that can be constructed from their knowledge bases.

When using the *pAct* strategy, agents prefer a move to open an inquiry dialogue over assert moves over close moves. This means that once the agents start asserting AOAs, they already know the truth value of all their propositions, due to performing an open *inq* dialogue move first and so all the AOAs presented in the subsequent *pAct* dialogue relate to the actual world state. Also, as the close move is the least preferred no agent will attempt to close the dialogue until it has run out of other moves and so the dialogue is exhaustive.

4 Implementation

The implementation of the framework detailed in this paper uses the *Java Agent DEvelopment Framework (JADE)*[4] to facilitate the storage, modelling and use of the agents' epistemic and normative knowledge at runtime. The user can inspect: the initial Value-based Argumentation Framework (VAF) that is used to evaluate the arguments produced by the agents following the protocol; the preferred extension of the VAF; and the final recommended action. Other elements that can be inspected include: the VATS of all the agents in the dialogue; further details on both the b-arguments, a-arguments and critical questions; and lastly the ability to view the complete resulting dialogue. In addition, the user can modify the value order after the dialogue has terminated, which may result in different recommended actions being generated.

---

[4] http://jade.tilab.com/

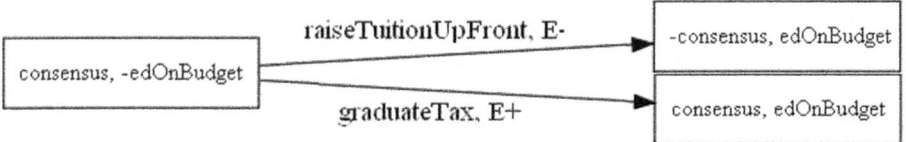

Fig. 10. The VATS for SU. **consensus** represents if the majority of the student union members are happy with the current fee level and **edOnBudget** represents if the educational system is currently on budget.

Agents are modelled within a closed environment, and communicate by broadcasting messages to other agents within the dialogue via a shared blackboard. The use of a blackboard to record communicative acts eliminates the need for agents to individually retain the dialogue history. Agents take turns to update the blackboard, to avoid the need for complex coordination strategies.

All attacks between arguments in the VAF are computed after the dialogue has terminated. As we assume a non-strategic approach, this allows agents to individually assert all relevant arguments prior to the VAF determining the arguments acceptability status. In future work we intend to investigate strategies.

Several issues were identified whilst developing a model to select a final recommended action from the asserted AOAs. Since several critical questions concern *problem formulation* issues [1], not all arguments have an associated value (e.g. those derived from CQ13, CQ15, CQ16); instead these arguments have been implemented to automatically defeat any other argument that they attack, as they are assigned the value 'truth' which always ranks higher than any other value, according to [2].

Finally, scenarios can arise whereby b-arguments may claim logical contradictions (e.g. p and $\neg p$). The current implementation resolves this issue by assuming that the assertion holds (i.e. p) and the contradiction is ignored (i.e. $\neg p$). This conflict strategy was selected because of time constraints but a conflict strategy should be chosen that suits the particular application the dialogue system is deployed in.

4.1 Implementation Example

Our system has so far been evaluated through the use of examples scenarios. One such scenario will now be detailed to show how our system works at runtime.

Consider three agents engaged in a dialogue about the recent UK tuition fee debate. These agents represent the student union SU (Figure 10), the University College Union UCU (Figure 11) and the government GOV (Figure 12). There are three values present in this simplified version of the debate: **E** representing the *equality* of future students when compared to previous ones; **JS** representing *job security* for public sector educational workers; and **NES** representing *national economic security*. Along with these values, there are two possible actions: **raiseTuitionFeesUpFront** representing raising the tuition fees for all new students; and **graduateTax** representing a tax for all workers who hold a degree.

As shown by their respective VATS each agent has different views on this scenario. The SU agent thinks raising tuition fees up front would unfairly affect future students

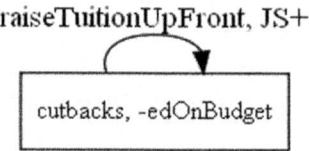

Fig. 11. The VATS for UCU. **cutbacks** represents if budget cutbacks are needed in the majority of universities. **edOnBudget** remains the same as in Fig. 11.

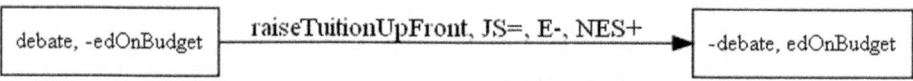

Fig. 12. The VATS for GOV. **debate** represents that there is currently a parliamentary review on how the educational budget can be improved. **edOnBudget** remains the same as in Fig. 11.

and thinks that a graduate tax would be a more appropriate alternative. SU's belief base is as follows:

$$\Sigma^{SU} = \{inFavourFees(X) > againstFees(Y) \rightarrow consensus,$$
$$inFavourFees(80\%), againstFees(20\%)\}$$

The UCU agent does not think that implementing a graduate tax is a workable alternative but on the other hand, it believes that no action will balance the educational budget. UCU's belief base is as follows:

$$\Sigma^{UCU} = \{\neg edOnBudget \rightarrow cutbacks\}$$

Lastly the GOV agent does not view a graduate tax as a possible action but realises that every value will be affected if it decides to raise the tuition fees. GOV's belief base is as follows:

$$\Sigma^{GOV} = \{budgetReview \wedge educationalReview \rightarrow debate,$$
$$budgetReview, educationalReview,$$
$$expenditure(X) \leq estimatedReturn(Y) + acceptableLosses(Z) \rightarrow edOnBudget,$$
$$expenditure(X) > estimatedReturn(Y) + acceptableLosses(Z) \rightarrow \neg edOnBudget,$$
$$expenditure(13.1), estimatedReturn(7), acceptableLosses(3)\}$$

The GOV agent tries to resolve the issue of what action (or non-action) should be recommended by starting a dialogue with the goal being to achieve $edOnBudget$ i.e. keeping education spending on budget. The value order that is used in this example is **NES** $\succ$ **JS** $\succ$ **E**. The dialogue order is SU followed by UCU and lastly GOV. For this example CQ16 is not asserted since it is assumed that all the values are recognised by all the agents.

Firstly the agents all open inquiry dialogues to find their correct state. Agent SU opens an inquiry dialogue to find the truth values of $consensus$ and $edOnBudget$.

The beliefs and b-arguments asserted are:

B1: (a defeasible rule asserted by GOV) $expenditure(X) \leq estimatedReturn(Y)$ $+ acceptableLosses(Z) \rightarrow edOnBudget$.

B2: (a defeasible rule asserted by SU) $inFavourFees(X) > againstFees(Y) \rightarrow$ $consensus$.

B3: (a defeasible fact asserted by GOV) $expenditure(13.1)$.

B4: (a defeasible fact asserted by GOV) $estimatedReturn(7)$.

B5: (a defeasible fact asserted by GOV) $acceptableLosses(3)$.

B6: (a defeasible fact asserted by SU) $inFavourFees(80\%)$.

B7: (a defeasible fact asserted by SU) $againstFees(20\%)$.

B8: (a b-argument asserted by SU) $\langle \{againstFees(20\%), inFavourFees(80\%),$ $inFavourFees(X) > againstFees(Y) \rightarrow consensus\}, consensus \rangle$.

With no other b-arguments possible, the conclusion of this inquiry dialogue is that only $consensus$ is true in the current state. As no b-arguments could be generated for $edOnBudget$ and our implementation operates under the closed world assumption then $edOnBudget$ is presumed to be false.

For the inquiry dialogues all shared propositions are only discussed once because the agents' protocols are exhaustive and so each shared proposition would always find the same truth value in each discussion.

The beliefs and b-arguments asserted in UCU's inquiry dialogue, which has been opened to find/confirm the truth values of $cutbacks$ and $edOnBudget$ are:

B9: (a defeasible rule asserted by UCU) $\neg edOnBudget \rightarrow cutbacks$.

B10: (a defeasible rule asserted by GOV) $expenditure(X) < estimatedReturn(Y)$ $+ acceptableLosses(Z) \rightarrow \neg edOnBudget$.

B11: (a b-argument asserted by GOV) $\langle \{acceptableLosses(3), estimatedReturn(7),$ $expenditure(13.1), expenditure(X) > estimatedReturn(Y) + acceptable$ $Losses(Z) \rightarrow \neg edOnBudget\}, \neg edOnBudget \rangle$.

B12: (a b-argument asserted by UCU) $\langle \{acceptableLosses(3), estimatedReturn(7),$ $expenditure(13.1), expenditure(X) > estimatedReturn(Y) + acceptable$ $Losses(Z) \rightarrow \neg edOnBudget, \neg edOnBudget \rightarrow cutbacks\}, cutbacks \rangle$.

The beliefs and b-arguments asserted in GOV's inquiry dialogue, which has been opened to find/confirm the truth values of $debate$ and $edOnBudget$ are:

B13: (a defeasible rule asserted by GOV) $budgetReview \wedge educationalReview \rightarrow$ $debate$.

B14: (a defeasible fact asserted by GOV) $budgetReview$.

B15: (a defeasible fact asserted by GOV) $educationalReview$.

B16: (a b-argument asserted by GOV) $\langle \{educationalReview, budgetReview,$ $budgetReview \wedge educationalReview \rightarrow debate\}, debate \rangle$.

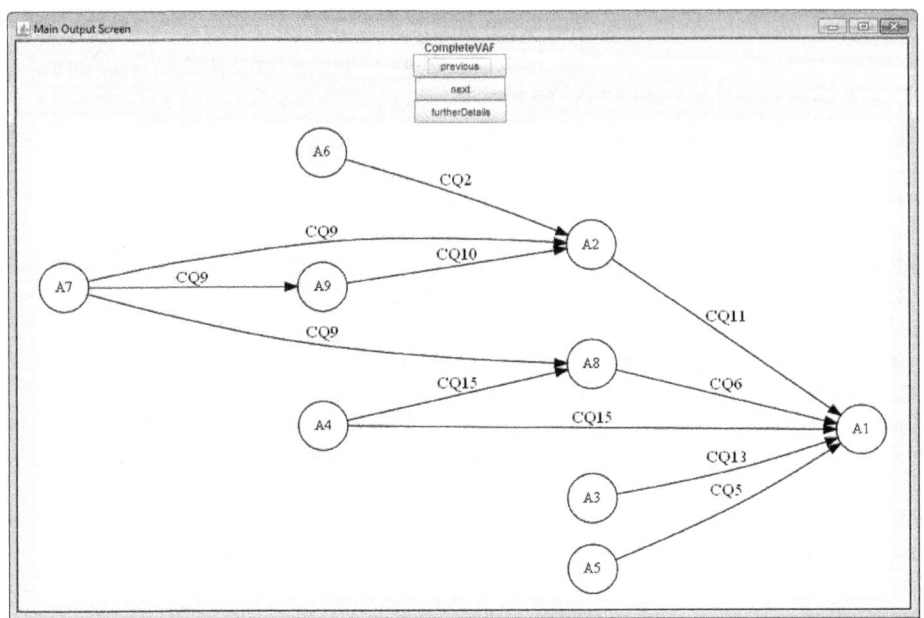

Fig. 13. VAF for the example dialogue. The nodes and edges represent the AOAs and attacks repectively. Each edge is labelled with its attack type.

After all the inquiry dialogues are complete, the agents will start to produce arguments over what actions to perform using the $pAct$ dialogue. The AOAs that are constructed in this example are [5]:

A1: (a-argument asserted by SU) As we are in state $[consensus, \neg edOnBudget]$, we should implement a **graduateTax**, which will achieve $edOnBudget$ and promote **equality**.

A2: (cq11-argument asserted by UCU) As we are in state $[cutbacks, \neg edOnBudget]$, we should **raiseTuitionFeesUpFront** which will promote **job security**.

A3: (cq13-argument asserted by UCU) A **graduateTax** is *not* a possible action.

A4: (cq15-argument asserted by UCU) Achieving $edOnBudget$ is *impossible*.

A5: (cq5-argument asserted by GOV) As we are in state $[debate, \neg edOnBudget]$, **raiseTuitionFeesUpFront**, would achieve the same state as **A1**.

A6: (cq2-argument asserted by GOV) As we are in state $[debate, \neg edOnBudget]$, **raiseTuitionFeesUpFront**, would achieve different state to **A2**.

A7: (cq9-argument asserted by GOV) As we are in state $[debate, \neg edOnBudget]$, **raiseTuitionFeesUpFront**, would demote **equality**.

A8: (cq-6 argument asserted by GOV) As we are in state $[debate, \neg edOnBudget]$, we should **raiseTuitionFeesUpFront** which will achieve $edOnBudget$ and promote **national economic security**.

---

[5] The formal characterisation of the dialogue is ommitted solely for reasons of space. The semantic meaning of the arguments can be found in Section 3.2 and Definition 6. The full list of attacks is visualised in Fig. 14.

A9: (cq10-argument asserted by *gov*) As we are in state [*debate,* ¬*edOnBudget*], **raiseTuitionFeesUpFront** will promote **national economic security**.

Now after argument **A9** the dialogue closes as the only further arguments that can be asserted would be repetitions of information already present in the CS. As discussed earlier, the full set of arguments put forward during the dialogue can now be organised into a VAF that shows the attack relations between them, as can be seen in Figure 13.

Evaluating the VAF to determine which arguments are defeated yields the preferred extension and the final recommended action as can be seen in Figure 14.

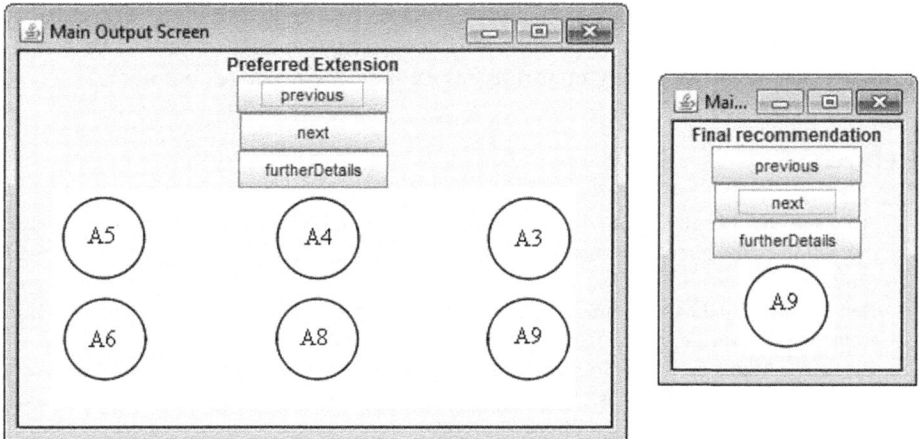

Fig. 14. The attack of **A7** on **A9** has not succeeded as **A9** promotes a higher value than **A7** according to the audience ranking of **NES** ≻ **JS** ≻ **E**. **A9** is then recommended as it is the only argument that promotes a value.

5   Discussion and Concluding Remarks

In this paper we have taken a formal specification of a dialogue system for inquiry and persuasion over action, extended it, and subsequently implemented it to produce a working agent dialogue system. Our contribution is in terms of the extended formalism and the implemented system itself, with both providing grounds for future work.

The inclusion of additional critical questions enables agents to better identify ambiguities within their shared models, and thus construct additional arguments when looking to find some consensus. However, the existence of these additional critical questions could also undermine the ability of agents forming some consensus. This is in part due to the fact that in some contexts, all existing arguments could be defeated by a carefully selected question which, when posed, may result in no recommended action. Thus, it may be possible that the efforts of other agents to arrive at consensus may be undermined by an agent that possesses flawed normative and/or epistemic knowledge. Implementing the stages of practical reasoning from [1] may eliminate this problem, as the first stage (*problem formulation*) would resolve representation issues, the second stage (*epistemic reasoning*) would be represented by the inquiry dialogue and the third stage (*action selection*) would be represented by the persuasion over action dialogue.

An interesting future extension to this work would be to see how this system could be modified to allow for each agent to have its own preference order, instead of the currently implemented single global preference order. This modification could lead to a multi-agent system with self-interested agents, which could be further explored by introducing aspects of coalition formation.

Acknowledgements. Luke Riley is grateful for support from the EPSRC. Katie Atkinson was partially supported by the FP7-ICT-2009-4 Programme, IMPACT Project, Grant Agreement No. 247228. Elizabeth Black is funded by the European Union Seventh Framework Programme (FP7/2007-2011), grant agreement 253911. The views expressed are those of the authors and are not necessarily representative of their project or funding body. We also thank the anonymous referees for their constructive comments.

References

1. Atkinson, K., Bench-Capon, T.J.M.: Practical reasoning as presumptive argumentation using action based alternating transition systems. Artificial Intelligence 171(10-15), 855–874 (2007)
2. Bench-Capon, T.J.M.: Persuasion in practical argument using value based argumentation frameworks. J. of Logic and Computation 13(3), 429–448 (2003)
3. Black, E., Atkinson, K.: Dialogues that account for different perspectives in collaborative argumentation. In: Proceedings of the Eighth International Conference on Autonomous Agents and Multi-Agent Systems, pp. 867–874 (2009)
4. Black, E., Hunter, A.: An inquiry dialogue system. JAAMAS 19(2), 173–209 (2009)
5. Dung, P.M.: On the acceptability of arguments and its fundamental role in nonmonotonic reasoning, logic programming and n-person games. Artificial Intelligence 77, 321–357 (1995)
6. García, A.J., Simari, G.R.: Defeasible logic programming an argumentative approach. Theory and Practice of Logic Prog. 4(1-2), 95–138 (2004)
7. McBurney, P., Parsons, S.: Games that agents play: A formal framework for dialogues between autonomous agents. J. of Logic, Language and Information 11(3), 315–334 (2002)
8. Hitchcock, D., McBurney, P., Parsons, S.: The eightfold way of deliberation dialogue. International Journal of Intelligent Systems 22(1), 95–132 (2007)
9. Prakken, H.: Formal systems for persuasion dialogue. KER 21(2), 163–188 (2006)
10. Rahwan, I., Ramchurn, S.D., Jennings, N.R., McBurney, P., Parsons, S., Sonenberg, L.: Argumentation-based negotiation. KER 18(4), 343–375 (2003)
11. Reed, C.: Dialogue frames in agent communications. In: 3rd Int. Conf. on Multi-Agent Systems, pp. 246–253 (1998)
12. Walton, D., Krabbe, E.: Commitment in dialogue: Basic concepts of interpersonal reasoning. State University of New York Press, Albany (1995)
13. Wooldridge, M., van der Hoek, W.: On obligations and normative ability: Towards a logical analysis of the social contract. J. of Applied Logic 3, 396–420 (2005)

An Argumentation Framework for Qualitative Multi-criteria Preferences

Wietske Visser, Koen V. Hindriks, and Catholijn M. Jonker

Man Machine Interaction Group, Delft University of Technology, The Netherlands
{Wietske.Visser,K.V.Hindriks,C.M.Jonker}@tudelft.nl

Abstract. Preferences between different alternatives (products, decisions, agreements etc.) are often based on multiple criteria. Qualitative Preference Systems (QPS) is a formal framework for the representation of qualitative multi-criteria preferences in which a criterion's preference is defined based on the values of attributes or by combining multiple subcriteria in a cardinality-based or lexicographic way. In this paper we present a language and reasoning mechanism to represent and reason about such qualitative multi-criteria preferences. We take an argumentation-based approach and show that the presented argumentation framework correctly models a QPS. Then we extend this argumentation framework in such a way that it can derive missing information from background knowledge, which makes it more flexible in case of incomplete specifications.

1 Introduction

In the context of practical reasoning, such as decision making and negotiation, preferences between the available alternatives play a key role. A system supporting a human user in such tasks should therefore have a representation of that user's preferences. In this paper we present an argumentation framework to represent and reason with *qualitative, multi-criteria* preferences. Preferences are modelled in a qualitative way because it is hard for humans to give exact numeric utilities. We use multiple criteria because it is a very natural thing to compare two alternatives on several criteria and base an overall preference on those comparisons. Criteria thus represent the *underlying interests*, or *reasons* for preferences. Moreover, the outcome space may be so large that it is infeasible to specify preference between outcomes directly.

We briefly present a framework for representing qualitative multi-criteria preferences, called Qualitative Preference Systems. In this framework, preferences between outcomes are determined by combining multiple criteria based on cardinality and lexicographic ordering. Ultimately, the criteria are based on preferences between the values of relevant variables. QPS is a framework that provides a formal definition of qualitative multi-criteria preferences. The aim of this paper is to provide a *language* and *reasoning mechanism* to reason about such qualitative preference systems. In addition, we provide the means of deriving information by default from background knowledge, which is useful when e.g. the outcomes are incompletely specified.

The approach we take is argumentation-based. Argumentation is a kind of defeasible reasoning, which allows for reasoning with incomplete information in a common-sense way, about things that are normally the case. Moreover, argumentation is a natural way

S. Modgil, N. Oren, and F. Toni (Eds.): TAFA 2011, LNAI 7132, pp. 85–98, 2012.

of reasoning for humans. As such, it is suitable for explaining the reasoning of a system to a human user. Finally, argumentation can be used in a persuasion dialogue, for example when multiple agents with different preferences have to agree on a common action.

Note that the argumentation framework presented here is *not* a preference-based argumentation framework (PAF) ins the sense of [1]. In a PAF, preference between arguments are used to determine the success of an attack between them. A similar approach, that considers preferences between rules in the logical language, has been taken in the specific context of decision making [2]. In contrast, the framework presented here aims to reason about preferences between objects outside of the argumentation framework ('outcomes') as opposed to preferences between arguments or logical rules.

The outline of the paper is as follows. In Section 2, we briefly recall qualitative preference systems. Section 3 presents the argumentation framework that provides the means to reason about a QPS. In Section 4 we extend the argumentation framework with background knowledge and the means to derive information by default. Finally, Section 5 concludes the paper.

2   Qualitative Preference Systems

In this section we briefly present qualitative preference systems. The main aim of a QPS is to determine preferences between *outcomes* (or *alternatives*). An outcome is represented as an assignment of values to a set of relevant variables. Every variable has its own domain of possible values. Constraints on the assignments of values to variables are expressed in a knowledge base. Outcomes are defined as variable assignments that respect the constraints in the knowledge base.

The preferences between outcomes are based on multiple *criteria*. Every criterion can be seen as a *reason* for preference, or as a preference from one particular *perspective*. A distinction is made between simple and compound criteria. Simple criteria are based on a single variable. Multiple (simple) criteria can be combined in order to determine an overall preference. In a QPS, this is done with compound criteria. There are two kinds of compound criteria: cardinality criteria and lexicographic criteria. The subcriteria of a cardinality criterion all have equal importance, and preference is determined by counting the number of subcriteria that support it. In a lexicographic criterion, the subcriteria are ordered by priority and preference is determined by the most important subcriteria.

Definition 1. (Qualitative preference system) A *qualitative preference system (QPS)* is a tuple $\langle Var, Dom, K, \Omega, \mathcal{C} \rangle$. *Var* is a finite set of *variables*. Every variable $X \in Var$ has a domain $Dom(X)$ of possible values. K is a set of constraints on the assignments of values to the variables in *Var*. Ω is the set of all outcomes. An *outcome* α is an assignment of a value $x \in Dom(X)$ to every variable $X \in Var$, such that no constraints in K are violated. α_X denotes the value of variable X in outcome α. $\mathcal{C} = \mathcal{C}_s \cup \mathcal{C}_c \cup \mathcal{C}_l$ is a set of criteria, where $\mathcal{C}_s$ contains simple criteria, $\mathcal{C}_c$ contains cardinality criteria and $\mathcal{C}_l$ contains lexicographic criteria. Weak preference between outcomes by a criterion c is denoted by the relation $\succeq_c$. $\succ_c$ denotes the strict subrelation, $\approx_c$ the indifference subrelation.

Definition 2. (Simple criterion) A *simple criterion* c is a tuple $\langle X_c, \geq_c \rangle$, where $X_c \in Var$ is a variable, and $\geq_c$, a preference relation on the possible values of X_c, is a preorder on $Dom(X_c)$. A simple criterion $c = \langle X_c, \geq_c \rangle$ *weakly prefers* an outcome α over an outcome β, denoted $\alpha \geq_c \beta$, iff $\alpha_{X_c} \geq_c \beta_{X_c}$.

Definition 3. (Cardinality criterion) A *cardinality criterion* c is a tuple $\langle C_c \rangle$ where C_c is a nonempty set of criteria (the *subcriteria* of c). A cardinality criterion $c = \langle C_c \rangle$ *weakly prefers* an outcome α over an outcome β, denoted $\alpha \geq_c \beta$, iff $|\{s \in C_c \mid \alpha >_s \beta\}| \geq |\{s \in C_c \mid \alpha \nsucceq_s \beta\}|$.

Definition 4. (Lexicographic criterion) A *lexicographic criterion* c is a tuple $\langle C_c, \rhd_c \rangle$, where C_c is a nonempty set of criteria (the *subcriteria* of c) and $\rhd_c$, a *priority relation* among subcriteria, is a strict partial order (a transitive and asymmetric relation) on C_c. A lexicographic criterion $c = \langle C_c, \geq_c \rangle$ *weakly prefers* an outcome α over an outcome β, denoted $\alpha \geq_c \beta$, iff $\forall s \in C_c (\alpha \geq_s \beta \vee \exists s' \in C_c (\alpha >_{s'} \beta \wedge s' \rhd_c s))$.

3 Argumentation Framework

In this section we present an argumentation framework for reasoning about qualitative multi-criteria preferences as defined in qualitative preference systems. The AF provides the logical language to represent facts about outcomes, criteria and preferences, and the means to construct arguments that infer preferences from certain input.

3.1 Abstract Argumentation Framework

Our argumentation framework is a concrete instantiation of an abstract argumentation framework as defined by Dung [3]. To define which arguments are justified, we use Dung's preferred semantics.

Definition 5. (Abstract argumentation framework) An *abstract argumentation framework (AF)* is a pair $\langle \mathcal{A}, \rightarrow \rangle$ where $\mathcal{A}$ is a set of *arguments* and $\rightarrow$ is a *defeat* relation among those arguments.

Definition 6. (Preferred semantics) A *preferred extension* of an AF $\langle \mathcal{A}, \rightarrow \rangle$ is a maximal (w.r.t. $\subseteq$) set $S \subseteq \mathcal{A}$ such that: $\forall A, B \in S : A \nrightarrow B$ and $\forall A \in S$: if $\exists B \in \mathcal{A} : B \rightarrow A$ then $\exists C \in S : C \rightarrow B$. An argument is credulously (resp. sceptically) *justified* w.r.t. preferred semantics if it is in some (resp. all) preferred extension(s). An argument is overruled if it is not in any extension. We also say that a formula is justified (resp. overruled) iff it is the conclusion of a justified (resp. overruled) argument.

An abstract AF can be instantiated by specifying the *structure of arguments* and the *nature of the defeat relation*. Prakken [4] presents such an instantiation that is itself still abstract: his *argumentation systems* define arguments as inference trees formed by applying inference rules and specify three kinds of defeat. We take the instantiation of an argumentation framework one step further and also define the *logical language* and the specific *inference schemes* that are used.

3.2   Arguments

Arguments are built from formulas of a logical language, that are chained together using inference steps. Every inference step consists of premises and a conclusion. Inferences can be chained by using the conclusion of one inference step as a premise in the following step. Thus a tree of chained inferences is created, which we use as the formal definition of an argument (cf. e.g. [5,4]).

Definition 7. (Argument) An *argument* is a tree, where the nodes are inferences, and an inference can be connected to a parent node if its conclusion is a premise of that node. Leaf nodes only have a conclusion (a formula from the knowledge base), and no premises. A subtree of an argument is also called a *subargument*. inf returns the last inference of an argument (the root node), and conc returns the conclusion of an argument, which is the same as the conclusion of the last inference.

3.3   Defeat

We define two different kinds of defeat: rebuttal and undercut (note that, unlike e.g. [4], in the current framework there is no distinction between *attack* and *defeat*). An argument *rebuts* another argument if its conclusion contradicts a conclusion of the other argument. Which conclusions contradict each other is defined below after the language is introduced. Defeat by rebuttal is mutual. The term undercut is used in different ways in the literature; we use it for the same concept as e.g. [4]. An *undercutter* is an argument for the inapplicability of an inference step made in another argument. Hence, it is a kind of meta-reasoning (the conlusion of an undercutting argument is not part of the object language). Undercut works only one way. Defeat is defined recursively, which means that rebuttal can attack an argument on all its premises and (intermediate) conclusions, and undercut can attack it on all its inferences.

Definition 8. (Defeat) An argument A *defeats* an argument B $(A \to B)$ if $\mathrm{conc}(A)$ and $\mathrm{conc}(B)$ are contradictory (*rebuttal*), or $\mathrm{conc}(A) = \mathrm{`inf}(B)$ is inapplicable' (*undercut*), or A defeats a subargument of B.

3.4   Language

The logical language provides the means to express statements about a the elements of a QPS. For a given QPS $S = \langle Var, Dom, K, \Omega, C \rangle$, the *domain of discourse* is $D = Var \cup \bigcup_{X \in Var} Dom(X) \cup \Omega \cup C$, i.e. variables and their possible values, outcomes and criteria.

We make a distinction between an *input* and *full* language. A knowledge base, which is the input for an argumentation framework, is specified in the input language. The input language allows us to express facts about the outcomes that are considered and details about the criteria that are used. With the full language we can also express preferences. Such statements can be *derived* from a knowledge base with the inference rules that will be introduced in the next section.

Basic expressions of the language (*atoms*) are built from predicates and terms. Let C be a set of *constants*. $i : C \mapsto D$ is an *interpretation function* that assigns an element

Table 1. The predicates in $\mathcal{P}_{in}$ and their interpretation

predicate	interpretation
$\texttt{val}(o,x,y)$	$i(o)_{i(x)} = i(y)$ where $i(o) \in \Omega, i(x) \in Var, i(y) \in Dom(i(x))$
	'the value of variable x in outcome o is y'
$\texttt{sc}(c,x)$	$i(c) \in \mathcal{C}_s, X_{i(c)} = i(x)$ where $i(x) \in Var$,
	'c is a simple criterion on variable x'
$\texttt{valpref}(c,y_1,y_2)$	$i(y_1) \geq_{i(c)} i(y_2)$ where $i(c) \in \mathcal{C}_s, i(y_1), i(y_2) \in Dom(X_{i(c)})$
	'simple criterion c weakly prefers value y_1 over value y_2'
$\texttt{cc}(c)$	$i(c) \in \mathcal{C}_c$
	'c is a cardinality criterion'
$\texttt{lc}(c)$	$i(c) \in \mathcal{C}_l$
	'c is a lexicographic criterion'
$\texttt{sub}(c,c_1)$	$i(c_1) \in C_{i(c)}$ where $i(c) \in \mathcal{C}_c \cup \mathcal{C}_l, i(c_1) \in C$
	'c_1 is a subcriterion of criterion c'
$\texttt{prior}(c,c_1,c_2)$	$i(c_1) \rhd_{i(c)} i(c_2)$ where $i(c) \in \mathcal{C}_l, i(c_1), i(c_2) \in C$
	'subcriterion c_1 has higher priority than subcriterion c_2
	according to lexicographic criterion c'

from the domain of discourse to every constant in C. There are two sets of *predicates*. $\mathcal{P}_{in}$ contains predicates that can be used in the input language. $\mathcal{P}_{out}$ contains predicates that cannot be used in the input language and can only be derived. The predicates in $\mathcal{P}_{in}$ and $\mathcal{P}_{out}$ and their interpretation are in Table 1 and 2.

Formulas of the input language are just atoms of the input language. Formulas of the full language are atoms (A) or weakly negated atoms ($\sim A$). Weak negation is negation as failure: $\sim A$ is justified if A is not. Strong negation is not needed to model qualitative preference systems, but it will be added in the extended version of the AF presented in Section 4 in order to reason with background knowledge.

Definition 9. (Language) The *input language* is defined as follows.
$atom_{in}$ ::= $p(t_1, \ldots, t_n)$ where p is an n-ary predicate $\in \mathcal{P}_{in}$
$literal_{in}$ ::= $atom_{in}$
$formula_{in}$ ::= $literal_{in}$
    The *full language* is defined as follows.
$atom_{out}$ ::= $p(t_1, \ldots, t_n)$ where p is an n-ary predicate $\in \mathcal{P}_{out}$
$literal$ ::= $literal_{in} \mid atom_{out}$
$formula$ ::= $literal \mid \sim literal$

Contradictory Formulas. Two arguments rebut each other if their conclusions are contradictory. There are two ways in which two formulas can be contradictory.

- The formulas specify different values for the same variable in the same outcome: $\texttt{val}(o,x,y)$ and $\texttt{val}(o,x,y')$ contradict each other if $y \neq y'$.
- $\texttt{prior}(c,c_1,c_2)$ and $\texttt{prior}(c,c_2,c_1)$ contradict each other, since priority is asymmetric.

Two other candidates for contradiction are not modelled as such because they are handled in a different way.

Table 2. The predicates in $\mathcal{P}_{out}$ and their interpretation

predicate	interpretation	
$\text{pref}(c,o_1,o_2)$	$i(o_1) \succeq_{i(c)} i(o_2)$	where $i(c) \in \mathcal{C}, i(o_1), i(o_2) \in \Omega$
	'criterion c weakly prefers outcome o_1 over outcome o_2'	
$\text{spref}(c,o_1,o_2)$	$i(o_1) \succ_{i(c)} i(o_2)$	where $i(c) \in \mathcal{C}, i(o_1), i(o_2) \in \Omega$
	'criterion c strictly prefers outcome o_1 over outcome o_2'	
$\text{epref}(c,o_1,o_2)$	$i(o_1) \approx_{i(c)} i(o_2)$	where $i(c) \in \mathcal{C}, i(o_1), i(o_2) \in \Omega$
	'criterion c equally prefers outcome o_1 and outcome o_2'	
$\text{sp}(c,o_1,o_2,n)$	$\|\{s \in C_{i(c)} \mid i(o_1) \succ_s i(o_2)\}\| = n$	where $i(c) \in \mathcal{C}_c, i(o_1), i(o_2) \in \Omega$
	'there are n subcriteria of cardinality criterion c	
	that strictly prefer outcome o_1 over outcome o_2'	
$\text{nwp}(c,o_1,o_2,n)$	$\|\{s \in C_{i(c)} \mid i(o_1) \not\succeq_s i(o_2)\}\| = n$	where $i(c) \in \mathcal{C}_c, i(o_1), i(o_2) \in \Omega$
	'there are n subcriteria of cardinality criterion c	
	that do not weakly prefer outcome o_1 over outcome o_2'	

One might argue that φ and $\sim \varphi$ are contradictory, and hence arguments concluding them should rebut each other. However, the status of these conclusions is not equal. φ has to be derived and is grounded in facts in the knowledge base. $\sim \varphi$ on the other hand is an assumption that can be made in the absence of evidence to the contrary. φ is such evidence to the contrary, and that is why an argument concluding φ undercuts the inference of $\sim \varphi$ instead of rebutting the conclusion (see the inference schemes for weak negation and its undercutter below).

Incompatible preference statements, such as e.g. $\text{spref}(c,o1,o2)$ and $\text{epref}(c,o1,o2)$ will resolve because $\text{epref}(c,o1,o2)$ can only be derived if $\text{pref}(c,o2,o1)$, in which case the $\sim\text{pref}(c,o2,o1)$ premise needed to derive $\text{spref}(c,o1,o2)$ will be undercut. Hence to have such arguments rebut each other would be superfluous.

Input Knowledge Base. An input knowledge base is a set of formulas of the input language. A knowledge base *KB corresponds to* a QPS $S = \langle Var, Dom, K, \Omega, \mathcal{C} \rangle$ if the following condition holds: a formula φ is in *KB* iff its interpretation holds in S. Note that a knowledge base corresponding to a QPS is conflict-free, i.e. does not contain contradictory formulas.

Example 1. *We will use a running example throughout the paper to illustrate the details of the argumentation framework. Anne is planning to go on holiday with a friend. Anne's overall preference is based on three simple criteria: c1: that someone (she or the accompanying friend) speaks the language (s1), c2: that it is sunny (su) and c3: that she has not been there before (bb). c1 and c2 have equal priority, so they are aggregated in a cardinality criterion c4. c3 and c4 are combined in a lexicographic criterion c5 where c3 has higher priority than c4. This information can be represented in the following knowledge base.*

Facts about two of the possible outcomes:

```
val(o1,s1,true)    val(o1,su,true)    val(o1,bb,true)
val(o2,s1,false)   val(o2,su,true)    val(o2,bb,false)
```

Information about the preferences:

```
lc(c5)          cc(c4)          sc(c1,sl)       valpref(c1,true,false)
sub(c5,c3)      sub(c4,c1)      sc(c2,su)       valpref(c2,true,false)
sub(c5,c4)      sub(c4,c2)      sc(c3,bb)       valpref(c3,false,true)
prior(c5,c3,c4)
```

3.5   Inference Rules

In this section we present the inference rules that are used in the argumentation framework to build arguments.

Weak Negation. The following two inference rules make sure that (i) a weakly negated formula can always be derived, but (ii) this inference will be undercut if the formula itself can be derived. So $\sim \varphi$ is sceptically justified iff φ is overruled.

$$\frac{}{\sim \varphi} \; asm(\sim \varphi) \qquad\qquad \frac{\varphi}{asm(\sim \varphi) \text{ is inapplicable}} \; asm(\sim \varphi)uc$$

Strict and Equal Preference. The following inference schemes are used to derive strict and equal preference from weak preference according to the common definitions.

$$\frac{\texttt{pref}(c,o_1,o_2) \quad \sim\texttt{pref}(c,o_2,o_1)}{\texttt{spref}(c,o_1,o_2)} \qquad\qquad \frac{\texttt{pref}(c,o_1,o_2) \quad \texttt{pref}(c,o_2,o_1)}{\texttt{epref}(c,o_1,o_2)}$$

Preference by a Simple Criterion. The following inference rule concludes that a simple criterion prefers one outcome over another if, for the variable that it is based on, it prefers the value of the first outcome over the value of the second. This is exactly the definition of preference by a simple criterion in a QPS.

$$\frac{\texttt{sc}(c,x) \quad \texttt{val}(o_1,x,y_1) \quad \texttt{val}(o_2,x,y_2) \quad \texttt{valpref}(c,y_1,y_2)}{\texttt{pref}(c,o_1,o_2)}$$

Example 2. *The following argument infers that simple criterion* c1 *prefers* o1 *over* o2. *Similar arguments can be constructed for* c2 *and* c3.

$$\frac{\texttt{sc}(c1,sl) \quad \texttt{val}(o1,sl,true) \quad \texttt{val}(o2,sl,false) \quad \texttt{valpref}(c1,true,false)}{\texttt{pref}(c1,o1,o2)}$$

Preference by a Cardinality Criterion. The next inference scheme derives preference by a cardinality criterion according to its definition in a QPS: an outcome o_1 is weakly preferred over an outcome o_2 if there are at least as many subcriteria that strictly prefer o_1 over o_2 as subcriteria that do not weakly prefer o_1 over o_2.

$$\frac{\texttt{cc}(c) \quad \texttt{sp}(c,o_1,o_2,l) \quad \texttt{nwp}(c,o_1,o_2,m) \quad l \geq m}{\texttt{pref}(c,o_1,o_2)}$$

Preference by a cardinality criterion is based on (i) the number of subcriteria that strictly prefer one outcome over the other, and (ii) the number of subcriteria that do not weakly prefer one outcome over the other. The following inference rules provide the required counting mechanism.

The next inference rules conclude that there are n subcriteria of c that strictly prefer o_1 over o_2, resp. that there are n subcriteria of c that do not weakly prefer o_1 over o_2.

$$\frac{\texttt{spref}(c_1,o_1,o_2) \quad \ldots \quad \texttt{spref}(c_n,o_1,o_2) \quad \texttt{sub}(c,c_1) \quad \ldots \quad \texttt{sub}(c,c_n)}{\texttt{sp}(c,o_1,o_2,n)} \; SP(c,o_1,o_2,n)$$

$$\frac{\texttt{\~pref}(c_1,o_1,o_2) \quad \ldots \quad \texttt{\~pref}(c_n,o_1,o_2) \quad \texttt{sub}(c,c_1) \quad \ldots \quad \texttt{sub}(c,c_n)}{\texttt{nwp}(c,o_1,o_2,n)} \; NWP(c,o_1,o_2,n)$$

If there are no subcriteria of c that strictly prefer o_1 over o_2, resp. that do not weakly prefer o_1 over o_2, no premises are needed to infer this.

$$\frac{}{\texttt{sp}(c,o_1,o_2,0)} \; SP(c,o_1,o_2,0) \qquad \frac{}{\texttt{nwp}(c,o_1,o_2,0)} \; NWP(c,o_1,o_2,0)$$

With these inference schemes, it is possible to derive a formula $\texttt{sp}(c,o_1,o_2,n)$ for any n between 0 and the actual number of subcriteria of c that strictly prefer o_1 over o_2. We want to make sure that only the formula that counts *all* subcriteria of c that strictly prefer o_1 over o_2 is justified. To this end, the following inference rules provide an undercutter for the previous schemes when they are non-maximal.

$$\frac{\texttt{spref}(c_1,o_1,o_2) \quad \ldots \quad \texttt{spref}(c_n,o_1,o_2) \quad \texttt{sub}(c,c_1) \quad \ldots \quad \texttt{sub}(c,c_n) \quad m<n}{SP(c,o_1,o_2,m) \text{ is inapplicable}} \; SP(c,o_1,o_2,m)uc$$

$$\frac{\texttt{\~pref}(c_1,o_1,o_2) \quad \ldots \quad \texttt{\~pref}(c_n,o_1,o_2) \quad \texttt{sub}(c,c_1) \quad \ldots \quad \texttt{sub}(c,c_n) \quad m<n}{NWP(c,o_1,o_2,m) \text{ is inapplicable}} \; NWP(c,o_1,o_2,m)uc$$

Example 3. *The following argument concludes that there is one subcriterion of* c4 *that strictly prefers* o1 *over* o2.

$$\frac{\dfrac{\vdots}{\texttt{pref}(c1,o1,o2) \quad \texttt{\~pref}(c1,o2,o1)}{\texttt{spref}(c1,o1,o2)} \qquad \texttt{sub}(c4,c1)}{\texttt{sp}(c4,o1,o2,1)}$$

It is also possible to construct an argument stating that there are two such criteria, but it will be undercut.

$$\frac{\dfrac{\vdots}{\texttt{pref}(c1,o1,o2) \quad \texttt{\~pref}(c1,o2,o1)}{\texttt{spref}(c1,o1,o2)} \qquad \dfrac{\vdots}{\texttt{pref}(c2,o1,o2) \quad \texttt{\~pref}(c2,o2,o1)}{\texttt{spref}(c2,o1,o2)} \quad ^* \qquad \texttt{sub}(c4,c1) \quad \texttt{sub}(c4,c2)}{\texttt{sp}(c4,o1,o2,2)}$$

$$\frac{\dfrac{\vdots}{\texttt{pref}(c2,o2,o1)}}{* \text{ is inapplicable}}$$

The following argument concludes that c4 *prefers* o1 *over* o2.

$$\frac{\overset{\vdots}{\text{sp}(c4,o1,o2,1)} \quad \text{nwp}(c4,o1,o2,0) \quad 1 \geq 0}{\text{pref}(c4,o1,o2)}$$

Preference by a Lexicographic Criterion. The following inference rule concludes that a lexicographic criterion c prefers an outcome o_1 over an outcome o_2 if o_1 is preferred over o_2 by a subcriterion of c. This inference is undercut by the next inference rule if there is a subcriterion of c with higher priority that does not prefer o_1 over o_2.

$$\frac{\text{lc}(c) \quad \text{sub}(c,c_1) \quad \text{pref}(c_1,o_1,o_2)}{\text{pref}(c,o_1,o_2)} \quad LC(c,c_1,o_1,o_2)$$

$$\frac{\text{lc}(c) \quad \text{sub}(c,c_2) \quad \sim\text{pref}(c_2,o_1,o_2) \quad \sim\text{prior}(c,c_1,c_2)}{LC(c,c_1,o_1,o_2) \text{ is inapplicable}} \quad LC(c,c_1,o_1,o_2)uc$$

According to its definition in a QPS, a lexicographic criterion c prefers o_1 over o_2 if every subcriterion either (weakly) prefers o_1 over o_2 or there is a higher priority subcriterion that strictly prefers o_1 over o_2. So if c prefers o_1 to o_2, all undominated (w.r.t. priority) subcriteria prefer o_1 to o_2. pref (c,o_1,o_2) can be derived based on any of those subcriteria, and there will be no justified undercutter. If c does not prefer o_1 to o_2, it may still be possible to construct an argument for pref (c,o_1,o_2), but it will be undercut because there is another subcriterion that does not prefer o_1 to o_2 and does not have lower priority. So together this pair of inference schemes correctly models the definition of preference by a lexicographic criterion in a QPS.

Example 4. *The following argument concludes that* c5 *prefers* o1 *to* o2 *based on its subcriterion* c4.

$$\frac{\text{lc}(c5) \quad \text{sub}(c5,c4) \quad \overset{\vdots}{\text{pref}(c4,o1,o2)}}{\text{pref}(c5,o1,o2)} \quad *$$

However, this argument is undercut by the following one stating that there is another subcriterion, c3, *that does not prefer* o1 *to* o2 *and does not have lower priority than* c4.

$$\frac{\text{lc}(c5) \quad \text{sub}(c5,c3) \quad \sim\text{pref}(c3,o1,o2) \quad \sim\text{sprior}(c5,c4,c3)}{* \text{ is inapplicable}}$$

The only justified argument for preference between o1 *and* o2 *by* c5 *is the following one.*

$$\frac{\text{lc}(c5) \quad \text{sub}(c5,c3) \quad \dfrac{\text{sc}(c3,bb) \quad \text{val}(o2,bb,\text{false}) \quad \text{val}(o1,bb,\text{true}) \quad \text{valpref}(c3,\text{false},\text{true})}{\text{pref}(c3,o2,o1)}}{\text{pref}(c5,o2,o1)}$$

3.6 Correspondence between QPS and AF

Theorem 1. Let $S = \langle Var, Dom, K, \Omega, \mathcal{C} \rangle$ be a QPS, KB a knowledge base that corresponds to S, and AF the argumentation framework built from KB. Then φ is a sceptically justified conclusion of AF iff its interpretation holds in S.

For every formula in KB, its interpretation holds in S (definition of correspondence). Every formula in the input language whose interpretation holds in S is in KB (definition of correspondence). All formulas in KB are justified since KB is conflict-free. For every inference rule, its conclusion is justified if and only if its premises are justified and all its undercutters (if any) are overruled. We have shown that every inference or pair of inference and its undercutter inference models the corresponding QPS definition: the interpretation of the conclusion holds in a QPS if and only if the interpretations of all premises hold and and the interpretations of the premises of all undercutters do not all hold.

4   Reasoning with Background Knowledge

The argumentation framework presented in the previous section models a QPS if the input is a knowledge base corresponding to that QPS. In order for a knowledge base to correspond to a QPS, it is necessary to specify the values of all variables for every outcome. This correpsonds to the formal (abstract) concept of an outcome as an assignment of a value to every variable in a given set of variables, as defined in the QPS framework.

In practice, an outcome is a concrete alternative (a decision, product, agreement etc.). The major difference is that not all attributes may be known. In a sense, such alternatives can be seen as partial outcomes (or sets of outcomes that share some attributes). Even though not all attributes may be specified beforehand, it is often possible to derive the values of some of the unspecified variables using background information. For example, if it is not specified whether someone speaks the language for a given holiday option, such information may be inferred if it is known that the destination is Barcelona which is in Spain, where the language is Spanish, Juan will accompany Anne, and he speaks Spanish.

In this section we introduce an extension of the argumentation framework in which it is possible to reason with such background knowledge. To this end, we extend the language and add one more inference scheme. This extension makes the system more flexible in case of incomplete specifications. If some attributes remain unknown even with reasoning with background knowledge, the argumentation framework still works correctly, it will just infer less preferences.

4.1   Language

Background knowledge is expressed using a set of predicates $\mathcal{P}_K$ which may differ per application domain. Atoms built with these predicates may also be negated (strong negation). Furthermore, a new construct is added to the input language: (defeasible) *rules* that consist of a set of (possibly weakly negated) antecedents and a consequent (the same kind of rules are used in [6]).

Definition 10. (Language) The input language is defined as follows.

$atom_{in}$ $::= p(t_1, \ldots, t_n)$ where p is an n-ary predicate $\in \mathcal{P}_{in}$

$atom_K$ $::= p(t_1, \ldots, t_n)$ where p is an n-ary predicate $\in \mathcal{P}_K$

$literal_{in}$ $::= atom_{in} \mid atom_K \mid \neg atom_K$

$rule$ $::= literal_{in}, \ldots, literal_{in}, \sim literal_{in}, \ldots, \sim literal_{in} \Rightarrow literal_{in}$

$formula_{in}$ $::= literal_{in} \mid rule$

The full language is defined as follows.

$atom_{out}$ $::= p(t_1, \ldots, t_n)$ where p is an n-ary predicate $\in \mathcal{P}_{out}$

$literal$ $::= literal_{in} \mid atom_{out}$

$formula$ $::= literal \mid \sim literal \mid rule$

Contradictory Formulas. Adding strong negation to the language also adds an additional way in which two formulas can be contradictory.

– A and $\neg A$ contradict each other.

Example 5. *Anne's criteria for a holiday are the same as before, but the information that she has about her options is different. The values of the variables* sl, su *and* bb *on which her preferences are based are not specified. Instead, for every outcome she only knows who of her friends is going with her (*fr*): Juan (*j*) or Mario (*m*), and the destination (*de*): Barcelona (*b*) or Rome (*r*). Besides, she has some relevant background information. All of this is specified in the following knowledge base.*

     Some facts from the background knowledge:

```
in(b,spain)                in(r,italy)
mediterranean(spain)       mediterranean(italy)
language(spain,spanish)    language(italy,italian)
speaks(j,spanish)          speaks(m,italian)
beenTo(b)
```

     Some rules from the background knowledge:

```
val(O,fr,X), val(O,de,C), in(C,Cn), language(Cn,L), speaks(X,L) => val(O,sl,true)
~val(O,sl,true) => val(O,sl,false)
val(O,de,C), in(C,Cn), mediterranean(Cn), ~val(O,su,false) => val(O,su,true)
val(O,de,C), beenTo(C) => val(O,bb,true)
~val(O,bb,true) => val(O,bb,false)
```

     Facts about some of the possible outcomes:

```
val(o1,fr,j)     val(o2,fr,j)     val(o3,fr,m)     val(o4,fr,m)
val(o1,de,b)     val(o2,de,r)     val(o3,de,b)     val(o4,de,r)
```

     Information about the preferences:

```
lc(c5)          cc(c4)        sc(c1,sp)     valpref(c1,true,false)
sub(c5,c3)      sub(c4,c1)    sc(c2,s)      valpref(c2,true,false)
sub(c5,c4)      sub(c4,c2)    sc(c3,n)      valpref(c3,false,true)
prior(c5,c3,c4)
```

4.2 Inferences

Defeasible Modus Ponens. This inference rule applies a rule $L_1, \ldots, L_k, \sim L_l, \ldots, \sim L_m \Rightarrow L_n$: when all its antecedents hold, the consequent is concluded.

$$\frac{L_1, \ldots, L_k, \sim L_l, \ldots, \sim L_m \Rightarrow L_n \quad L_1 \quad \ldots \quad L_k \quad \sim L_l \quad \ldots \quad \sim L_m}{L_n} \; DMP$$

Note the difference between a rule in the language and an inference rule. Defeasible modus ponens is an inference rule that applies a rule from the language. We reserve inference rules for domain-independent inferences, and provide the possibility to specify domain-specific rules in the language. Instead of possible undercutters of an inference rule, it is possible to have weakly negated antecedents for the same purpose.

Example 6. *Below are some of the arguments that can be built with the knowledge base from Example 5. The values for the variables* su *and* bb *can be derived in a similar way.*

$$\frac{r \quad \text{val(o1,fr,j)} \quad \text{val(o1,de,b)} \quad \text{in(b,spain)} \quad \text{lang(spain,spanish)} \quad \text{speaks(j,spanish)}}{\text{val(o1,s1,true)}}$$

where r is val(O,fr,X), val(O,de,C), in(C,Cn), lang(Cn,L), speaks(X,L) => val(O,s1,true).

$$\frac{\text{~val(O,s1,true)} => \text{val(O,s1,false)} \quad \text{~val(o2,s1,true)}}{\text{val(o2,s1,false)}}$$

The argument deriving a preference for o1 *over* o2 *by criterion* c5 *is the same as in Example 4, except that* val(o2,bb,false) *and* val(o1,bb,true) *are derived instead of taken directly from the knowledge base (for reasons of space, the argument is cut in three).*

$$\frac{\text{lc(c5)} \quad \text{sub(c5,c3)} \quad \dfrac{\text{sc(c3,bb)} \quad A \quad B \quad \text{valpref(c3,false,true)}}{\text{pref(c3,o2,o1)}}}{\text{pref(c5,o2,o1)}}$$

$$A: \quad \frac{\text{~val(O,bb,true)} => \text{val(O,bb,false)} \quad \text{~val(o2,bb,true)}}{\text{val(o2,bb,false)}}$$

$$B: \quad \frac{\text{val(O,de,C), beenTo(C)} => \text{val(O,bb,true)} \quad \text{val(o1,de,b)} \quad \text{beenTo(b)}}{\text{val(o1,bb,true)}}$$

5   Conclusion

In this paper we presented an argumentation framework for representing and reasoning about qualitative multi-criteria preferences. We showed that this argumentation framework models the preferences as defined by qualitative preference systems. Qualitative preference systems use both cardinality and lexicographic ordering to combine multiple criteria, which are ultimately based on the attributes of the outcomes. In an extension of the base argumentation framework we added the means to reason with background knowledge, which adds expressivity and flexibility in case of incomplete specifications.

Argumentation about preferences has been studied extensively in the context of *decision making* [7,8]. The aim of decision making is to choose an action to perform. The quality of an action is determined by how well its consequences satisfy certain criteria. For example, [8] present an approach in which arguments of various strengths in favour of and against a decision are compared. However, it is a two-step process in which argumentation is used only for epistemic reasoning. Also in [9,10], preferences are based on arguments, but not themselves derived using argumentation. In our approach, we combine reasoning about knowledge, criteria and preferences between outcomes in a single argumentation framework.

Within the context of argumentation, an approach that is related to criteria is *value-based argumentation* [11,12]. Values are used in the sense of 'fundamental social or personal goods that are desirable in themselves' [12], and are used as the basis for persuasive argument in practical reasoning. A value can be seen as a binary criterion that is satisfied if the value is promoted. In value-based argumentation, arguments are associated with values that they promote. Values are ordered according to importance to a particular audience. An argument only defeats another argument if it attacks it and the value promoted by the attacked argument is not more important than the value promoted by the attacker. In this framework, every argument is associated with only one value, while in many cases there are multiple values or interests at stake. [13] define so-called *value-specification argumentation frameworks*, in which arguments can support multiple values, and preference statements about values can be given. However, the preference between arguments is not derived from the preference between the values promoted by the arguments. Besides, there is no guarantee that a value-specification argumentation framework is consistent, i.e., some sets of preference statements do not correspond to a preference ordering on arguments.

In value-based argumentation, we cannot argue *about* what values are promoted by the arguments or the ordering of values; this mapping and ordering are supposed to be given. But these might well be the conclusion of reasoning, and might be defeasible. Therefore, it would be natural to include this information at the object level. [14] describe some argument schemes regarding the influence of certain perspectives on values. However, for the aggregation of multiple values, they assume a given order on sets of values, whereas we want to derive such an order from an order on individual values.

In our future work we would like to look into the possibilities that the presented framework offers to not only derive missing information about the attributes of outcomes, but also information about e.g. the criteria that are used and their preferences between attribute values, or priority between subcriteria. This would be especially useful when modelling other agents' preferences, e.g. the opponent in negotiation or someone you have to make a joint decision with. Often, another person's preferences are not (completely) known, but some of them may be inferred by default.

Acknowledgements. This research is supported by the Dutch Technology Foundation STW, applied science division of NWO and the Technology Program of the Ministry of Economic Affairs. It is part of the Pocket Negotiator project with grant number VICI-project 08075.

References

1. Amgoud, L., Cayrol, C.: Inferring from inconsistency in preference-based argumentation frameworks. Journal of Automated Reasoning 29, 125–169 (2002)
2. Kakas, A., Moraïtis, P.: Argumentation based decision making for autonomous agents. In: Autonomous Agents and Multiagent Systems (AAMAS 2003), pp. 883–890 (2003)

3. Dung, P.M.: On the acceptability of arguments and its fundamental role in nonmonotonic reasoning, logic programming and *n*-person games. Artificial Intelligence 77, 321–357 (1995)
4. Prakken, H.: An abstract framework for argumentation with structured arguments. Argument and Computation 1(2), 93–124 (2010)
5. Vreeswijk, G.A.W.: Abstract argumentation systems. Artificial Intelligence 90(1-2), 225–279 (1997)
6. Prakken, H., Sartor, G.: Argument-based extended logic programming with defeasible priorities. Journal of Applied Non-Classical Logics 7, 25–75 (1997)
7. Ouerdane, W., Maudet, N., Tsoukiàs, A.: Argumentation theory and decision aiding. In: Ehrgott, M., Figueira, J.R., Greco, S. (eds.) New Trends in Multiple Criteria Decision Analysis. Springer, Heidelberg (2010)
8. Amgoud, L., Prade, H.: Using arguments for making and explaining decisions. Artificial Intelligence 173(3-4), 413–436 (2009)
9. Bonnefon, J.F., Fargier, H.: Comparing sets of positive and negative arguments: Empirical assessment of seven qualitative rules. In: 17th European Conference on Artificial Intelligence (ECAI 2006), pp. 16–20. IOS Press (2006)
10. Dubois, D., Fargier, H., Bonnefon, J.F.: On the qualitative comparison of decisions having positive and negative features. Journal of Artificial Intelligence Research 32, 385–417 (2008)
11. Bench-Capon, T.J.M.: Persuasion in practical argument using value based argumentation frameworks. Journal of Logic and Computation 13(3), 429–448 (2003)
12. Bench-Capon, T., Atkinson, K.: Abstract argumentation and values. In: Rahwan, I., Simari, G.R. (eds.) Argumentation in Artificial Intelligence, pp. 45–64. Springer, Heidelberg (2009)
13. Kaci, S., van der Torre, L.: Preference-based argumentation: Arguments supporting multiple values. International Journal of Approximate Reasoning 48(3), 730–751 (2008)
14. van der Weide, T.L., Dignum, F., Meyer, J.-J.C., Prakken, H., Vreeswijk, G.A.W.: Practical reasoning using values. In: McBurney, P., Rahwan, I., Parsons, S., Maudet, N. (eds.) ArgMAS 2009. LNCS, vol. 6057, pp. 79–93. Springer, Heidelberg (2010)

Modeling and Solving AFs
with a Constraint-Based Tool: ConArg[*]

Stefano Bistarelli[1,2] and Francesco Santini[1,3,**]

[1] Dipartimento di Matematica e Informatica, Università di Perugia, Italy
{bista,francesco.santini}@dmi.unipg.it
[2] Istituto di Informatica e Telematica (CNR), Pisa, Italy
stefano.bistarelli@iit.cnr.it
[3] Centrum Wiskunde & Informatica, Amsterdam, The Netherlands
F.Santini@cwi.nl

Abstract. *ConArg* is a tool based on *Constraint Programming* which is able to model and solve different problems related to Argumentation Frameworks (AFs). To practically implement the tool, we have used *JaCoP*, a Java library which provides the user with a *Finite Domain Constraint Programming* paradigm. *Constraint Satisfaction Problems* (CSPs) offer a wide number of efficient techniques (as inference and search algorithms) that can tackle the complexity in finding all the possible Dung's conflict-free, admissible, complete and stable extensions in AFs. Moreover, we can use the tool to solve some of the preference-based problems presented in literature. ConArg is able to randomly generate networks with small-world properties in order to find Dung's extensions on such interaction graphs. We present the main features of ConArg and we report the performance in time.

1 Introduction

Interactions are a core part of all multi-party systems (e.g. multi-agent systems). *Argumentation* [14] is based on the exchange and the evaluation of interacting arguments which may represent information of various kinds, especially beliefs or goals. Argumentation can be used for modeling some aspects of reasoning, decision making, and dialogue. For instance, when an agent has conflicting beliefs (viewed as arguments), a (nontrivial) set of plausible consequences can be derived through argumentation from the most acceptable arguments for the agent.

---

[*] Research partially supported by MIUR PRIN 20089M932N project: "Innovative and multi-disciplinary approaches for constraint and preference reasoning", by CCOS FLOSS project "Software open source per la gestione dell'epigrafia dei corpus di lingue antiche", and by INDAM GNCS project "Fairness, Equità e Linguaggi".

[**] This work was carried out during the tenure of an ERCIM "Alain Bensoussan" Fellowship Programme. This Programme is supported by the Marie Curie Co-funding of Regional, National and International Programmes (COFUND) of the European Commission.

S. Modgil, N. Oren, and F. Toni (Eds.): TAFA 2011, LNAI 7132, pp. 99–116, 2012.

Argumentation has become an important subject of research in Artificial Intelligence and it is also of interest in several disciplines, such as Logic, Philosophy and Communication Theory [20].

Many theoretical and practical developments build on Dung's seminal theory of argumentation. A *Dung Argumentation Framework* (*AF*) is a directed graph consisting of a set of arguments and a binary conflict based *attack relation* among them. The sets of arguments to be considered are then defined under different semantics, where the choice of semantics equates with varying degrees of scepticism or credulousness. The main issue for any theory of argumentation is the selection of acceptable sets of arguments, based on the way arguments interact. Intuitively, an acceptable set of arguments must be in some sense coherent and strong enough (e.g. able to defend itself against all attacking arguments).

Constraint Programming [25] (CP) is a powerful paradigm for solving combinatorial search problems that draws on a wide range of techniques from artificial intelligence, computer science, databases, programming languages, and operations research. CP is currently applied with success to many domains, such as scheduling, planning, vehicle routing, configuration, networks, and bioinformatics [25]. The basic idea in constraint programming is that the user states the constraints and a general purpose constraint solver is used to solve them. Constraints are just relations, and a *Constraint Satisfaction Problem* [25] (*CSP*) states which relations should hold among the given decision variables.

Constraint solvers search the solution space either systematically, as with backtracking or branch and bound algorithms, or use forms of local search which may be incomplete. Systematic method often interleave search and inference, where inference consists of propagating the information contained in one constraint to the neighboring constraints [25].

In this paper we present *ConArg* [7,9] (*Arg*umentation with *Con*straints), a tool that can practically find all Dung's classical extensions [14] (i.e. conflict-free, admissible, complete and stable extensions) by defining the properties of these extensions with constraints and solving the related CSP; we program these constraints in JaCoP [19]. To show the feasibility of such solution, in the paper we test the tool on different randomly generated small-world networks and we report the performance in time; since the number of these extensions, which in practice are subsets of the $\mathcal{A}_{rgs}$ set of arguments, may explode for large $\mathcal{A}_{rgs}$ (i.e., the powerset of $\mathcal{A}_{rgs}$), it is important to use techniques to tackle this inherent complexity, as CP. This is particularly true for conflict-free extensions, which represent the least constrained extensions.

Moreover, ConArg can solve different hard problems in literature related to weighted AFs (see Sec. 6), as the ones presented in [15]. We consider weighted AFs as AFs where attacks are associated with a weight capturing their relative strength [15,8]. For example, given a weighted argument system, a set of arguments $S \subseteq \mathcal{A}_{rgs}$ and an inconsistency budget β, to find if β is minimal w.r.t. S represents a co-NP-complete problem [15]. This and other hard problems have been highlighted in [15].

This paper continues the research line on connecting AFs to CP opened in [6,8,10]. The remainder is organized as follows: in Sec. 2 we report the theory behind Dung Argumentation (and weighted AFs [15]), while in Sec. 3 we summarize the background on constraints and weighted constraints. In Sec. 4 we show a mapping from AFs to CSPs, which is used in ConArg to find the Dung's extensions by representing their properties with constraints and solving the obtained CSP. In Sec. 5 we show the tests on *Barabasi* [22,3] and *Kleinberg* [22] small-world networks, which prove the feasibility of a constraint-based tool. In Sec. 6 we show how to model hard problems related to AFs (e.g. preferred extensions or the problems presented in [15]) in JaCoP [19]. A comparison with related work is given in Sec. 7. Finally, Sec. 8 presents conclusions and ideas about future work.

2 Dung Argumentation

In [14], the author has proposed an abstract framework for argumentation in which he focuses on the definition of the status of arguments. For that purpose, it can be assumed that a set of arguments is given, as well as the different conflicts among them. An argument is an abstract entity whose role is solely determined by its relations to other arguments.

Definition 1. *An Argumentation Framework (AF) is a pair $\langle \mathcal{A}_{rgs}, R \rangle$ of a set $\mathcal{A}_{rgs}$ of arguments and a binary relation R on $\mathcal{A}_{rgs}$ called the attack relation. $\forall a_i, a_j \in \mathcal{A}_{rgs}$, $a_i R a_j$ means that a_i attacks a_j. An AF may be represented by a directed graph (the interaction graph) whose nodes are arguments and edges represent the attack relation. A set of arguments $\mathcal{B}$ attacks an argument a if a is attacked by an argument of $\mathcal{B}$. A set of arguments $\mathcal{B}$ attacks a set of arguments $\mathcal{C}$ if there is an argument $b \in \mathcal{B}$ which attacks an argument $c \in \mathcal{C}$.*

The "acceptability" of an argument [14] depends on its membership to some sets, called *extensions*. These extensions characterize collective "acceptability".

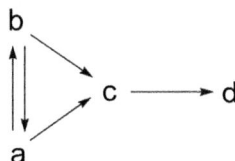

Fig. 1. An example of Dung Argumentation Framework; e.g. c attacks d

In Fig. 1 we show an example of AF represented as an *interaction graph*: the nodes represent the arguments and the directed arrow from c to d represents the attack of c towards d, that is $c R d$. Dung [14] gave several semantics to "acceptability". These various semantics produce none, one or several acceptable sets of arguments, called extensions. In Def. 2 we define the concepts of conflict-free and stable extensions:

Definition 2. *A set $\mathcal{B} \subseteq \mathcal{A}_{rgs}$ is conflict-free iff no two arguments a and b in $\mathcal{B}$ exist such that a attacks b. A conflict-free set $\mathcal{B} \subseteq \mathcal{A}_{rgs}$ is a stable extension iff for each argument which is not in $\mathcal{B}$, there exists an argument in $\mathcal{B}$ that attacks it.*

The other semantics for "acceptability" rely upon the concept of defense:

Definition 3. *An argument b is defended by a set $\mathcal{B} \subseteq \mathcal{A}_{rgs}$ (or $\mathcal{B}$ defends b) iff for any argument $a \in \mathcal{A}_{rgs}$, if a attacks b then $\mathcal{B}$ attacks a.*

An admissible set of arguments according to Dung must be a conflict-free set which defends all its elements. Formally:

Definition 4. *A conflict-free set $\mathcal{B} \subseteq \mathcal{A}_{rgs}$ is admissible iff each argument in $\mathcal{B}$ is defended by $\mathcal{B}$.*

Besides the stable semantics, three semantics refining admissibility have been introduced by Dung [14]:

Definition 5. *A preferred extension is a maximal (w.r.t. set inclusion) admissible subset of $\mathcal{A}_{rgs}$. An admissible $\mathcal{B} \subseteq \mathcal{A}_{rgs}$ is a complete extension iff each argument which is defended by $\mathcal{B}$ is in $\mathcal{B}$. The least (w.r.t. set inclusion) complete extension is the grounded extension.*

A stable extension is also a preferred extension and a preferred extension is also a complete extension. Stable, preferred and complete semantics admit multiple extensions whereas the grounded semantics ascribes a single extension to a given argument system.

In the paper we want to to deal also with hard problems related to *weighted AFs* [15]. Formally, a weighted AF is a triple $\langle \mathcal{A}_{rgs}, R, w \rangle$ where $\langle \mathcal{A}_{rgs}, R \rangle$ is a Dung-style abstract argument system, and $w : \mathcal{A}_{rgs} \to \mathbb{R}^+$ is a function assigning real valued weights to attacks.

A key idea presented in [15] is the *inconsistency budget*, $\beta \in \mathbb{R}^+$, which the authors use to characterise how much inconsistency they are prepared to tolerate. The intended interpretation is that, given an inconsistency budget β, *we would be prepared to disregard attacks up to a total weight of β* [15]. Conventional AFs implicitly assume an inconsistency budget of 0. Weighted AFs have been already modeled also in [8], by considering a semiring-based constraint programming framework.

3   Constraint Programming

The classic definition of a *Constraint Satisfaction Problem* [25] (*CSP*) is as follows. A *CSP* P is a triple $P = \langle X, D, C \rangle$ where X is an n-tuple of variables $X = \langle x_1, x_2, \ldots, x_n \rangle$, D is a corresponding n-tuple of domains $D = \langle D_1, D_2, \ldots, D_n \rangle$ such that $x_i \in D_i$, C is a t-tuple of constraints $C = \langle C_1, C_2, \ldots, C_t \rangle$. A constraint

C_j is a pair $\langle R_{S_j}, S_j \rangle$ where R_{S_j} is a relation on the variables in $S_i = scope(C_i)$. In other words, R_i is a subset of the Cartesian product of the domains of the variables in S_i. A solution to the CSP P is an n-tuple $A = \langle a_1, a_2, \ldots, a_n \rangle$ where $a_i \in D_i$ and each C_j is satisfied in that R_{S_j} holds on the projection of A onto the scope S_j. In a given task one may be required to find the set of all solutions, $sol(P)$, to determine if that set is non-empty or just to find any solution, if one exists. If the set of solutions is empty the CSP is unsatisfiable.

This simple but powerful framework captures a wide range of significant applications in fields as diverse as artificial intelligence, operations research, scheduling, supply chain management, graph algorithms, computer vision and computational linguistics [25].

Constraint solvers search the solution space either systematically, as with backtracking or branch and bound algorithms, or use forms of local search which may be incomplete. Systematic method often interleave search and inference, where inference consists of propagating the information contained in one constraint to the neighboring constraints. Such inference (usually called constraint propagation) is useful since it may reduce the parts of the search space that need to be visited.

A backtracking search algorithm, used also by JaCoP [19], performs a depth-first traversal of a search tree, where the branches out of a node represent alternative choices that may have to be examined in order to find a solution, and the constraints are used to prune subtrees containing no solutions. Backtracking search algorithms usually come with a guarantee that a solution will be found if one exists (i.e., it is a complete search method), and can be used to show that a CSP does not have a solution or to find a provably optimal solution. Many techniques for improving the efficiency of a backtracking search algorithm have been suggested and evaluated including constraint propagation, nogood recording, backjumping, heuristics for variable and value ordering, and randomization and restart strategies [25]. In this paper we use also an incomplete backtracking search, that is the *Limited Discrepancy Search* (*LDS*) [18] (see Sec. 5).

One of the main reasons why constraint programming quickly found its way into applications has been the early availability of usable constraint programming systems, as JaCoP, which we will use in this paper to find a solution for AFs [19].

Various generalizations of the classic CSP model have been developed subsequently. One of the most significant is the *Constraint Optimization Problem* (*COP*) for which there are several significantly different formulations, and the nomenclature is not always consistent [25]. Perhaps the simplest COP formulation retains the CSP limitation of allowing only hard Boolean-valued constraints but adds a cost function over the variables, that must be minimized. A *weighted constraint* $\langle c, w \rangle$ is just a classical constraint c, plus a weight w (over natural, integer, or real numbers). The cost of an assignment t of the variable is the sum of all $w(c)$, for all constraints c which are violated by t [25].

In this paper we use weighted constraints to model and solve weighted AFs [8], and also the hard problems presented in [15], which are solved in Sec. 6.

4   Mapping AFs to CSPs in ConArg

In this section we propose a mapping from AFs to CSPs, which is used inside ConArg [7,9] as the core of the tool. Given an $AF = \langle A_{rgs}, R \rangle$, we define a variable for each argument $a_i \in A_{rgs}$ ($V = \{a_1, a_2, \ldots, a_n\}$) and each of these argument can be taken or not, i.e. the domain of each variable is $D = \{1, 0\}$.

In the following explanation, notice that "b attacks a" means that b is a parent of a in the interaction graph, and "c attacks b attacks a" means that c is a grandparent of a. To compute the extensions of Dung [14] we need to define different sets of constraints:

1. **Conflict-free constraints.** In order to find conflict-free sets, if $R(a_i, a_j)$ is in the graph we need to prevent the solution to include both a_i and a_j in the considered extension: $\neg(a_i = 1 \wedge a_j = 1)$. The other possible assignment of the variables $(a = 0 \wedge b = 1)$, $(a = 1 \wedge b = 0)$ and $(a = 0 \wedge b = 0)$ are permitted: in these cases we are choosing only one argument between the two (or none of the two) and thus, we have no conflict.

2. **Admissible constraints.** For the admissibility, we need that, if child argument a_i has a parent node a_f but a_i has no grandparent node a_g (parent of a_f), then we must avoid to take a_i in the extension because it is attacked and cannot be defended by any ancestor: expressed with a unary constraint, $a_i = 0$.
   Moreover, if a_i has several grandparents $a_{g1}, a_{g2}, \ldots, a_{gk}$ and only one parents a_f (child of $a_{g1}, a_{g2}, \ldots, a_{gk}$), we need to add a $k + 1$-ary constraint $\neg(a_i = 1 \wedge a_{g1} = 0 \wedge \cdots \wedge a_{gk} = 0)$. The explanation is that at least a grandparent must be taken in the admissible set, in order to defend a_i from one of his parents a_f. Notice that, if a node is not attacked (i.e. he has no parents), it can be taken or not in the admissible set.

3. **Complete constraints.** To compute a complete extension $\mathcal{B}$, we impose that each argument a_i which is defended by $\mathcal{B}$ is in $\mathcal{B}$, except those a_i that, in such case, would be attacked by $\mathcal{B}$ itself [5]. This can be enforced by imposing that for each a_i taken in the extension, also all its $a_{s1}, a_{s2}, \ldots, a_{sk}$ grandchildren (i.e. all the arguments defended by a_i), whose fathers are not taken in the extension, must be in $\mathcal{B}$. Formally, $(a_i = 1 \wedge a_{s1} = 1 \wedge \cdots \wedge a_{sk} = 1)$ only for those a_{si} for which it stands that $(a_{fs_1} = 0 \wedge a_{fs_2} = 0 \wedge \cdots \wedge a_{fs_z} = 0)$, where $a_{fs_1}, a_{fs_2}, \ldots, a_{fs_z}$ are the fathers of a_{si}.

4. **Stable constraints.** If we have a child node a_i with multiple parents $a_{f1}, a_{f2}, \ldots, a_{fk}$, we need to add the constraint $\neg(a_i = 0 \wedge a_{f1} = 0 \wedge \cdots \wedge a_{fk} = 0)$. In words, if a node is not taken in the extension (i.e. $a_i = 0$), then it must be attacked by at least one of the taken nodes, that is at least a parent of a_i needs to be taken in the stable extension (that is, $a_{fj} = 1$).
   Moreover, if a node a_i has no parent in the graph, it has to be included in the stable extension (notice a_i cannot be attacked by nodes inside the extension, since he has no parent). The corresponding unary constraint is $\neg(a_i = 0)$.

The following proposition states the equivalence between solving an AF_S and its related CSP.

Proposition 1 (Solution equivalence [6]). *Given an AF = $\langle \mathcal{A}_{rgs}, R \rangle$, the solutions of the related CSP obtained with the mapping corresponds to find over AF all the*

- *conflict-free extensions by using conflict-free constraint classes.*
- *admissible extensions by using conflict-free and admissible constraint classes.*
- *complete extensions by using conflict-free, admissible and complete constraint classes.*
- *stable extensions by using conflict and stable constraint classes.*

Notice that the presented soft constraint framework can be easily used to solve argumentation problems with additional constraints, as proposed in [13]. We can find further requirements on the sets of arguments which are expected as extensions, like "extensions must contain argument a when they contain b" or "extensions must not contain one of c or d when they contain a but do not contain b".

In Fig. 2 and Fig. 3 we show two screenshots of ConArg: both the randomly generated networks have 12 nodes (i.e. arguments) although they have different edges (i.e., attacks among arguments); they have been created according to the principles of Barabasi defined in [3,22] (see Sec. 5).

In Fig. 2 we show one of the 78 conflict-free extensions (i.e. in particular, the 70th solution). The arguments in the considered extension are highlighted in gray: $\{a_1, a_3, a_8\}$. The attacks from these arguments are associated with a ticker edge. All the 78 solutions can be navigated by using the scrolling arrows of the application (see Fig. 2).

In Fig. 3 we show the only stable extension that can be found over the considered interaction graph: the nodes in the stable extension are: $\{a_4, a_9, a_{10}, a_{11}\}$.

5   Testing ConArg

In this section we test the implementation with constraints of the mapping given in Sec. 4 and we test how ConArg [7,9] is efficient in finding all the possible conflict-free, admissible, complete and stable extensions over a randomly generated small-world network. Therefore, we show how ConArg is able to deal with real-world cases with several arguments and attacks (e.g. discussion fora) where the topology of interaction graphs follows the properties of a social network [24,17].

For the implementation we use two Java libraries, the *Java Constraint Programming* solver [19] (JaCoP) and the *Java Universal Network/Graph Framework (JUNG)* [22]. JaCoP [19] is a Java library which provides the Java user with *Finite Domain Constraint Programming* paradigm [25]. It provides different type of constraints: most commonly used primitive constraints, such as arithmetical constraints, equalities and inequalities, logical, reified and conditional constraints, combinatorial (global) constraints. It provides a significant number of (global) constraints to facilitate an efficient modeling. It also provides a modular design of search to help the user on specific characteristics of the problem being addressed.

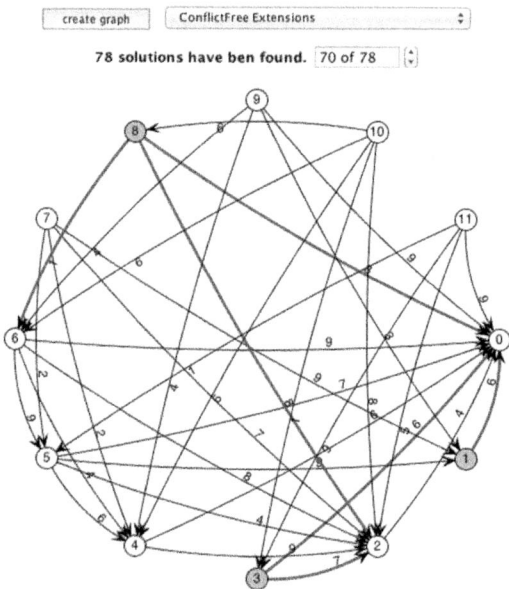

Fig. 2. A screenshot of ConArg, showing the 70th conflict-free extension found over the considered interaction graph: the arguments belonging to the extension are highlighted in gray

To practically develop and test our model, we also adopted JUNG [22], a software library for the modeling, generation, analysis and visualization of graphs. We suppose that interaction graphs, where nodes are arguments and edges are attacks (see Sec. 2), represent in this case a kind of social network and consequently show the related small-world properties [24]. A practical example can be the study of discussion fora, where the users post their arguments that can attack other users' arguments [24,17].

Therefore, for the following tests we use the *BarabasiAlbertGenerator* class [22,3], which is an evolving small-world random graph generator. At each time step, a new vertex is created and is connected to existing vertices according to the principle of "preferential attachment", whereby vertices with higher degree have a higher probability of being selected for attachment. At a given step, the probability p of creating an edge between an existing vertex v and the newly added vertex is $p = (degree(v) + 1)/(|E| + |V|)$. $|E|$ and $|V|$ are, respectively, the number of edges and vertices currently in the network (counting neither the new vertex nor the other edges that are being attached to it). An example of such random graphs with 40 nodes is shown in Fig. 4.

In the first tests we use a *Depth First Search (DFS)* algorithm [19,25]: this algorithm searches for a possible solution by organizing the search space as a search tree. In every node of this tree a value is assigned to a domain variable and a decision whether the node will be extended or the search will be cut in this node is made. The search is cut if the assignment to the selected domain

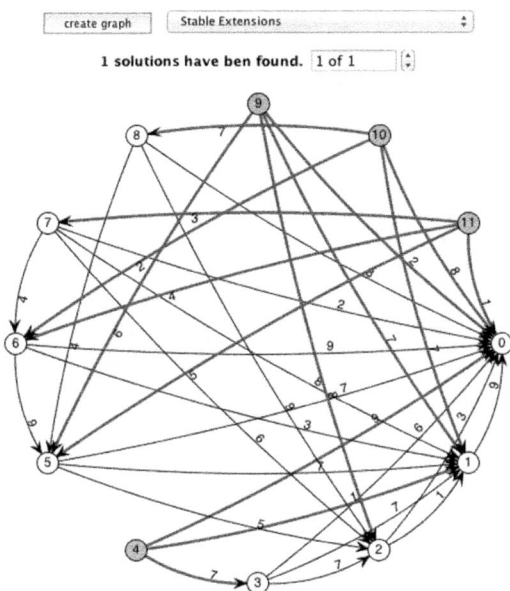

Fig. 3. A screenshot of ConArg, showing the only stable extension over an interaction graph: the arguments in the extension are $\{a_4, a_9, a_{10}, a_{11}\}$

variable does not fulfill all constraints. Each time during the search, we select the variable which has most constraints assigned to it and we assign to it a random value from its current domain: we use *MostConstrainedStatic()* as the variable selection method and *IndomainSimpleRandom()* as the value selection method, offered by JaCoP. Moreover, we set a timeout of 180 sec. to interrupt the search procedure and to report the number of solutions found only in that interval; we run our experiments over 4 different sets of random graphs with 10, 40, 60 and 100 nodes. The performance in Tab. 1 reports the average results for each set consisting in 5 different random graphs each. Each row in Tab. 1 shows the number of nodes and edges for the graphs and the average number of found conflict-free, admissible, complete and stable extensions; the time is measured in milliseconds.

Table 1. The test small-world network generated with JUNG [22] and the corresponding statistics; time is in milliseconds

Nodes	Edges	Conf-free (time)	Admissible (time)	Compl. (time)	Stable (time)
10	23	73 (317)	32 (307)	29 (99)	1 (161)
40	142	421.697 ($\simeq$3min)	30.720 (3.193)	3.108 (938)	1 (64)
60	240	320.828 ($\simeq$3min)	320.466 ($\simeq$3min)	3.104 (1.035)	2 (98)
100	451	219.194 ($\simeq$3min)	220.528 ($\simeq$3min)	377.610 ($\simeq$3min)	1 (112)

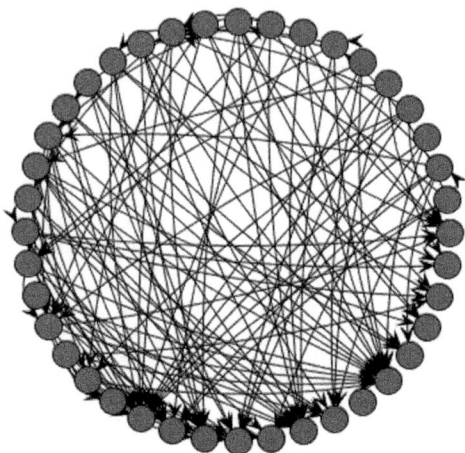

Fig. 4. A small-world network with 40 nodes generated with JUNG by using the *BarabasiAlbertGenerator* class [22,3]

The implementation easily finds all the admissible extensions up to 40 nodes and complete extensions (which do not scale between 40 and 60 nodes) up to 60 nodes. Stable extensions, due to the characteristics of this kind of small-world network (i.e., *Barabasi* [3]), are very few. The main problem is represented by finding all the conflict-free extensions, since we experienced problems already with 40 nodes: in fact, they represent the "less" constrained extension w.r.t. the others and, therefore, we have a large number of them. However, we remind that an AF contains at most $2^{|\mathcal{A}_{rgs}|}$ extensions, considering all the possible subsets of $\mathcal{A}_{rgs}$, thus the problem explodes very quickly.

However, the constraint framework comes with different performant solving techniques: to show how the performance can be improved, we also used a partial method, the *Limited Discrepancy Search (LDS)*, which is a kind of *Depth First Search* procedure adopting the method proposed in [18]. If a given number of different decisions along a search path is exhausted, then backtracking is initiated [19,18]. In Tab. 2 we show the improved results only for conflict-free and admissible extensions. With this method we can find up to five times more the number of extensions w.r.t. DFS, except for conflict-free extensions for graphs with 100 nodes: in this case the number of extensions is so huge that the LDS method performs as plain DFS within 3 minutes.

In order to study our implementation on networks with distinct properties, we have repeated the same tests over a different kind of small-world network, the *KleinbergSmallWorldGenerator* [19]. In this is graph generator, the model is an $m \times n$ (optionally toroidal) lattice. Each node u has four local connections, one to each of its neighbors, and in addition one long range connection to some node v, where v is chosen randomly according to probability proportional to d^α where d is the lattice distance between u and v and α is the clustering exponent. An example of such graph with 36 nodes is shown in Fig. 5.

Table 2. The test small-world network generated with JUNG [22] and the corresponding statistics by using the LDS search [18]: time is in milliseconds

Nodes	LDS n	Conf-free (time)	Admissible (time)
40	30	2.455.079 ($\simeq$3min)	30.720 (3.120)
60	40	1.716.880 ($\simeq$3min)	1.562.289 ($\simeq$3min)
100	451	215.362 ($\simeq$3min)	843.927 ($\simeq$3min)

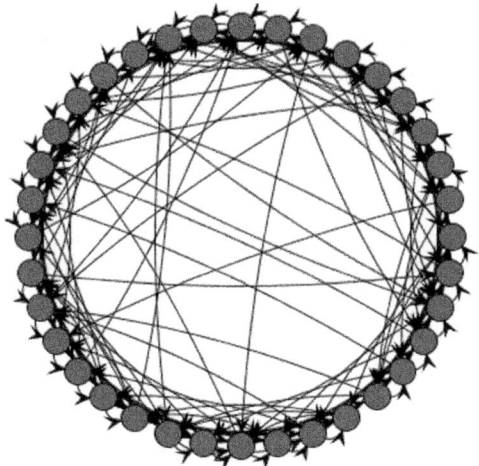

Fig. 5. A small-world network with 36 nodes generated with JUNG by using the *KleinbergSmallWorldGenerator* class [22,3]

In Tab. 3 we report the performance collected with the same methodology as for Tab 1. The higher number of found stable extensions w.r.t. complete ones in case of 100 nodes can be justified with the fact that stable extensions are more constrained and therefore they are easier to find within the timeout of the search procedure (i.e. still 180 sec.).

The performance in this section have been collected using a MacBook with 2.4Ghz Core Duo and 4Gb 1067Mhz DDR3 of RAM. Notice that the coalition structure generation problem is extremely challenging due to the number of possible solutions that need to be examined. Other works in literature, finding different kinds of constrained (and optimized according to some criteria) coalitions are usually tested over networks of 15-30 nodes [23].

6 Hard Problems Solved with ConArg

In this section we propose some hard problems related to AFs; in particular, on finding preferred extensions and weighted grounded extensions. Then we show how to implement the corresponding constraints in JaCoP.

Table 3. The test small-world network generated with JUNG [22] and the corresponding statistics: time is in milliseconds

Nodes	Edges	Conf-free (time)	Admissible (time)	Compl. (time)	Stable (time)
9	45	22 (83)	12 (71)	8 (80)	6 (73)
36	180	411.317 (121.371)	5.1412 (8.511)	525 (1.249)	449 (717)
64	320	290.910 ($\simeq$3min)	292.480 ($\simeq$3min)	92.725 (134.896)	63.878 (15.601)
100	500	219.194 ($\simeq$3min)	215.273 ($\simeq$3min)	49.787 ($\simeq$3min)	58.728 ($\simeq$3min)

Preferred Extensions. The first interesting problem is determining whether a set of arguments is a preferred extension, which is a co-NP-complete [5] problem. Since a preferred extension S^* is a maximal (w.r.t. set inclusion) admissible subset of $\mathcal{A}_{rgs}$ (see Sec. 2), we can implement the search by finding an admissible extension S' such that $S^* \subset S'$; before we check if S^* is at least admissible. If S' exists, then S^* is not a preferred extension, otherwise it is. The problem can be solved by assigning to 1 the variables representing the arguments in S^*, given as the input of the decidability problem. This ensures that the search procedure will try to find a superset of S^*. As we meet a S' solution during the search that includes the input set, then we can say that S^* is not a preferred extension, otherwise, it is.

Weighted Grounded Extensions. As shown in [15], while the the problem of finding the weighted version of the classical extensions (e.g. stable or admissible) is not computationally harder than the original problem, there are some important problems related to *weighted grounded extensions* (*wge*) that are very difficult to solve. The weighted AFs and the concept of inconsistency budget β have been introduced in Sec. 2. In ConArg we are able to find all the β-grounded extensions given a random small-world interaction graph, by selecting also the desired amount for the β budget.

With ConArg we can also solve some hard problems proposed in [15], which are related to β grounded extensions. We now define one of these problems in Prop. 2, and then we describe how ConArg models β-grounded extensions and solves the problem Prop. 2 with constraints.

Proposition 2 ([15]). *Given a weighted argument system $\langle X, A, w \rangle$, an inconsistency budget β and argument $a \in X$, the problem of checking whether $\exists S \in wge(X, A, w, \beta)$ such that $a \in S$ is NP-complete.*

Implementation in ConArg. To compute all the weighted grounded extensions, first of all we need to extend our representation to include weighted constraints, as defined in Sec. 3. The weighted constraints represent the costs associated to the attacks among the arguments in weighted AFs. To do so, we need to redefine the conflict-free constraints as proposed also in [8], thus leading to a weighted AF as defined in Sec. 2, i.e. $\langle \mathcal{A}_{rgs}, R, w \rangle$: if $w(R(a_i, a_j)) = s$ is in the graph we need assign a s consistency budget to the solution that includes both a_i and a_j in the considered conflict-free extension, that is $(a_i = 1 \land a_j = 1) = s$. For the other possible assignment of the variables we have $(a = 0 \wedge b = 1) = 0, (a = 1 \wedge b = 0) = 0$ and $(a = 0 \wedge b = 0) = 0$, since these assignments are permitted with no cost also in the classical semantics: in these cases we are choosing only one argument between the two (or none of the two) and thus, we have no conflict.

The easy way (if memory is not an issue) is to use *ExtensionalConflict* constraints [19]. In our implementation we can specify, for example, an assignment $[1, 1, 10]$ for three variables: the first two values states that if we take a_1 and a_2 in the same extension (i.e. $a_1 = 1$ and $a_2 = 1$), then the cost to be paid is represented by the third value (i.e. 10). In our model, this cost represents the cost associated to the attack between a_1 and a_2.

Since with this representation we need to specify different costs for any assignment of the considered variables, in general this extensional form could cause the system to run out of memory. However, for the weighted AF case it is easy and not memory-consuming to express all the costs of attacks in this way, since the variables can be assigned only to 0,1 to represent the fact an argument is taken into the extension or not; therefore, there are only four cases to define for each attack. In the following piece of JaCoP code we show how to fully express a constraint that defines an attack with cost 10 between a_1 and a_2.

```
store.impose(new ExtensionalConflictVA(new IntVar[]{a1, a2, cost},
new int[][]
{{1, 1, 10},
{0, 1, 0},
{1, 0, 0},
{0, 0, 0}}));
```

Therefore, to check Prop. 2 we impose $a = 1$ (i.e. a must be present in the extension) and we have also to constrain the sum of the β inconsistency budget to be not worse than the given β (i.e. $cost \leq \beta$). As soon as we find a β-grounded extension containing a, we can (successfully) stop the search.

ConArg can also solve the following hard problems defined in [15]:

Proposition 3 ([15]). *Given a weighted argument system* $\langle X, A, w \rangle$, *an inconsistency budget* β *and an argument* $a \in X$, *the problem of checking whether,* $\forall Y \in wge(X, A, w, \beta)$, *we have* $a \in Y$ *is co-NP-complete.*

Proposition 4 ([15]). *Given a weighted argument system* $\langle X, A, w \rangle$, *a set of arguments* $S \subseteq X$ *and an inconsistency budget* β, *checking whether* β *is minimal w.r.t.* $\langle X, A, w \rangle$ *and* S *is co-NP-complete.*

To check Prop. 3 we can impose a constraint on the inconsistency budget to be not worse than the given β (i.e. $cost \leq \beta$), and then we solve the problem with a self-implemented solution listener [19] (a plug-in in JaCoP that is called by search when a solution is found) that exits from the search when $a \notin X$, that is when the given a does not belong to a solution. In this case we can reply that a does not belong to every solution with a β inconsistency budget.

To solve the problem expressed by Prop. 4, we simply solve the CSP by minimizing *cost* and then to check if the found solution(s) have a *cost* value equal to (or less than) the desired β inconsistency budget. JaCoP offers methods for finding a solution that minimizes a given cost function. A very simple way is to find all the solutions by invoking the following method.

```
IntVar cost;
...
boolean result = label.labeling(store, select, cost);
```

In Fig. 6 we show a screenshot of the ConArg application: in this case, we show the solution of the search of all the weighted grounded extensions [15] over the interaction graph shown in Fig. 6. The same example with a budget β of 3 is proposed also in [15]. ConArg finds 4 different β-grounded extensions: $\{\{\emptyset\}, \{a_0, a_1, a_3, a_5\}, \{a_2, a_4, a_6, a_7\}, \{a_0, a_1, a_3, a_4, a_6, a_7\}\}$, respectively obtained by removing no attacks (the $\emptyset$ case) and, for the other three solutions, by removing the attacks (a_4, a_3), (a_3, a_4) and both $\{(a_4, a_3), (a_3, a_4)\}$. In particular, Fig. 6 shows the second solution, i.e. $\{a_0, a_1, a_3, a_5\}$, obtained by removing the attack (a_4, a_3) with a weight of 2. The corresponding edge in Fig. 6 is dotted.

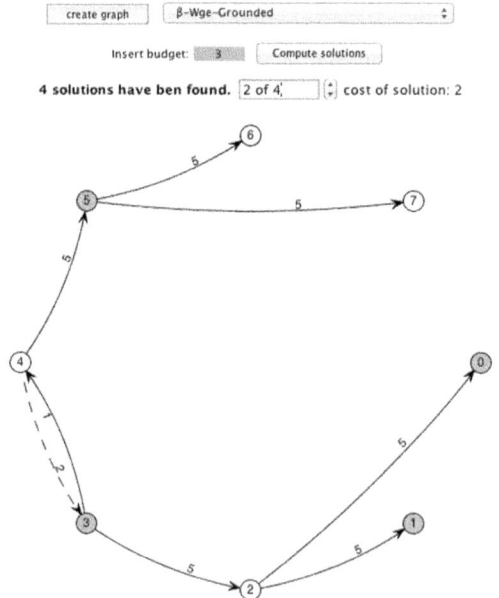

Fig. 6. An example of a β-grounded extension, with $\beta = 3$, obtained by removing the attack with cost 2 from a_4 to a_3. The same interaction graph is given in [15].

7   Related Work

In this work we propose ConArg, a constraint-based tool as an ideal framework where to solve (weighted) AFs. ConArg [7,9] can find all the classical extensions (see Sec. 4) and hard problems related to weighted AFs (as shown in Sec. 6).

In [15], one of the main inspiration sources for this work, no solving mechanism is proposed to solve the problems presented in the paper: the focus of the authors is mainly in defining the computationally hard problems and proposing their complexity proofs. We considered constraint programming techniques because they can tackle this complexity in practice.

In [5] the authors associates to each subset S of arguments a formula in propositional logic; then, S is an extension under a given semantics if and only if the formula is satisfiable (i.e. they solve the problem with SAT [11]). An extensive survey of the difference between SAT and CP can be found in [11]: summarizing, CP is more expressive for the modeling phase in order to find more complex semantics (e.g. grounded or semi-stable [12] ones) and further user-defined constraints on classical semantics [13]. In addition, in CP the user has the possibility to inform the solver about problem specific information and then to appropriately tune it, while in SAT there is usually little room and need for this parametrization. The modeling in [5] does not include preferred, grounded or weighted extensions [8,15]; furthermore, the encoding presented in [5] has no practical implementation and performance tests.

In a very recent paper [2], the authors present how to encode AFs as CSPs too; they show how to represent preferences over arguments as a (partial or total) preorder. In this work, and also in the past ones [6,8], we choose to model quantitative preferences instead of qualitative ones, even if qualitative preferences can be clearly cast also in our semiring-based framework. Moreover, while in [6] some of the authors of this paper model the preferences over arguments, in this paper we associate weights with attacks instead, as proposed in [8,15]. In addition to [2], we provide a practical implementation of the constraint modeling (in JaCoP [19]) and performance tests.

The are some frameworks based on Logic Programming-like languages. For example, the system *ASPARTIX* [16] is a tool for computing acceptable extensions for a broad range of formalizations of Dung's argumentation framework and generalizations thereof, e.g. value-based AFs [4] or preference-based [1]. *ASPARTIX* relies on a fixed disjunctive datalog program which takes an instance of an argumentation framework as input, and uses the answer-set solver DLV for computing the type of extension specified by the user. However, *ASPARTIX* does not solve any quantitative argumentation case, as well as other Answer Set Programming systems [21]. As far as we know, all these implemented systems are not tested on large interaction graphs as we do in this paper, thus it is not possible to have an idea on how these systems scale over the number of arguments (for example).

In [6] the authors have proposed a semiring-based constraint framework to model AFs with weights on arguments, i.e. withe the nodes of the interaction graph. In [8] the authors have extended [6] in order to solve over-constrained weighted AF problems, where weights are instead associated with arcs and represent the cost of the attack between two arguments. to relax the notion of conflict-free extensions to α-conflict-free ones (and also for the other extensions of Dung), in order to include in the same set also attacking arguments, whose attack costs are not worse than a threshold α.

In [10] the authors extend the classical AFs [14] in order to deal with coalitions of arguments. The initial set of arguments is partitioned into subsets, or coalitions. Each coalition represents a different *line of thought*, but all the coalitions show the same property inherited by Dung, e.g. all the coalitions in the partition are admissible (or conflict-free, complete, stable). Also in [10] JaCoP is the tool used to find the coalitions of arguments.

8   Conclusions and Future Work

In the paper we have tested how Dung's classical extensions (conflict-free, admissible, complete and stable ones) [14] can be found by using a constraint-based tool we named ConArg [7,9]. We have presented the mapping from AFs to CSPs implemented in ConArg and solved the obtained CSP with JaCoP libraries [19]. We have proposed an unifying computational framework with strong mathematical foundations and solving techniques.

Moreover, we have tested ConArg over two different kinds of small-world networks and we have reported the performance in Sec. 5. The performance show the efficiency of our framework and the paper proposes the first computational tests of Dung's AFs applied to interaction graphs with small-world properties. The generation of coalitions given a set of entities and their relationships (in this case according to the constraints defined by Dung) is a challenging problem in literature, due to the rapid explosion of the solutions [23]. We used small-world networks in order to study the possible application of ConArg over real-world cases with several arguments and attacks, as, for example, in discussion fora or in social networks: the relationships in this kind of graphs show small-world properties [24,17,3].

As a further result we have shown how to define and program in ConArg the constraints to model and solve hard problems related to AFs (e.g. the preferred extension) and weighted AFs, as, for example, the problem of finding the weighted ground extensions presented in [15]. Constraint solving techniques prove to be able to deal with large scale problems, even if the treated problems are difficult [25].

For the future we have many open issues. First of all, we would like to investigate the properties of interaction graphs, in order to reproduce the tests we have presented in this paper on more appropriate graphs (not randomly generated). In particular we would like to find the small-world properties in real interaction graphs. Moreover, we would like to compare the performance with other systems, as *ASPARTIX* [16]. To do so, we would like to find common benchmarks with the same properties in order to test the two systems on the same different problems.

A further intent is to optimize the performance in order to speed up the search of conflict-free and admissible extensions. We would like to further investigate the search plug-ins provided by JaCoP: the search-plugin is an object, which is informed about the current state of the search and may influence the behavior of the search [19]. Moreover, we would like to directly program ad-hoc methods that can improve the performance during the search in the JaCoP solver.

At last, we would like to extend ConArg in order to be able to solve other kinds of extensions, as the semi-stable one [12], and also the coalition extension presented in [10]. Moreover, we can easily extend ConArg to include also user-defined constraint [13], as "extensions must contain argument a when they contain b" or "extensions must not contain one of c or d when they contain a but do not contain b".

Acknowledgements. We would like to thank the Davide Diosono, Valerio Egidi and Francesco Vicino, who developed the first version of ConArg as the final project of the exam "Constraint Systems", for their Master's degree in Computer Science at the University of Perugia.

References

1. Amgoud, L., Cayrol, C.: Inferring from inconsistency in preference-based argumentation frameworks. J. Autom. Reasoning 29(2), 125–169 (2002)
2. Amgoud, L., Devred, C.: Argumentation Frameworks as Constraint Satisfaction Problems. In: Benferhat, S., Grant, J. (eds.) SUM 2011. LNCS, vol. 6929, pp. 110–122. Springer, Heidelberg (2011)
3. Barabasi, A.L., Albert, R.: Emergence of scaling in random networks. Science 286(5439), 509–512 (1999)
4. Bench-Capon, T.J.M.: Persuasion in practical argument using value-based argumentation frameworks. J. Log. Comput. 13(3), 429–448 (2003)
5. Besnard, P., Doutre, S.: Checking the acceptability of a set of arguments. In: Workshop on Non-Monotonic Reasoning, pp. 59–64 (2004)
6. Bistarelli, S., Pirolandi, D., Santini, F.: Solving Weighted Argumentation Frameworks with Soft Constraints. In: Larrosa, J., O'Sullivan, B. (eds.) CSCLP 2009. LNCS, vol. 6384, pp. 1–18. Springer, Heidelberg (2011)
7. Bistarelli, S., Santini, F.: ConArg: ARGumentation with CONstraint, http://www.dmi.unipg.it/francesco.santini/argumentation/conarg.zip
8. Bistarelli, S., Santini, F.: A common computational framework for semiring-based argumentation systems. In: ECAI 2010 - 19th European Conference on Artificial Intelligence. Frontiers in Artificial Intelligence and Applications, vol. 215, pp. 131–136. IOS Press (2010)
9. Bistarelli, S., Santini, F.: Conarg: A constraint-based computational framework for argumentation systems. In: 23rd IEEE International Conference on Tools with Artificial Intelligence. IEEE (to appear 2011)
10. Bistarelli, S., Campli, P., Santini, F.: Finding partitions of arguments with Dung's properties via SCSPs. In: ACM Symposium on Applied Computing (SAC), pp. 913–919. ACM (2011)
11. Bordeaux, L., Hamadi, Y., Zhang, L.: Propositional satisfiability and constraint programming: A comparative survey. ACM Comput. Surv., 38 (December 2006)

12. Caminada, M.: Semi-stable semantics. In: Computational Models of Argument: Proceedings of COMMA 2006. Frontiers in Artificial Intelligence and Applications, vol. 144, pp. 121–130. IOS Press (2006)
13. Coste-Marquis, S., Devred, C., Marquis, P.: Constrained argumentation frameworks. In: Knowledge Representation and Reasoning (KR), pp. 112–122. AAAI Press (2006)
14. Dung, P.M.: On the acceptability of arguments and its fundamental role in non-monotonic reasoning, logic programming and n-person games. Artif. Intell. 77(2), 321–357 (1995)
15. Dunne, P.E., Hunter, A., McBurney, P., Parsons, S., Wooldridge, M.: Inconsistency tolerance in weighted argument systems. In: Conf. on Autonomous Agents and Multiagent Systems, pp. 851–858. IFAAMS (2009)
16. Egly, U., Alice Gaggl, S., Woltran, S.: ASPARTIX: Implementing Argumentation Frameworks using Answer-Set Programming. In: Garcia de la Banda, M., Pontelli, E. (eds.) ICLP 2008. LNCS, vol. 5366, pp. 734–738. Springer, Heidelberg (2008)
17. Gordon, T.F., Karacapilidis, N.I.: The zeno argumentation framework. KI 13(3), 20–29 (1999)
18. Harvey, W.D., Ginsberg, M.L.: Limited discrepancy search. In: IJCAI (1), pp. 607–615 (1995)
19. Kuchcinski, K., Szymanek, R.: Jacop - java constraint programming solver (2001), http://jacop.osolpro.com/
20. Modgil, S.: Reasoning about preferences in argumentation frameworks. Artif. Intell. 173(9-10), 901–934 (2009)
21. Nieves, J.C., Cortés, U., Osorio, M.: Possibilistic-based argumentation: An answer set programming approach. In: Mexican International Conference on Computer Science(ENC), pp. 249–260. IEEE Computer Society (2008)
22. O'Madadhain, J., Fisher, D., White, S., Boey, Y.: The JUNG (Java Universal Network/Graph) framework. Technical report, UC Irvine (2003)
23. Rahwan, T., Ramchurn, S.D., Jennings, N.R., Giovannucci, A.: An anytime algorithm for optimal coalition structure generation. J. Artif. Int. Res. 34, 521–567 (2009)
24. Ravid, G., Rafaeli, S.: Asynchronous discussion groups as small world and scale free networks. First Monday 9(9) (2004)
25. Rossi, F., van Beek, P., Walsh, T.: Handbook of Constraint Programming (Foundations of Artificial Intelligence). Elsevier Science Inc., New York (2006)

Resource Boundedness and Argumentation

Nicolás D. Rotstein, Nir Oren, and Timothy J. Norman

dot.rural Digital Economy Hub,
University of Aberdeen, United Kingdom
{nico.rotstein,n.oren,t.j.norman}@abdn.ac.uk

Abstract. In this paper we extend the traditional Dung argumentation framework with cardinality constraints over the set of warranted arguments. This results in a new definition for argumentation semantics wherein arguments within an extension are both in some sense consistent and compliant with the constraints imposed on the system. After discussing the theoretical aspects of such a *resource-bounded argumentation framework* we describe its utility via an application to a concrete application domain: the scheduling of demand responsive transport.

1 Introduction

A common subthread of recent work in argumentation has concentrated on applying abstract argumentation semantics to problems from other domains. Examples of such work include addressing normative conflict [15], trust [14], practical reasoning [3], amongst others. In this paper, we follow this tradition, and apply ideas from argumentation theory to the domain of demand responsive transport (DRT) [2]. This domain (described in more detail in Section 4) can be seen as an instantiation of a more general scheduling problem. Here, a set of passengers with certain requirements must be allocated to a set of vehicles, with each vehicle able to contain only a fixed number of passengers and traverse only a certain route during a single period of time. Now the semantics for abstract argument frameworks might allow us to identify some alternatives, but such semantics do not take the *resource bounds* (such as a vehicle being able to hold only a certain number of passengers) of the domain into account. This limitation suggests the possibility of enhancing Dung's argumentation framework [11] to cater for such resource bounds, and this paper investigates this process.

We modify Dung's seminal argument framework (AF) with the addition of constraints over different properties of arguments. For example, as we discuss in more depth in Section 4, we can associate a possible trip for a passenger with an argument. Now, if different passengers have different weights, and different vehicles have different maximum loads, we can constrain the legal combinations of trips based on the total weights the vehicles would have to carry. This constraint depends on a relatively complex attribute associated with an argument, which could be interpreted as consumption of resources. Taking this into account we then place some sort of limit (*e.g.*, less than, equal to or greater than) over the total number of arguments that can appear within an extension.

The remainder of this paper is structured as follows: we begin with Section 2 explaining *why* traditional AFs are ill-suited for representing resource bounds, in Section 3 we

S. Modgil, N. Oren, and F. Toni (Eds.): TAFA 2011, LNAI 7132, pp. 117–131, 2012.

formalise the notion of a *resource-bounded argumentation framework* (RAF), and then describe two possible semantics of such argumentation frameworks. Section 4 then examines our application domain in more detail, and finally we discuss related work and future extensions to our current approach in Section 5, before concluding in Section 6.

2   Resource Boundedness and Traditional Conflicts

Constraints and resource bounds add a new source of conflict to arguments which is not adequately captured by standard Dung argumentation frameworks. In order to illustrate this assertion, let us assume the existence of two undefeated arguments a_1, a_2 within some argument framework. Since both are undefeated, most semantics would have them warranted. Now consider the situation where each of these arguments represent the use of some resource, and only a single instance of this resource is available. This additional constraint does not make a_1 or a_2 any less acceptable but simply states that both arguments cannot be warranted together. The simplest representation for this situation is a mutual defeat. However, not every RBness[1] situation can be modelled with such simplicity.

Now consider the case of three undefeated arguments a_1, a_2, a_3 requiring resource r in order to be applicable. Again, any semantics would have these three arguments warranted. Assume that there are 2 available instances (or *tokens*) of r, and that the RB on r requires maximising its consumption; hence, exactly two of these arguments should be warranted together. One possible approach to representing this scenario utilising the previous idea is to set a mutual defeat between every pair of arguments, as shown in Figure 1(a). In this case, the preferred semantics would yield the following set of extensions: $\{\{a_1\}, \{a_2\}, \{a_3\}\}$. Intuitively, the desired set of extensions is $\{\{a_1, a_2\}, \{a_2, a_3\}, \{a_1, a_3\}\}$, *i.e.*, every combination of three arguments with no repetition, taken two at a time. This indicates that there are two problems that must be solved: devising a framework yielding the desired set of extensions, and dealing with the combinatorial explosion brought about by the construction of this set. This combinatorial explosion should be avoided both in the representation, as well as in the computation of warranted arguments.

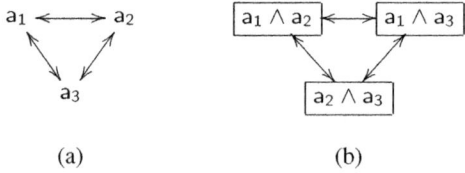

(a)   (b)

Fig. 1. (a) Three arguments, two tokens (b) Clusters of arguments

A potential solution for the representation problem generates a combinatorial problem: the creation of *clusters of arguments* attacking one another, as depicted in

---

[1] "RBness" will be used as the short form for "resource boundedness".

Figure 1(b). For n arguments competing for m ($m < n$) tokens of the same resource, the amount of clusters is $\frac{n!}{m!(n-m)!}$, which grows factorially. This is applicable to every set of arguments in the framework that is involved in RBness.

3   The Resource-Bounded Argumentation Framework

There is a need to add a new element to the argumentation framework to best represent the kind of conflicts brought by RBs, *i.e.*, conflicts at the level of extensions. RBs indicate that certain arguments, competing for a certain resource, cannot be warranted together. This *resource boundedness relation* will be a new element in the new framework we are proposing.

Definition 1 (Resource-Bounded Argumentation Framework (RAF))
A RAF is a tuple $\langle A, D, R \rangle$, where A is a set of arguments, $D \subseteq A \times A$ is the defeat relation over arguments[2], and $R \subseteq 2^A \times f$ is the resource boundedness relation over arguments, where $f : 2^A \rightarrow \{true, false\}$ is a boolean function.

We refer to elements of R, of the form (ρ, f) as resource bounds, abbreviated as RB. Here, ρ is a set of arguments, and f is the boolean function found in the resource boundedness relation.

The RB relation could be also represented with a single boolean function. We have chosen to explicitly specify each set of arguments along with its particular bounding function for the sake of clarity. Both approaches are equivalent.

The previous example concerning three arguments competing for two tokens of the same resource can now be represented by the RAF $\langle \{a_1, a_2, a_3\}, \{\}, \{(\{a_1, a_2, a_3\}, \Sigma = 2)\} \rangle$, as illustrated in Figure 2, where the sum symbol is a shortcut for the constraint over the amount of tokens allowed. Hence, $f(\{a_1, a_2, a_3\}) \equiv \sum(a_1, a_2, a_3) = 2 \equiv false$ and the RB needs to be taken into account.

In order to simplify our presentation, in the remainder of the paper we will only consider one type of resource bound, which operates over the summation of the number of arguments appearing in the extension and referred to by the resource bound. While we refer only to this type of RB, our results are intended to be applicable to any type of resource bound of the form described in the definition above.

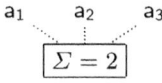

Fig. 2. Three arguments and a resource bound

Definition 2 (RB Compliance). *A set of arguments E **complies** (or is **compliant**) with an RB (ρ, f) iff $f(E \cap \rho) = true$. Given a set of RBs R, if E complies with every RB in R, we say that E **complies** (or is **compliant**) with R.*

---

[2] An argument a defeating an argument b will be written as a $\rightarrow$ b.

The combinatorial explosion over the representation is now solved[3], as there is no need to explicitly state all the sets of arguments compliant with all RBs, as illustrated in Figure 1(b). However, the semantic problem of how to compute extensions compliant with all RBs still remains. The rest of this section is focused on this, providing two different approaches.

RBs may not apply to all arguments found in a RAF. Arguments unaffected by RBs are referred to as *unbounded arguments*.

Definition 3 (Unbounded Argument). *An argument* a *is **unbounded** wrt. a set of RBs* R *iff* a $\notin \rho$, *for any* $(\rho, \cdot) \in$ R.

Before introducing the notion of a *resource-bounded extension* we need the definition for an *admissible set of arguments*. In this article, we restrict the study to the admissibility based semantics, and concentrate on the preferred semantics.

Definition 4 (Acceptable Arg., Admissible Set of Args., Preferred Extension [11]). *Given an AF* (A, *attacks*), *two arguments* a, b $\in$ A *and* $S \subseteq$ A:

1. a *is attacked by* S *iff there is an argument in* S *attacking* a;
2. a *is **acceptable** wrt.* S *iff for each* b: *if* b *attacks* a *then* b *is attacked by* S.
3. *A conflict-free set of args.* S *is **admissible** iff each arg. in* S *is acceptable wrt.* S.
4. *A maximal admissible set (wrt.* $\subseteq$) S *is the **preferred extension** of* (A, *attacks*).

Definition 5 (AF Extension). *Given a RAF* $F = \langle$A, D, R$\rangle$ *and an argumentation semantics* S, *a set of arguments* $E_{af} \subseteq$ A *is an **AF extension** of* F *following* S *iff* E_{af} *is an extension of the associated AF* $\langle$A, D$\rangle$ *following* S.

3.1 An Approach for RB-Compliant Extensions: Partitioning

Once the set of extensions of a framework has been computed, a straightforward solution for making extensions compliant with RBs is to partition them. Next, we define how such partitioning should be performed, and what the set of extensions looks like.

Definition 6 (RB Extension by Partitioning). *Let* $F = \langle$A, D, R$\rangle$ *be a RAF and* E_{af}, *an AF extension of* F *following argumentation semantics* S. *An **RB extension** $E_{rb} \subseteq$ E_{af} for* F *following* S *is a set of arguments such that all of the following hold:*

1. E_{rb} *complies with* R
2. E_{rb} *is admissible*
3. *if* a $\in$ E_{af} *and* a *is unbounded wrt.* R, *then* a $\in E_{rb}$

An RB extension denotes a set of arguments that is warranted in terms of the chosen semantics while being compliant with all RBs. Note that there is no maximality requirement for these extensions, as we only rely on compliance with RBs. Sometimes, a set of arguments qualifying as an RB extension could have several subsets that are also RB extensions, which differs with traditional argumentation semantics. However, as stated

---

[3] As long as representing f does not require an exponential amount of space wrt. the set of arguments, *e.g.*, by listing all the valid subsets.

by condition (3), every unbounded argument in an AF extension E_{af} must belong to an RB extension E_{rb}, and $E_{rb} \subseteq E_{af}$. This condition ensures some sort of partial maximality, *i.e.*, maximality only for those arguments that are not actually constrained by any resource limitation.

Consider the RAF depicted in Figure 2, changing the RB to $\Sigma \leq 2$. In this case, under the preferred semantics, every set of two arguments would be an RB extension, but also every singleton set, and even the empty set. All of these sets are admissible and compliant with the RB. Adding an unbounded argument a_4 to the example would make a_4 appear in all the RB extensions. Choosing one of these extensions is, as always, dependent on the application domain.

Returning to traditional AFs, an informal interpretation of their semantics is that an extension describes a maximal set of arguments that can be believed in despite having information in opposition. Hence, sometimes we will be interested in *maximal RB extensions*, with no regard to their subsets. Again, this is totally dependent on the application domain. Another question worth asking is whether there is a relation between the number of AF extensions and the number of RB extensions of a given RAF. The answer is that there is not. For instance, the RAF $\langle \{a\}, \{\}, \{(\{a\}, \Sigma \leq 1)\}\rangle$ has only one AF extension: $\{a\}$, but two RB extensions: $\{\{\}, a\}$. On the other hand, if we consider the RAF $\langle \{a, b\}, \{a \rightarrow b\}, \{(\{a, b\}, \Sigma = 2)\}\rangle$, it has one AF extension ($\{a\}$) but no RB extensions. Furthermore, even when considering single-extension semantics (such as the grounded semantics) the partitioning approach could yield multiple extensions.

Now it should be clear that there are some situations where no RB extension is compliant with some resource bound. We refer to such a case as a resource bound truncated RAF.

Definition 7 (RB Truncated RAF). *Given a set of AF extensions X for a RAF $F = \langle A, D, R\rangle$ following argumentation semantics S and an RB $\rho \in R$, F is **RB truncated** iff there is no $E' \subseteq E, E \in X$ such that E' complies with ρ.*

In other words, a RAF is RB truncated iff it has an empty set of RB extensions and a non-empty set of AF extensions. A simple example of RB truncation is a RAF with an argument a that is a member of all extensions and is the only one associated with an RB wherein $f = \Sigma \geq 2$ Such a RAF is truncated since the RB cannot be complied with. Truncated RAFs yield no extensions, and can thus not be solved. An external mechanism could be defined to discover the source of truncation, and even be used as a tool to understand what constraints are preventing the framework from yielding extensions. Dealing with truncated RAFs in this way is out of the scope of the current paper. While RB truncated RAFs may occur, we can also guarantee the existence of RB extensions in some cases, as formalised by the following proposition:

Proposition 1. *Given a RAF $F = \langle A, D, R\rangle$ and an argum. semantics S, it holds that:*

1. *E_{af} is both an AF extension and an RB extension for F following S iff E_{af} complies with R;*
2. *the set of AF extensions and RB extensions for F following S coincide iff every AF extension for F following S complies with every R.*

Due to space constraints, we do not provide algorithms for computing RB extensions in this paper. However, the following examples give some indication of this process. Here, and in the rest of the paper, examples will utilise the preferred semantics, unless stated otherwise.

Example 1. *Consider the following* RAF $F_1 = \langle A_1, D_1, R_1 \rangle$, *where* $A_1 = \{a, b, c\}$, $D_1 = \{a \rightarrow b, b \rightarrow c\}$, *and* $R_1 = \{(\{a, c\}, \Sigma \leq 1)\}$.

$$a \longrightarrow b \longrightarrow c$$
$$\boxed{\Sigma \leq 1}$$

The only AF extension of F is $\{a, c\}$, *which is not compliant with the RB* $\Sigma \leq 1$. *The subsets of this extension that verify Definition 6 are* $\emptyset$ *and* $\{a\}$, *since* $\{c\}$ *is not defended by its own set* (i.e., *is inadmissible*) *and thus does not meet condition* (2).

Example 2. *Let* $\langle A_2, D_2, R_2 \rangle$ *be a* RAF, *where* $A_2 = \{a, w, x, y, z\}$, $D_2 = \{x \rightarrow y, y \rightarrow x, z \rightarrow w\}$, *and* $R_2 = \{(\{a, z\}, \Sigma \leq 1), (\{a, x, z\}, \Sigma \leq 2)\}$, *representing that* a *and* z *require a resource whose only token has to be consumed, and* a, x, z *require a resource with 2 tokens available. Figure 3 shows* F_2 *along with the set of preferred AF extensions* $\{\{a, x, z\}, \{a, y, z\}\}$.

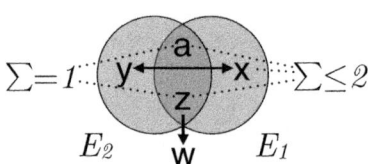

Fig. 3. AF extensions and resource bounds in F_2

Extension $E_1 = \{a, x, z\}$ *cannot be taken as a whole due to RB* $\Sigma \leq 2$. *Therefore, the seven RB-compliant subsets are:* $E_{11} = \{a, x\}$; $E_{12} = \{a, z\}$; $E_{13} = \{x, z\}$; $E_{14} = \{a\}$, $E_{15} = \{x\}$, $E_{16} = \{z\}$, $E_{17} = \emptyset$. *Note that some of these subsets are also tied to RB* $\Sigma = 1$. *Those that do not contain* a *and/or* z *will not be RB extensions, namely,* E_{12}, E_{15} *and* E_{17}. *Regarding extension* $E_2 = \{a, y, z\}$, *the subsets complying with RB* $\Sigma = 1$ *are* $E_{21} = \{a, y\}$; $E_{22} = \{z, y\}$. *Both comply with RB* $\Sigma \leq 2$. *Finally, the set of RB extensions is* $\{\{a\}, \{z\}, \{a, x\}, \{x, z\}, \{a, y\}, \{z, y\}\}$. *Maximal RB extensions wrt.* $\subseteq$ *are shown in Figure 4.*

The process of computing RB extensions can be seen to resemble a *meta-argumentation* process [5,7], up to one level. The difference between our approach and those based on meta-argumentation is the nature of these *meta-conflicts*. In the RAF, these are not traditional pairwise conflicts, but more complex structures relating sets of potentially warranted arguments to the resources they require and their availability.

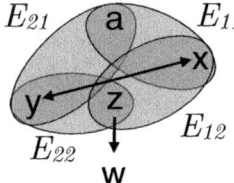

Fig. 4. Maximal RB extensions considering resource bounds in F_2

3.2   An Approach for RB-compliant Extensions: Modifying the RAF

There is an alternative for obtaining RB-compliant extensions. The previous approach partitions an extension whenever it encounters RBs that it does not comply with. Once every extension complies with all RBs, the process stops, and an appropriate criterion should be used to choose which extension is taken. In this section we introduce a different method, which removes enough arguments in every extension in order to make them compliant with all RBs, and we will show that this method can result in different extensions than the approach described in Definition 6.

Now one question that arises is whether this method is intuitively correct. More specifically, is it appropriate to remove arguments from an extension that have been deemed as warranted by the argumentation semantics? In order to answer this question, consider a framework with only one extension; every argument in that extension would be sceptically warranted. Now assume that a subset of these warranted arguments is not compliant with a certain RB, and that this situation could be resolved by removing some of the (previously) warranted arguments. The absence of those arguments could lead to other arguments becoming warranted. These newly warranted arguments could be seen to be "sub-optimally warranted". In some domains, accepting such arguments allows one to obtain a solution where no solution would have otherwise been computable, and we thus claim that there is a dependency on the application domain regarding whether taking these options instead of the (discarded) best ones makes sense.

Example 3. *Consider the RAF in Example 2, depicted in Figure 3. At least one argument has to be removed in order for the extensions to comply with* R_2. *Note that deleting z allows both RBs to be complied with. Now, since z no longer exists, argument w becomes undefeated, thus it is included in both extensions. This is shown in Figure 5.*

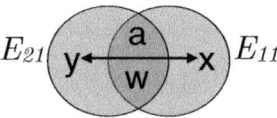

Fig. 5. Framework F_2 after an RB-driven deletion

If we compare extension E_{21} from Example 2 with extension E_{22} from Example 3, the difference is the inclusion of argument w in E_{22}. This occurred due to the modification of the framework caused by withdrawing x, the sole defeater of w.

In most domains, it makes little sense to withdraw arguments at random. Instead, some *selection criteria*, often based on preferences, can be used to identify which arguments should be withdrawn.

Definition 8 (Selection Criterion). *Given a set of arguments A, a **selection criterion** $\succ \subset A \times A$ determines a total order over A. Argument a being **preferred** to argument b is denoted as* a$\succ$b.

Now given a resource bound that needs to be satisfied, these selection criteria can guide us as to which arguments must be deleted in order to comply with the RB. We encapsulate this concept within the definition of an RB deletion, as follows.

Definition 9 (RB Deletion). *Let $F = \langle A, D, R \rangle$ be a RAF and $\succ$, a selection criterion over A. An **RB deletion** for a set of arguments γ wrt. an RB $(\rho, \mathfrak{f}) \in R$ is a set of arguments $\delta \subseteq \rho$ such that both of the following hold:*

1. $\mathfrak{f}(\rho \cap \gamma \setminus \delta) = true$.
2. *for every* e $\in \delta$ *and every* f $\in (\rho \setminus \delta)$ *it holds that* f$\succ$e.

The first condition requires the set including those arguments that are both in γ and the RB (*i.e.*, in ρ), minus those in the RB deletion δ, to comply with the RB. The second condition specifies that those arguments that are to be left out are the least preferred wrt. the selection criterion. Note that unbounded arguments are never included into RB deletions, as these are subsets of the RB set.

Since the selection criterion has to make a decision for every pair of arguments, an RB deletion will leave out the least preferred arguments in the set in order to satisfy some RB. However, there could be many RB deletions associated with each RB, and thus a choice has to be made. For instance, consider the RB deletion for set {a, b} wrt. RB ({a, b}, $\Sigma \leq 1$), where a$\succ$b. The two valid RB deletions are {b} and {a, b}, as both satisfy the RB. If we choose to delete the least number of arguments, we would drop only b, the least preferred argument. The choice is up to the application domain.

We are now in a position to define how an RB extension can be computed through RB deletions. Informally, given an extension of the underlying argument framework wrt. some semantics, and a subset of R which this extension does not comply with, we identify those RB deletions that, when applied to the original argument framework, will allow all RBs to be complied with. We then compute the RB extension by creating a new argumentation framework that does not contain those arguments found in the RB deletions. This is formalised by the following definition.

Definition 10 (RB Extension by Deletions). *Let $F = \langle A, D, R \rangle$ be a RAF. An **RBD extension** E_{rbd} for F following argumentation semantics S is an extension of the AF $\langle A', D' \rangle$ following S, where:*

1. $A' = A \setminus \{\delta \mid$ *where δ is an RB deletion for an AF extension for F following S wrt. an RB in* R$\}$;

2. $D' = \{a \rightarrow b \in D \mid a \in A', b \in A'\}$;

3. E_{rbd} complies with R.

The process of computing RBD extensions can be seen as operating in two steps. First, given an RB unaware solution (the AF extension), arguments that are not compliant with the RBs are removed. Following this, the argumentation process is repeated in order to discover new solutions. Deleting a warranted argument allows for the consideration of those arguments that were defeated by it. Therefore, RBD extensions may include arguments that were not present in any AF extension.

Now it is important to note that different combinations of RB deletions could yield different RBD extensions, and our definition permits any of these combinations to be applied. Thus a single RAF may yield multiple RBD extensions for each extension computable from the original AF according to some semantics.

Example 4. *Consider the* RAF $\langle \{a, b\}, \{\}, \{(\{a, b\}, \Sigma \leq 2)\} \rangle$ *and the selection criterion* a≻b. *The AF extension* $\{a, b\}$ *is already compliant with the RB. However, the possible RB deletions are:* $\emptyset, \{b\}, \{a, b\}$. *It is up to the application domain to choose whether to maximise resource consumption, minimise it, or do something in between.*

Example 5. *Consider* RAF F_2 *as in Example 3, with the selection criterion defined as* a≻x≻z. *The least preferred argument in* $(\{a, x, z\}, \Sigma \leq 2)$ *is z and its deletion renders both RB functions true. Hence, the only RB deletion is* $\{z\}$. *Finally, the RBD extensions of* F_2 *are the extensions of the AF* $\langle A'_2, D'_2 \rangle$, *where* $A'_2 = \{a, w, x, y\}$ *and* $D' = \{y \rightarrow x, x \rightarrow y\}$.

The following example illustrates a situation where RB extensions and RBD extensions can overlap.

Example 6. *Consider the* RAF $\langle A_6, D_6, R_6 \rangle$, *as depicted in Figure 6. Set of arguments* $\{c, e\}$ *is not compliant with the RB* $\Sigma = 1$. *In the partitioning approach, extension* $E = \{b, c, e, f\}$ *should be split in two:* $E_1 = \{b, e, f\}$ *and* $E_2 = \{b, c, f\}$, *but* E_2 *does not comply with the RB* $\Sigma > 1$, *nor does any subset, since the RB specifies a lower bound. Hence the only RB extension is* E_1.

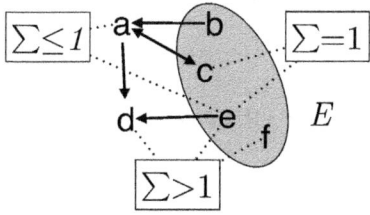

Fig. 6. RAF for Example 6

Regarding the approach by deletions, let assume the selection criterion establishes that c is the least preferred argument within E. In this case, the resulting AF would yield E_1 *again. On the other hand, if e were the least preferred argument and hence removed, the resulting AF would yield the extension* $\{b, c, d, f\}$.

Finally, to illustrate the utility of RBD extensions, consider the following example, which presents an RB truncated RAF that possesses a non-empty RBD extension.

Example 7. *Consider the RAF F_7 depicted in Figure 7. The only AF extension for F_7 is $\{a, c, d\}$. The partitioning approach would attempt to shrink that extension until it gets RB-compliant subsets. Note that this is not possible, as there is no partition compliant with RB $\Sigma = 3$.*

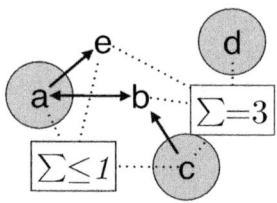

Fig. 7. RAF for Example 7

On the other hand, the approach by deletions would consider one of the RB deletions $\{c\}$ and $\{a, c\}$, given that both of them yield the RBD extension $\{b, d, e\}$, which is compliant with both RBs. In this case, we could choose to perform a minimal RB deletion, removing only c.

It is important to note that the application of one RB deletion, which deletes arguments from the AF, can cause other RBs in the RAF to no longer be complied with, requiring additional RB deletions over those previously applied in order to comply with an RB. This suggests that the order in which RB deletions are applied is important when computing extensions, and that an implemented system might have to utilise backtracking if properties such as the maximality of an RBD extension are required.

As with the partitioning approach, there is no general relation between the cardinality of the set of RBD extensions wrt. the set of AF extensions of a given RAF.

Proposition 2. *Given a RAF $F = \langle A, D, R \rangle$ and an argum. semantics S, it holds that:*

1. *E_{af} is both an AF extension and an RBD extension for F following S iff the empty set is an RBD deletion for E_{af} wrt. an RB in R;*
2. *the set of AF extensions and RBD extensions for F following S coincide iff the empty set is an RB deletion for every AF extension for F following S wrt. every RB in R.*

Now, given a RAF, it could be the case that there is no RB deletion for one or more members of its RB set R. We call such a RAF an *RBD truncated* RAF. However, a RAF may not be truncated but still have no RBD extensions, as it is possible that a feasible set of RB deletions does not exist.

Having formalised various concepts around the idea of resource bounded argumentation, we now describe its use in the FITS project, wherein it is used to assess the alternative solutions provided by a transport scheduler.

4   The FITS **Project: An Application for the RAF**

The FITS project (standing for "flexible integrated transport services") falls within the Rural Digital Economy Hub (*dot.rural*[4]) agenda, funded by Research Councils UK. FITS aims at providing a *virtual transport marketplace* to improve the existing connection between transport demand and supply in rural areas of the United Kingdom. The engine of this system is argumentation-based.

Flexible transportation systems (FTS) provide services to users based on demand, attempting to maximise ubiquitousness without compromising cost and quality. FTSs involve several elements, namely: acquisition of data (passengers, traffic, environment, fleet, *etc.*), evaluation of plausible options and journey planning. In this research the main focus is set on assisting the decision making process of passengers when selecting transport mode and route, while taking into account the operators' preferences, who provide resources. The objective of this research is the implementation of an argumentation-based expert system built upon a multi-agent system. Each passenger would have an associated agent, in charge of collecting relevant data, either permanently or triggered by trip requests. Similarly, agents will act on behalf of operators, imposing their restrictions and preferences regarding the usage of vehicles.

The aim of this effort is not to emulate (or replace) already existing FTS solutions for scheduling [2,8]. Instead, our approach looks to add value to the user's choice, taking the plausible options given as an output by some scheduler. We contend that the nature of argumentation for decision making allows for a clear presentation of how the process is carried out. Moreover, an interesting interface challenge lies in how to show an incrementally complex argument graph backing the suggestions made by the system.

Including RBs into the system permits a more concise representation, while accurately reflecting how the decision is being made. The following are the different elements that come into play in the FITS project domain:

- each possible journey for any passenger will constitute an **argument**;
- each sub-graph containing the alternatives for one passenger will be star-connected with **conflicts**;
- **defeat** is determined upon conflict, relying on passengers' preferences;
- the number of available seats in a given vehicle at a certain stop will determine a **resource bound**;
- the **selection criterion** for arguments is designed to ensure that passengers with fewer options get seats.

Additional RBs can also be taken into account, such as operator-side constraints controlling the minimum number of seats or minimum revenue. However, these have not yet been integrated into the system.

Our current focus involves how to determine the best journey for each passenger, while taking into account other passengers through RBs. This means that passenger's choices are not always weighted in isolation; a global balance must be sought in order to achieve a certain degree of fairness. We have implemented a prototype system, which we are currently evaluating over randomly generated scenarios (maps, passengers

---

[4] http://www.dotrural.ac.uk

and vehicles). The system computes possible journeys (*i.e.*, arguments), conflicts, RBs, and then makes a decision using the RBD approach on top of the preferred semantics. Within FITS, individual passengers' arguments associated with different journey options form separate subgraphs linked by RBs, and all RBs indicate a single upper bound (representing seat availability within a vehicle). When an RB is not met, we choose to maximise resource consumption, and therefore consider minimal RB deletions.

Figure 8 illustrates an argument graph generated by a sample scenario from our system. Rectangles representing individual journeys (and thus arguments) contain a passenger name, the journey's total cost and distance, the sequence of stops, the sequence of vehicles used and at which stops, and the passenger's preferences. For example, argument (6) states that one of Ruth's possible journeys has a cost of $52; a length of 86 kilometres; goes from $s4$ to $s5$ through $s1$ and $s3$, and is travelled by taking vehicles $v7$ at $s4$ and then $v5$ at $s1$; finally, she has a preference for shorter trips. If distances are equal, she looks for as few changes as possible, and if these are equal she looks to minimise the trip's cost. If multiple trips equally satisfy all these requirements, one of them is randomly selected. Solid arrows within the graph indicate defeat based on preferences, with two-way arrows linking equivalently preferred options. Elliptical nodes indicate RBs due to seat restrictions; for instance, $v5/s3(1)$ means that vehicle $v5$ at stop $s3$ has only one seat available.

Treating Figure 8 as a standard AF, we see that trip (6) would appear in all preferred extensions. This indicates that this trip is the most preferred one by Ruth. However, when considering resource bounds, our system does not allow Ruth to undertake this trip. To see why, consider, for example, the resource bound $v5/s3(1)$, stating that vehicle 5 has only one free seat at location s3. Utilising RBD extensions, our seat allocation strategy

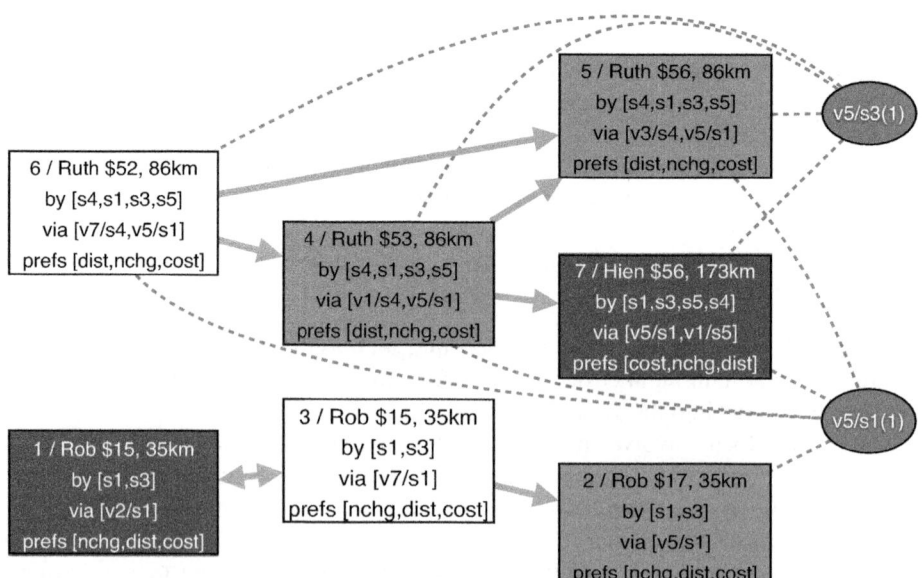

Fig. 8. Screenshot

(allocating seats to passengers with fewer options) means that $(7) \succ (6)$[5]. Therefore, argument (7) will appear in the RBD extension, while argument (6) will not, meaning that Hien will obtain this vehicle's seat. This resource bound also means that arguments (4) and (5) cannot appear in the RBD extension, and neither does argument (2). Since arguments (1) and (3) are equally preferred, the system randomly allocates to Rob the trip represented by argument (1). Within Figure 8, the different statuses of these arguments is indicated by different colours and shadings.

5   Discussion and Future Work

As hinted at in Section 4, selection criteria can be used in the process of computing RBD extensions, to provide us with certain desirable properties for those extensions. Such properties could, for example, include minimising the number of arguments removed from an extension, maximising the size of some (or the largest) extension, and so on. Such properties have an analogy with the concept of *minimal change* in the area of belief revision [1]. More specifically, since the RB deletion approach fundamentally modifies the knowledge base, it is desirable to ensure that this change is performed with the smallest possible representational impact.

One (computationally intensive) approach for minimising the representational impact of RB deletions would involve considering all possible RB deletions, examining the resultant extension/s, and selecting the one that meets our (domain specific) requirements. As an example, consider yet another possible metric of minimal change: preservation of arguments within an extension. Referring to Example 1, we see that the AF extension $\{a, c\}$ does not comply with RB $\Sigma \leq 1$. There are three possible RB deletions: $\{a, c\}, \{a\}, \{c\}$. The first and second options yield RBD extension $\{b\}$, while the third one yields $\{a\}$. The latter is the one that best preserves the AF extension. By having $\{c\}$ as the least preferred extension wrt. the selection criterion, we could ensure that the desired extension is obtained. One interesting piece of future work involves determining the conditions required to guarantee that RBD extensions will meet some representational impact requirements. One possible inspiration for this work might arise from the recent interest in change over argumentation frameworks [4,6,17].

The formal properties of RAFs provide fertile ground for additional future work. For example, it would be useful to identify the situations under which RB deletions exist but would not yield an RBD extension. In the short term, our focus lies in investigating whether a mapping exists between RAFs and traditional AFs. The propositions in this paper identify some situations where such a mapping exists, and a more complex conversion process, following the ideas of [16], has allowed us to map between RAFs and Dung AFs in many situations. However, this mapping yields incorrect results in some cases, and we intend to investigate this further.

Regarding related work, in [10] a *constrained argumentation framework* is presented. A propositional formula is added to the framework to place a constraint on admissible sets of arguments. In contrast, we propose linking arguments directly to resource bounds, which is more natural and compact as a representation. They redefine the grounded, preferred and stable semantics to make extensions compliant with the

---

[5] The rest of the ordering determined by the selection criterion is not relevant for this example.

constraint. The way in which the latter is done resembles our partitioning approach. However, they do not present an alternative similar to our approach by deletions.

Within the argumentation domain, recent work on weighted argument frameworks [12] bears some relation to our approach. Here, the authors define an abstract argument framework that assigns different weights to attacks. An inconsistency budget permits conflicts to exist within an extension only when the total weight of attacks is below this budget. This can be viewed as a single type of constraint on attacks. In our current work we constrain arguments, rather than attacks, and allow for arbitrary types of constraints. We believe that there is no mapping between weighted frameworks and RAFs, unless attacks are associated to resource consumption, but further theoretical investigation is needed to verify this intuition. More generally, RBs within a RAF can be seen as a type of constraint. Hence, we will study what is the relation behind RAFs and constraint satisfaction problems [13] and constraint optimisation [9].

6   Conclusions

In this article we have presented a novel approach to formal argumentation that takes into account limitation of resources. The introduction of resource bounds (RB) calls for a redefinition on how to compute the set of warranted arguments. To this end, we propose two different approaches: one of them considers partitions of pre-calculated extensions; the other one considers the removal of those arguments that make extensions not to be compliant with certain RBs.

Resource-bounded argumentation frameworks are based on Dung's standard argument framework, but augment it by labelling arguments with additional parameters, and placing constraints on the combinations of labels that are permitted to exist within an extension. We concentrated on arguments with a single label consisting of the number 1, with our constraints requiring that the sum of these for some set of arguments to be equal, less than, or greater than some value. Naturally, our formalism allows for the implementation of any other constraint.

Finally, we detailed how RAFs can be used to attack an important real world problem, namely that of scheduling dynamic transport provision. We believe that RAFs are applicable to a large variety of other domains where constraints exist, and can be used to bring the strong reasoning processes found in argumentation theory to tackle such problems, something which current work in argumentation has been unable to do.

Acknowledgments. The authors would like to thank Nagendra R. Velaga and John D. Nelson (*dot.rural Digital Economy Hub, University of Aberdeen*) for their contribution on the transportation domain application.

References

1. Alchourrón, C., Gärdenfors, P., Makinson, D.: On the logic of theory change: Partial meet contraction and revision functions. The Journal of Symbolic Logic 50, 510–530 (1985)
2. Ambrosino, G., Nelson, J.D., Romanazzo, M.: Demand responsive transport services: Towards the flexible mobility agency. ENEA Publications (2004)

3. Atkinson, K., Bench-Capon, T.J.M.: Practical reasoning as presumptive argumentation using action based alternating transition systems. Artif. Intell. 171(10-15), 855–874 (2007)
4. Baumann, R., Brewka, G.: Expanding argumentation frameworks: Enforcing and monotonicity results. In: COMMA, pp. 75–86 (2010)
5. Boella, G., Gabbay, D., van der Torre, L., Villata, S.: Meta-Argumentation Modelling I: Methodology and Techniques. Studia Logica 93, 297–355 (2009)
6. Boella, G., Kaci, S., van der Torre, L.: Dynamics in argumentation with single extensions: Abstraction principles and the grounded extension. In: Sossai, C., Chemello, G. (eds.) ECSQARU 2009. LNCS, vol. 5590, pp. 107–118. Springer, Heidelberg (2009)
7. Boella, G., van der Torre, L., Villata, S.: On the acceptability of meta-arguments. In: IAT, pp. 259–262 (2009)
8. Brake, J., Mulley, C., Nelson, J.D., Wright, S.: Key lessons learned from recent experience with flexible transport services. Transport Policy 14(6), 458–466 (2007)
9. Chua, L.O., Lin, G.N.: Non-linear optimization with constraints: A cook-book approach. International Journal of Circuit Theory and Applications 11(2), 141–159 (1983)
10. Coste-Marquis, S., Devred, C., Marquis, P.: Constrained argumentation frameworks. In: KR, pp. 112–122 (2006)
11. Dung, P.M.: On the acceptability of arguments and its fundamental role in nonmonotonic reasoning, logic programming and n-person games. Artif. Intell. 77(2), 321–358 (1995)
12. Dunne, P.E., Hunter, A., McBurney, P., Parsons, S., Wooldridge, M.: Weighted argument systems: Basic definitions, algorithms, and complexity results. Artif. Intell. 175(2), 457–486 (2011)
13. Kumar, V.: Algorithms for constraint-satisfaction problems: a survey. AI Mag. 13, 32–44 (1992)
14. Matt, P.A., Morge, M., Toni, F.: Combining statistics and arguments to compute trust. In: AAMAS, pp. 209–216 (2010)
15. Oren, N., Panagiotidi, S., Vázquez-Salceda, J., Modgil, S., Luck, M., Miles, S.: Towards a Formalisation of Electronic Contracting Environments. In: Hübner, J.F., Matson, E., Boissier, O., Dignum, V. (eds.) COIN@AAMAS 2008. LNCS, vol. 5428, pp. 156–171. Springer, Heidelberg (2009)
16. Oren, N., Reed, C., Luck, M.: Moving between argumentation frameworks. In: Computational Models of Argument, Proceedings of COMMA 2010, pp. 379–390 (2010)
17. Rotstein, N.D., Moguillansky, M., Falappa, M.A., García, A.J., Simari, G.R.: Argument Theory Change: Revision Upon Warrant. In: COMMA, pp. 336–347 (2008)

An Empirical Study of a Deliberation Dialogue System

Elizabeth Black[1] and Katie Bentley[2]

[1] Intelligent Systems Group, Universiteit Utrecht, De Uithof, 3584 CC Utrecht, NL
lizblack@cs.uu.nl
[2] Vascular Biology Lab, London Research Institute, Cancer Research UK, Lincoln's Inn Fields,
WC2A 3LY, UK
katie.bentley@cancer.org.uk

Abstract. We present an empirical simulation-based study of the use of value-based argumentation in two-party deliberation dialogues, investigating the impact that argumentation can have on the quality of the outcome reached. Our simulation allows us to vary the number of values, actions and arguments that appear in the system; we investigate how the behaviour of the system changes as these parameters vary. This parameter sensitivity analysis tells us whether a value-based deliberation dialogue system may be useful for a particular real-world application. We measure the quality of the dialogue outcome (i.e. the action that the agents agree to) against a global view of whether that action would be agreeable to each agent if all of the agents' knowledge were taken into account. We compare the deliberation outcome with a simple consensus forming procedure (where no arguments are exchanged). Our results show that the deliberation dialogue system we present outperforms consensus forming.

ACM Category: I.2.11 Multiagent systems.

General terms: performance, experimentation.

Keywords: dialogue, value-based argumentation, simulation, agreement, deliberation.

1 Introduction

There is little work on evaluating whether an argumentation-based approach to a problem is a good approach to take. Most works assume that the decision to use argumentation has already been made and disregard the question of whether there is a better approach to take. We present what we believe to be the first simulation-based study of an argumentation-based deliberation dialogue system, which allows us to start addressing this question and allows us to investigate the effect of varying the parameters of the system.

Simulation is an imperative next step for bridging the gap between argumentation theory and real-world agent applications. Given the complexity of argumentation-based dialogue systems, it is very hard to theoretically investigate their properties without making many restrictive assumptions. In order to gain a full understanding of the behaviour of such systems, theoretical investigations need to be complemented with empirical simulation-based studies. Simulation provides a unique opportunity to generate

S. Modgil, N. Oren, and F. Toni (Eds.): TAFA 2011, LNAI 7132, pp. 132–146, 2012.
© Springer-Verlag Berlin Heidelberg 2012

large, complex scenarios and analyse their results across thousands of iterations and permutations.

There are few existing works that take a simulation approach to investigating the performance of argumentation-based dialogue systems. Two notable examples are [1,2]. Each of these focus on a form of argumentation-based negotiation (ABN), where arguments providing reasons for an agent's position are shared; this exchange of information allows the negotiation space to change. In [1], the information exchanged relates to the influence of social commitments between roles, whilst [2] focusses on interest-based negotiation where agents exchange information about their underlying goals and different ways to achieve these.

In [3], ABN is used to address the distributed constraint satisfaction problem. Importantly, the authors have performed experiments with their model to investigate the performance of their argument-based approach. Agents in the system use arguments in the sense that they put forward a proposal and provide a justification for this by giving their local constraints, these constraints are propagated by the receiving agent.

Our deliberation context differs from ABN (which is generally concerned with the allocation of scarce resources), as agents in our system have a shared goal and wish to come to an agreement on how to act in order to achieve that goal. Similarly to the systems discussed above, our agents also share arguments regarding actions to achieve the goal and this allows the set of actions that an agent finds agreeable to change. In our system, however, these arguments are value-based (relating to various social values that may be promoted or demoted by performing an action) and, unlike in [1,2,3], our agents also use argumentation as the reasoning mechanism with which they determine which actions they find agreeable.

We specifically investigate two questions:

– Do our deliberation dialogues perform better across the entire parameter space than a simple consensus forming approach, where agents try to find an action they each find agreeable without sharing any arguments?
– How does the behaviour of both the dialogue system and the consensus forming mechanism change as the number of arguments, actions and values present in the system varies?

Our results clearly show that the deliberation dialogue system outperforms consensus forming across all parameter combinations. Further, we have identified particular parameter settings that optimise dialogue performance in terms of quality of outcome and length of dialogue. This detailed parameter sensitivity analysis allows a designer of an agent system to evaluate whether value-based deliberation dialogues are useful for their particular application domain.

2 Model

In this section we describe the model that we are simulating. We give details of the value-based argumentation model, the dialogue system, the consensus forming mechanism, the evaluation metric that we use and our experimental set up. The model was written in c++ on a standard workstation. A complete parameter sensitivity analysis of 1.8 million runs took less than an hour to complete.

2.1   Argumentation Model

We are investigating the performance of the system formally specified in [4], which is based on the popular argument scheme and critical question approach [5]. Arguments are generated by an agent instantiating a **scheme for practical reasoning** [6]: In the current circumstances R, we should perform action A, which will result in new circumstances S, which will achieve goal G, which will promote value V.

The scheme is associated with a set of characteristic critical questions (CQs) that can be used to identify challenges to proposals for action that instantiate the scheme. An unfavourable answer to a CQ will identify a potential flaw in the argument. Since the scheme makes use of what are termed as 'values', this caters for arguments based on subjective preferences as well as more objective facts. Such values represent qualitative social interests that an agent wishes to uphold by realising the goal stated [7].

An agent has a **Value-based Transition System** (VATS), that it uses to instantiate the scheme for practical reasoning. This transition system represents the agent's knowledge about the effect of actions and the values that are promoted or demoted. (For brevity, we omit the definition here; the reader is referred to [4].) Given its VATS, an agent can instantiate the practical reasoning argument scheme in order to construct arguments for (or against) actions to achieve a particular goal because they promote (or demote) a particular value. Note that here we are focussing on the **choice of action** stage (as defined in [6]), we assume that any discrepancies between the agents in either the problem formulation or epistemic reasoning stages have been resolved (perhaps with some other type of dialogue); thus, for example, agents do not need to question here whether an action in question does achieve the desired goal or whether a certain set of circumstances hold.

Definition 1. An **argument** *constructed by an agent x from its VATS is a 4-tuple $A = \langle a, p, v, s \rangle$ where:*
$s = +$ *iff a is an* **action** *that will achieve* **goal** *p and will* **promote** *value v;*
$s = -$ *iff a is an* **action** *that will achieve* **goal** *p but will* **demote** *value v.*

We define the functions: $\mathsf{Act}(A) = a$; $\mathsf{Goal}(A) = p$; $\mathsf{Val}(A) = v$; $\mathsf{Sign}(A) = s$. *If* $\mathsf{Sign}(A) = +(-resp.)$, *then we say A is a* **positive (negative** *resp.) argument* **for (against** *resp.) action a. We denote the* **set of all arguments an agent** *x* **can construct from its VATS** *as $Args^x$; we let $Args^x_p = \{A \in Args^x \mid \mathsf{Goal}(A) = p\}$. The set of* **values** *for a set of arguments $\mathcal{X}$ is defined as* $\mathsf{Vals}(\mathcal{X}) = \{v \mid A \in \mathcal{X} \text{ and } \mathsf{Val}(A) = v\}$.

If we take a particular argument for an action, it is possible to generate attacks on that argument by posing the various CQs related to the practical reasoning argument scheme. The relevant CQs are used to generate a set of arguments for and against different actions to achieve a particular goal, where each argument is associated with a motivating value. To evaluate the status of these arguments we use a Value Based Argumentation Framework (VAF) (introduced in [7]), an extension of the argumentation frameworks (AF) of Dung [8]. In an AF an argument is admissible with respect to a set of arguments S if all of its attackers are attacked by some argument in S, and no argument in S attacks an argument in S. In a VAF an argument succeeds in defeating an argument it attacks if its value is ranked higher than or at least as high as the value of the argument attacked;

a particular ordering of the values is characterised as an **audience**. Arguments in a VAF are admissible with respect to an audience A and a set of arguments S if they are admissible with respect to S in the AF which results from removing all the attacks which are unsuccessful given the audience A. A maximal admissible set of a VAF is known as a **preferred extension**.

Although VAFs are often considered abstractly, here we give an instantiation in which we define the attack relation between the arguments. This attack relation is derived from the CQs, for details the reader is referred to [4].

Definition 2. *An* **instantiated value-based argumentation framework** *(iVAF) is defined by a tuple* $\langle \mathcal{X}, \mathcal{A} \rangle$ *s.t.* $\mathcal{X}$ *is a finite set of arguments and* $\mathcal{A} \subset \mathcal{X} \times \mathcal{X}$ *is the* **attack relation**. *A pair* $(A_i, A_j) \in \mathcal{A}$ *is referred to as* "A_i *attacks* A_j" *or* "A_j *is attacked by* A_i". *For two arguments* $A_i = \langle a, p, v, s \rangle$, $A_j = \langle a', p', v', s' \rangle \in \mathcal{X}$, $(A_i, A_j) \in \mathcal{A}$ *iff* $p = p'$ *and either: (1)* $a = a'$, $s = -$ *and* $s' = +$; *or (2)* $a = a'$, $v \neq v'$ *and* $s = s' = +$; *or (3)* $a \neq a'$ *and* $s = s' = +$.

An **audience** *for an agent* x *over the values* V *is a binary relation* $\mathcal{R}^x \subset V \times V$ *that defines a* total order *over* V *where exactly one of* (v, v'), (v', v) *are members of* $\mathcal{R}^x$ *for any distinct* $v, v' \in V$. *If* $(v, v') \in \mathcal{R}^x$ *we say that* v *is* **preferred to** v', *denoted* $v \succ_{\mathcal{R}^x} v'$. *We say that an argument* A_i *is* **preferred to** *the argument* A_j *in the audience* $\mathcal{R}^x$, *denoted* $A_i \succ_{\mathcal{R}^x} A_j$, *iff* $\mathtt{Val}(A_i) \succ_{\mathcal{R}^x} \mathtt{Val}(A_j)$. *If* R^x *is an audience over the values* V *for the iVAF* $\langle \mathcal{X}, \mathcal{A} \rangle$, *then* $\mathtt{Vals}(\mathcal{X}) \subseteq V$.

We use the term 'audience' to be consistent with the literature. Note, however, audience does not refer to the preference of a *set* of agents; rather, it represents a particular agent's preference over values.

Given an iVAF and a particular agent's audience, we can determine acceptability of an argument as follows. Note that (as in [4]) if an attack is symmetric, then an attack only succeeds in defeat if the attacker is more preferred than the argument being attacked; however, if an attack is asymmetric, then an attack succeeds in defeat if the attacker is at least as preferred as the argument being attacked. Asymmetric attacks occur only when an argument against an action attacks another argument for that action; in this case, if both arguments are equally preferred then we do not wish the argument for the action to withstand the attack. If we have a symmetric attack where the arguments attacking one another are equally preferred, then we must have arguments for two different actions that promote the same value; here, the defeat is not successful, since it is reasonable to choose either action.

Definition 3. *Let* $\mathcal{R}^x$ *be an audience and let* $\langle \mathcal{X}, \mathcal{A} \rangle$ *be an iVAF.*
For $(A_i, A_j) \in \mathcal{A}$ *s.t.* $(A_j, A_i) \notin \mathcal{A}$, A_i **defeats** A_j *under* $\mathcal{R}^x$ *if* $A_j \nsucc_{\mathcal{R}^x} A_i$.
For $(A_i, A_j) \in \mathcal{A}$ *s.t.* $(A_j, A_i) \in \mathcal{A}$, A_i **defeats** A_j *under* $\mathcal{R}^x$ *if* $A_i \succ_{\mathcal{R}^x} A_j$.
An argument $A_i \in \mathcal{X}$ *is* **acceptable w.r.t** S *under* $\mathcal{R}^x$ $(S \subseteq \mathcal{X})$ *if: for every* $A_j \in \mathcal{X}$ *that defeats* A_i *under* $\mathcal{R}^x$, *there is some* $A_k \in S$ *that defeats* A_j *under* $\mathcal{R}^x$.
A subset S *of* $\mathcal{X}$ *is* **conflict-free** *under* $\mathcal{R}^x$ *if no argument* $A_i \in S$ *defeats another argument* $A_j \in S$ *under* $\mathcal{R}^x$.
A subset S *of* $\mathcal{X}$ *is* **admissible** *under* $\mathcal{R}^x$ *if: S is conflict-free in* $\mathcal{R}^x$ *and every* $A \in S$ *is acceptable w.r.t* S *under* $\mathcal{R}^x$.

A subset S of $\mathcal{X}$ is a **preferred extension** *under $\mathcal{R}^x$ if it is a maximal admissible set under $\mathcal{R}^x$.*

An argument A is **acceptable** *in the iVAF $\langle \mathcal{X}, \mathcal{A} \rangle$ under audience $\mathcal{R}^x$ if there is some preferred extension containing it.*

We have defined a mechanism with which an agent can determine attacks between arguments for and against actions; it can then use an ordering over the values that motivate such arguments (its audience) in order to determine their acceptability. Next, we define our dialogue system.

2.2   Dialogue System

The dialogue system investigated here is formally defined in [4]. For readability and brevity, we omit the formal definitions here but informally describe the dialogue system. The communicative acts in a dialogue are called **moves**. We assume that there are always exactly two agents (**participants**) taking part in a dialogue, each with its own identifier taken from the set $\mathcal{I} = \{Ag1, Ag2\}$ and each with a knowledge base of arguments that it knows about (those it can construct from its VATS). Each participant takes it in turn to make a move to the other participant. We refer to participants using the variables x and $\overline{x}$ such that: x is $Ag1$ if and only if $\overline{x}$ is $Ag2$; x is $Ag2$ if and only if $\overline{x}$ is $Ag1$.

We assume that the participants have agreed to partake in a deliberation dialogue whose **topic** is the joint goal in question. During the dialogue, agents can either:

- assert a positive argument (an argument *for* an action);
- assert a negative argument (an argument *against* an action);
- agree to an action;
- indicate that they have no arguments that they wish to assert (with a pass).

The agents take it in turn to make a single move. A dialogue terminates under one of two conditions: **failure**, when two pass moves appear one immediately followed by the other in the dialogue; **success** with **outcome** a, when two moves each agreeing to the action a appear one immediately followed by the other in the dialogue.

In order to evaluate which actions it finds agreeable at a point in a dialogue with topic p, an agent x considers the iVAF that it constructs from all the arguments that it currently has available to it relating to p; this consists of the arguments from its own VATS, as well as the arguments that the other agent has asserted thus far. We call this agent x's **dialogue iVAF**, which is the iVAF $\langle \mathcal{X}, \mathcal{A} \rangle$ where $\mathcal{X} = Args_p^x \cup \{A \mid \overline{x}$ has previously asserted A during the dialogue$\}$. An action is **agreeable** to an agent x if and only if there is some argument *for* that action that is acceptable in x's dialogue iVAF under the audience that represents x's preference over values. Note that the set of actions that are agreeable to an agent may change over the course of the dialogue, due to its dialogue iVAF changing as arguments asserted by $\overline{x}$ are added to it.

The **protocol** defines which moves an agent x (whose turn it is) is allowed to make at any point in a deliberation dialogue with topic p as follows:

- It is permissible to assert an argument A iff $\text{Goal}(A) = p$ (i.e. the argument is for or against an action to achieve the topic of the dialogue) and A has not been asserted previously during the dialogue.

- It is permissible to `agree` to an action a iff either:
  - the immediately preceding move was an `agree` to the action a, or
  - the other participant $\overline{x}$ has at some point previously in the dialogue asserted a positive argument A for the action a.
- It is always permissible to `pass`.

We have thus defined a protocol that determines which moves it is permissible to make during a dialogue; however, an agent still has considerable choice when selecting which of these permissible moves to make. In order to select one of the permissible moves, an agent uses a particular strategy. The **strategy** that our agents use is as follows:

- If it is permissible to `agree` to an action that the agent finds *agreeable*, then make such an `agree` move; else
- if it is permissible to `assert` a positive argument *for* an action that the agent finds *agreeable*, then assert some such argument; else
- if it is permissible to `assert` a negative argument *against* an action and the agent finds that action *not agreeable* then assert some such argument; else
- make a `pass` move.

We have now defined how our dialogue system regulates the moves that agents may make, and the strategy that the agents use to select one of the permissible moves to make. (For an example of a dialogue produced by this system, please refer to [4].) Next, we define a method with which two agents may form a consensus without exchanging any arguments.

2.3 Consensus Forming

In order to start investigating the question of whether it is worth using argumentation-based deliberation dialogues to decide how to act to achieve a shared goal, we compare outcomes produced by our dialogue system with those produced by a simple consensus forming method. For two agents $x, \overline{x}$ who are about to enter into a deliberation dialogue with topic p, the outcome produced by **consensus forming** is simply the *intersection* of the following two sets:

- the set of actions to achieve p that agent x finds agreeable at the start of the dialogue;
- the set of actions to achieve p that agent $\overline{x}$ finds agreeable at the start of the dialogue.

That is to say, the consensus set contains all the actions that each agent finds agreeable, given the arguments they can construct from their VATS and without any exchange of arguments. If consensus forming returns a non-empty set, then we say that a **consensus was found** and that the consensus forming was **successful**.

This gives us a non-argumentative approach to which we can compare our dialogue system. We next discuss how we compare these systems, namely on the quality of outcome.

2.4 Measuring Quality of Outcome

Unless they exchange all arguments, agents in our system only ever have a partial view of all of the available knowledge. We can, however, take a global view of which potential outcomes are best for each of the agents. For this purpose, we define for a dialogue the **omniscient argumentation framework** (OAF), which is the iVAF constructed from the union of the arguments that each participant can construct from its VATS that relate to the topic of the dialogue. For a dialogue with participants $x, \overline{x}$ and topic p, the associated OAF is thus the iVAF $\langle \mathcal{X}, \mathcal{A} \rangle$ where $\mathcal{X} = Args_p^x \cup Args_p^{\overline{x}}$. We say that an action is **globally agreeable** to an agent x if and only if there is some *positive* argument *for* that action that is acceptable in the OAF under the audience that represents x's value preference.

We can now measure the quality of a particular outcome (i.e. an action to achieve the goal p) by considering whether it is globally agreeable to each agent. Such a quality measure can be applied to both the outcome produced by a dialogue and the outcome produced by consensus forming.

For a particular outcome a, we assign an **outcome quality score** as follows:

- if a is **globally agreeable to both** x and $\overline{x}$, score 3;
- if a is **globally agreeable to only one** of x or $\overline{x}$, score 2;
- if a is **not globally agreeable to either** x or $\overline{x}$, score 1.

If there is **no successful outcome** (i.e. dialogue terminates in failure or consensus forming returns an empty set) then the outcome quality **score is 0**. Where the consensus forming returns a set of more than one action, we assign the outcome quality score to be that of the action from the set which receives the lowest score (since this is the best that the consensus forming method can guarantee to do, given that only one action can be selected).

Our simple scoring metric reflects the intuition that any outcome is better than no outcome, but an outcome that is globally agreeable to an agent is better than one that is not. We plan to study more sophisticated scoring metrics in future work.

2.5   Experimental Set Up

The dialogue system and consensus forming mechanism were implemented as described in the previous sections. We also implemented a **random scenario generator**; this generates **scenarios** that initialise the agents' knowledge bases (i.e. the arguments known to each agent at the start of the dialogue, which all relate to the joint goal which the agents wish to achieve) and their audiences. The generator takes three parameters (Args, Vals, Acts), where

- Args is the number of distinct arguments to appear in the union of the agents' knowledge bases;
- Vals is the number of distinct values that may motivate those arguments;
- Acts is the number of distinct actions that the arguments may relate to.

The generator randomly constructs without replacement (i.e. does not allow duplicate arguments) the required number of arguments from the allowed values and actions and the symbols $\{+, -\}$ (where each combination is equally likely). For example, when given parameters $(8, 2, 2)$, the generator will construct the following set of arguments:

$$\{\langle a1, p, v1, + \rangle, \langle a1, p, v1, - \rangle, \langle a1, p, v2, + \rangle, \langle a1, p, v2, - \rangle,$$
$$\langle a2, p, v1, + \rangle, \langle a2, p, v1, - \rangle, \langle a2, p, v2, + \rangle, \langle a2, p, v2, - \rangle\}.$$

(Note, it is not possible for the generator to construct a set of arguments from parameters (Args, Vals, Acts) if Args > Vals × Acts × 2. For a particular number of values and a particular number of actions, the total **possible arguments** is Vals × Acts × 2.)

The generator randomly assigns each agent an audience over the allowed values and it randomly allocates exactly half of the constructed arguments to one agent, and the other half to the other agent. Our generator is therefore simulating the construction of arguments from the agents' VATS. It allows us to run experiments over all possible combinations of the parameters (Args, Vals, Acts). In the experiments reported here we consider all possible parameter combinations where:

- Vals ∈ {2, 4, 6, 8, 10},
- Acts ∈ {2, 4, 6, 8, 10},
- 2 ≤ Args ≤ Vals × Acts × 2.

Our experiments investigate how the outcome quality scores of the dialogue system and the consensus forming mechanism compare across the space of possible parameter combinations. We performed 1000 runs of our simulation for each possible parameter combination. In each run, a random scenario is generated. We first calculate the consensus set of the scenario and then simulate a dialogue from the same scenario; we compare the quality scores assigned to the outcomes produced by these two approaches.

3 Results

3.1 Dialogue Is Significantly More Likely to Be Successful Than Consensus Forming

Figure 1 shows strikingly across all parameter combinations that the frequency of successful consensuses is never as great as the frequency of successful dialogues. There is a significant difference between these two frequencies: across all parameters, consensus forming fails more than 50% of the time, whilst up to 90% of dialogues are successful.

We also found that, across all runs for each possible parameter combination (a total of 1.8 million runs), for every run in which a consensus was found the dialogue produced was also successful. It is not immediately clear whether the converse situation (i.e. a consensus is found but the dialogue produced is not successful) is theoretically impossible, but this result strongly suggests that this may be the case and so identifies a property worthy of theoretical investigation.

Consensus forming is relatively robust to the number of values present in the system; however there is a marked difference when Vals = 2, in which case the frequency of successful consensuses is approximately half that of when Vals ∈ {4, 6, 8, 10}.

When Acts = 2, the highest frequency of consensuses found is seen when Args is equal to approximately 50% of the total arguments possible. A higher number of arguments present in the system leads to a higher frequency of successful consensuses; in contrast, the frequency of successful dialogues drops as the number of arguments present in the system increases (although the number of successful dialogues is still greater than the number of consensuses found).

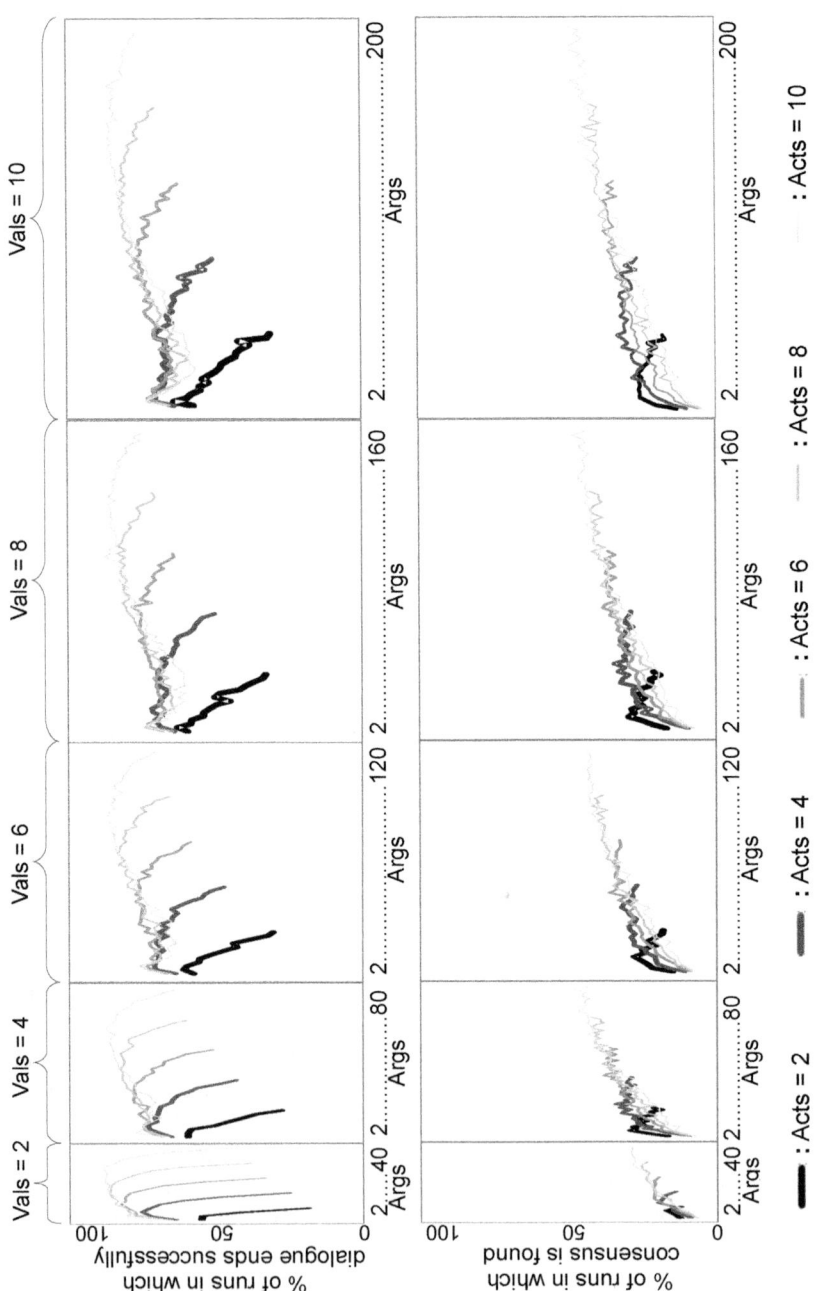

Fig. 1. Top: Percentage of dialogues that ended successfully out of 1000 runs across each possible parameter combination. Bottom: Percentage of 1000 runs across each possible parameter combination in which a consensus was found.

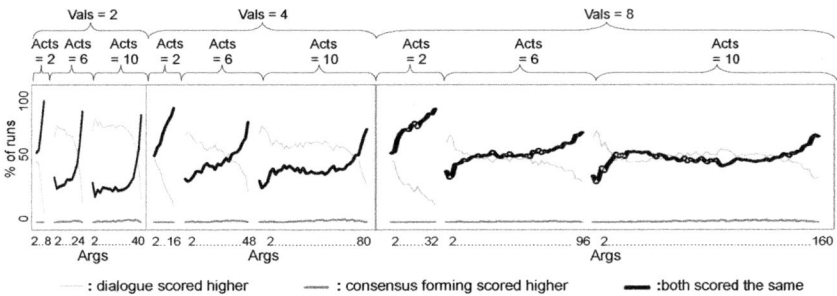

Fig. 2. Percentage of 1000 runs across each possible parameter combination where Acts $\in$ $\{2, 6, 10\}$ and Vals $\in \{2, 4, 8\}$ in which: dialogue outcome quality score was higher than consensus outcome quality score; consensus outcome quality score was higher than dialogue outcome quality score; dialogue outcome quality score was the same as consensus outcome quality score

3.2   Successful Dialogues Are More Likely with Higher Numbers of Actions and Values

Looking at the top of Figure 1 in depth, we can see how sensitive the dialogue system is to the parameters. The dialogue system appears to be most sensitive to the parameter settings Acts = 2 and Vals = 2.

Across all parameter settings, the frequency of successful dialogues is closely related to the percentage of the total possible arguments present in the system: if Acts $\in$ $\{4, 6, 8, 10\}$ and Vals $\in \{4, 6, 8, 10\}$, this frequency peaks when Args is around 75% of the total possible; if Acts = 2, this frequency peaks when Args $\approx$ 4; if Acts $\in$ $\{4, 6, 8, 10\}$ and Vals = 2, this frequency peaks when Args is around 50% of the total possible.

When Acts = 2, the highest frequency of successful dialogues seen is lower than the highest frequencies seen for the other settings of Acts. Both the maximum and the minimum frequency of successful dialogues recorded is greater when more actions are under consideration, and the minimum frequency of successful dialogues is greater when more values are present in the system.

Generalising these results, we can say that the dialogue system performs better (i.e. reaches agreement more often) when Acts $\neq$ 2. The more values and the more actions present in the system the better the system performs, with the frequency of successful dialogues dependent on the percentage of the total possible arguments present in the system.

3.3   Quality of Dialogue Outcome Is Very Rarely Worse Than Quality of Consensus Outcome

We next consider for each run whether the dialogue system or consensus forming resulted in a higher outcome quality score. We investigated this across all possible parameter combinations; since we found a trend that repeats across the whole parameter space, we present in Figure 2 only the results for when Acts $\in \{2, 6, 10\}$ and Vals $\in \{2, 4, 8\}$.

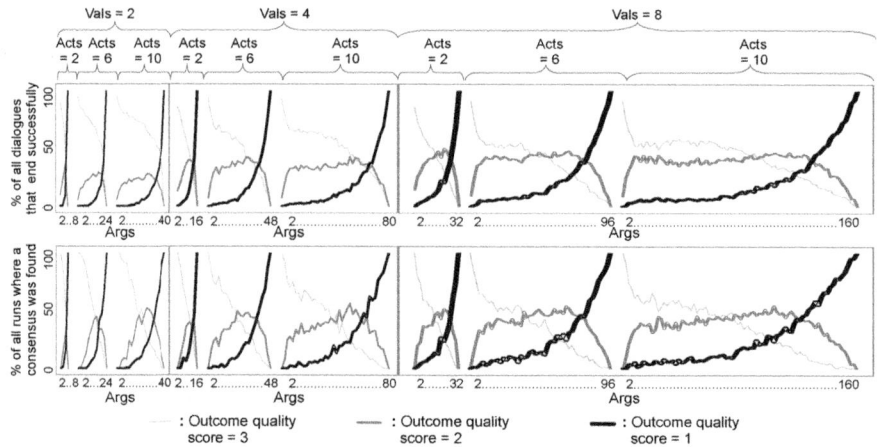

Fig. 3. Top: across all possible parameter settings where Acts $\in \{2, 6, 10\}$ and Vals $\in \{2, 4, 8\}$, percentage *of the dialogues that ended successfully* that received each outcome quality score. Bottom: across all possible parameter settings where Acts $\in \{2, 6, 10\}$ and Vals $\in \{2, 4, 8\}$, percentage of the runs *in which a consensus was found* that received each outcome quality score.

Figure 2 shows clearly that only very rarely (in less than 3% of the runs across all possible parameter settings) does consensus forming produce a higher quality outcome than the dialogue system. However, if there are only two actions, then the two methods produce the same quality outcome more often than the dialogue system produces a higher quality outcome. This is a useful observation, particularly considering the higher computational overheads associated with the dialogue system.

3.4 Successful Dialogue Outcomes Are More Likely to Be Globally Agreeable to Both Agents Than Successful Consensus Outcomes

We now consider how the outcome quality score varies for successful outcomes produced by both the dialogue system and consensus forming across the parameter space. We performed this analysis across all possible parameter settings and found a trend that occurs across the entire parameter space; hence we present in Figure 3 only those results where where Acts $\in \{2, 6, 10\}$ and Vals $\in \{2, 4, 8\}$. The top of this figure shows what percentage *of the dialogues that ended in agree* received which outcome quality score. The bottom of this figure shows what percentage *of the runs in which a consensus was found* received which outcome quality score. (Recall the outcome quality score metric: 3 - outcome is globally agreeable to both agents; 2 - outcome is globally agreeable to only of the agents; 1 - outcome is not globally agreeable to either of the agents.)

As discussed earlier, Figure 1 shows that the frequency of dialogues that end successfully is considerably higher than the frequency of consensuses found, and that each of these frequencies vary as the parameters change; thus, it is important to bear in mind here that the percentages denoted on the y-axes of the graphs in Figure 3 relate to different sized sets depending on the particular parameter settings and on whether dialogue outcome or consensus outcome is being considered. Considering only the proportion of *successful* dialogues and consensuses that receive the different outcome quality scores (as seen in Figure 3) allows us to clearly see the following points.

Of the successful outcomes produced by both methods (consensus forming and the dialogue system), a higher proportion of those produced by the dialogue system are globally agreeable to each agent (i.e. outcome quality score = 3). The difference between the proportion of successful dialogues that receive outcome quality score 3 and the proportion of consensuses that receive outcome quality score 3 is bigger the more actions and the fewer values that are present in the system.

It is interesting to note that the points on the graphs in Figure 3 where the green line (i.e. outcome quality score = 1) and the red line (i.e. outcome quality score = 2) intersect occur at the same position on the x-axis for both the dialogue outcome and the consensus outcome. If Vals = 2, this occurs when Args is equal to approximately 95% of the total possible arguments, otherwise this occurs when Args is equal to approximately 80% of the total possible arguments. Thus, if a successful outcome is produced either by consensus forming or by the dialogue system and there are more than 80% of the total possible arguments present in the system (95% if Vals = 2), it is likely that this outcome is not globally agreeable to either agent.

The quality of successful outcomes produced by both the dialogue system and consensus forming is most sensitive to the number of arguments present in the system, and is little affected by changes to the number of values or actions under consideration. Consensus forming is more sensitive than the dialogue system to the number of arguments.

3.5 Average Dialogue Outcome Quality Score Is Higher Than Average Consensus Outcome Quality Score

Figure 4 shows the average outcome quality score produced by both the dialogue system and the consensus forming mechanism across all parameter settings where Args = 25%, 50% and 75% of the total possible arguments. It is very clear from these results that, on average, the dialogue system outperforms consensus forming.

Looking at Figure 4 in more depth, we see that the highest outcome quality score averages for the dialogue system are seen when Vals = 2, whilst this parameter setting produces the lowest outcome quality score averages for consensus forming. For all settings of Acts and Vals, the smallest difference between the outcome quality score averages of the two methods is seen when Args = 75% of the total possible arguments. For all settings of Vals and Args, the smallest difference between the two outcome quality score averages is seen when Acts = 2. We can conclude that if Vals = 2 and Acts $\neq$ 2, it is likely that the outcome produced by the dialogue system will be higher quality than that produced by consensus forming.

3.6 Dialogue Length Grows Exponentially with Increasing Arguments

Figure 5 shows that the time it takes to complete dialogues increases exponentially with the number of arguments. However as the number of values increases this trend flattens and increases are more linear. Indeed as values and actions increase the curve becomes

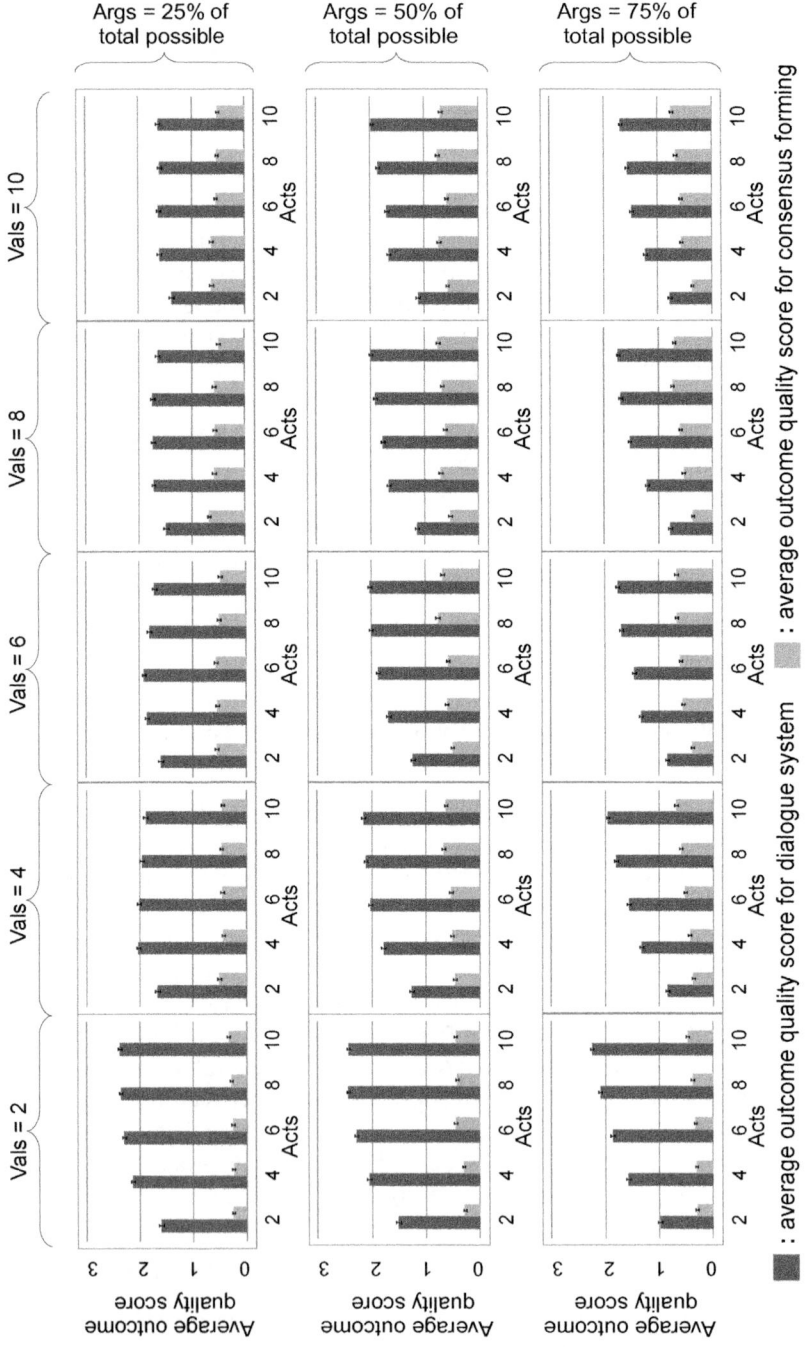

Fig. 4. Average quality outcome score over 1000 runs for both the dialogue system and consensus forming, across every parameter combination where Args = 25%, 50% or 75% of the total possible arguments. The error bars show the standard errors of the means.

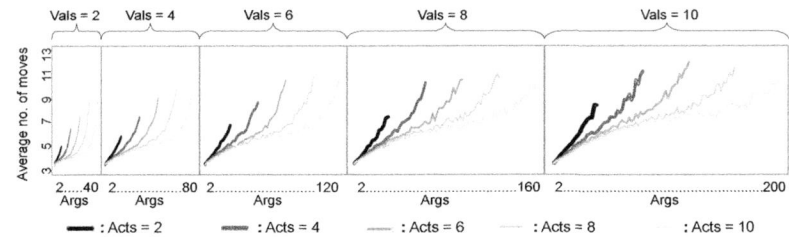

Fig. 5. Average number of moves in a terminated dialogue

almost sigmoidal. This indicates that if speed is a key factor for an applied dialogue system, deliberation dialogues are most useful when either the number of arguments is low or the number of values and actions is high.

4   Discussion

We have presented empirical results from what we believe is the first simulation-based study of a deliberation dialogue system, where the agents involved used value-based argumentation to determine agreeable actions. Our results show that the dialogue system we present outperforms a simple consensus forming mechanism. We provide an in-depth analysis of the behaviour that can be expected from the system based on the number of actions, values and arguments that are present. For instance, the dialogue system reaches agreement more frequently when there is a higher number of actions and values under consideration; the quality of a successful dialogue outcome is more likely to be higher when there are less than 80% of the total possible arguments present.

These results take a significant step towards demonstrating the applicability of value-based deliberation dialogue systems, as well as demonstrating the importance of complementing theoretical evaluations with simulation-based studies. Our specific quantitative results can be compared against the parameters derived from a particular domain in order to determine the suitability of value-based deliberation dialogues.

Our simulation facilitates many avenues of future work, for example it is simple to adapt it to allow multiple agents and we are particularly interested in investigating different strategies that the agents might use and seeing how these compare with one another. Our next step is to analyse why the system behaves as it does. We have already begun to investigate how the topology of the OAF (which is itself determined by the combination of parameters) affects the dialogue behaviour, and it is clear that they are closely linked. Here, we have restricted the system so that the agents each get exactly half of the arguments present in the system; certainly altering this split will have a marked effect of the behaviour of the system and this is something we are keen to investigate. We also intend to extend our dialogue model to take into account the other stages of practical reasoning (problem formulation and epistemic reasoning [6]).

It would be very interesting to see how an argumentative agent would perform against a non-argumentative agent, such as one that uses classical decision theory to determine the actions it finds agreeable. There is a large body of work on computational social choice (see e.g. [9]), which considers mechanisms with which group decisions can be made. Although beyond the scope of this paper, we plan to compare deliberation dialogues with social choice mechanisms (more sophisticated that the simple consensus forming method presented here). Such comparisons of an argumentation-based approach with approaches from other fields are of vital importance if we are to demonstrate the value of argumentation theory to the wider field of Artificial Intelligence.

Our investigation here takes a fundamental first step towards evaluating the potential benefit of a value-based deliberation dialogue system; however, it is not clear whether the scenarios that our simulation randomly generates are reflected in any real world setting. For example: Are there any real applications where more than 75% of all possible arguments are present in the system? Is it realistic that negative arguments are as likely to appear within the system as positive arguments? In order to be sure that the results are useful beyond a randomised setting, it is important to test argumentation-based approaches using real world data. This presents a challenge for the community, since it is hard to get access to such data that can be represented as arguments. We plan to collaborate with researchers working on real applications in order to validate our approach.

This simulation has been invaluable in identifying areas of future work that have the potential to be of benefit to real world applications, and in providing us with an implemented framework that we can adapt to investigate these areas.

Acknowledgements. E. Black funded by the European Union Seventh Framework Programme (FP7/2007-2011) under grant agreement 253911. K. Bentley funded by the Fondation Leducq.

References

1. Karunatillake, N.C., Jennings, N.R., Rahwan, I., McBurney, P.: Dialogue games that agents play within a society. Artificial Intelligence 173(9-10), 935–981 (2009)
2. Pasquier, P., Hollands, R., Rahwan, I., Dignum, F., Sonenberg, L.: An empirical study of interest-based negotiation. Autonomous Agents and Multi-Agent Systems 22(2), 249–288 (2011)
3. Jung, H., Tambe, M.: Towards argumentation as distributed constraint satisfaction. In: Proc. of AAAI Fall Symposium on Negotiation Methods for Autonomous Cooperative Systems (2001)
4. Black, E., Atkinson, K.: Choosing persuasive arguments for action. In: Proc. of the 10th Int. Conf. on Autonomous Agents and Multi-Agent Systems, pp. 905–912 (2011)
5. Walton, D.N.: Argumentation Schemes for Presumptive Reasoning. Lawrence Erlbaum Associates, Mahwah (1996)
6. Atkinson, K., Bench-Capon, T.J.M.: Practical reasoning as presumptive argumentation using action based alternating transition systems. Artificial Intelligence 171(10–15), 855–874 (2007)
7. Bench-Capon, T.J.M.: Agreeing to differ: Modelling persuasive dialogue between parties without a consensus about values. Informal Logic 22(3), 231–245 (2002)
8. Dung, P.M.: On the acceptability of arguments and its fundamental role in nonmonotonic reasoning, logic programming and n-person games. Artificial Intelligence 77, 321–357 (1995)
9. Chevaleyre, Y., Endriss, U., Lang, J., Maudet, N.: A Short Introduction to Computational Social Choice. In: van Leeuwen, J., Italiano, G.F., van der Hoek, W., Meinel, C., Sack, H., Plášil, F. (eds.) SOFSEM 2007. LNCS, vol. 4362, pp. 51–69. Springer, Heidelberg (2007)

Selective Revision by Deductive Argumentation

Patrick Krümpelmann[1], Matthias Thimm[1],
Marcelo A. Falappa[2], Alejandro J. García[2],
Gabriele Kern-Isberner[1], and Guillermo R. Simari[2]

[1] Technische Universität Dortmund, Germany
[2] Universidad Nacional del Sur, Bahía Blanca, Argentina

Abstract. The *success* postulate of classic belief revision theory demands that after revising some beliefs with by information the new information is believed. However, this form of *prioritized* belief revision is not apt under many circumstances. Research in *non-prioritized* belief revision investigates forms of belief revision where *success* is not a desirable property. Herein, *selective revision* uses a two step approach, first applying a transformation function to decide if and which part of the new information shall be accepted, and second, incorporating the result using a prioritized revision operator. In this paper, we implement a transformation function by employing deductive argumentation to assess the value of new information. Hereby we obtain a non-prioritized revision operator that only accepts new information if believing in the information is justifiable with respect to the beliefs. By making use of previous results on selective revision we prove that our revision operator satisfies several desirable properties. We illustrate the use of the revision operator by means of examples and compare it with related work.

1 Introduction

Belief revision [4,12] is concerned with changing beliefs in the light of new information. Usually, the beliefs of an agent are not static but change when new information is available. In order to be able to act reasonably in a changing environment the agent has to integrate new information and give up outdated beliefs. In particular, if the agent learns that some beliefs have been misleadingly assumed to be true its beliefs have to be *revised*. The research field of belief revision distinguishes between *prioritized* and *non-prioritized* belief revision. In prioritized belief revision [12] new information is always assumed to represent the most reliable and correct information available and revising the agent's beliefs by the new information is expected to result in believing the new information. This is a reasonable assumption for many imaginable situations and there are many technical challenges in realizing prioritized belief revision, cf. e. g. [12]. However, many circumstances demand that new information is not blindly accepted but weighted against the current beliefs. The field of non-prioritized belief revision [11] investigates change operations where revising some beliefs by new information may not result in believing the new information. Imagine a

S. Modgil, N. Oren, and F. Toni (Eds.): TAFA 2011, LNAI 7132, pp. 147–162, 2012.

multi-agent system where agents exchange information. In general, agents may be cooperative or competitive. Information that is passed from one agent to another may be intentionally wrong, mistakenly wrong, or correct. It is up to the receiver of the information to evaluate whether it should be integrated into the beliefs or not. In particular, in non-prioritized belief revision the satisfaction of the *success* postulate—which demands that new information is believed after revision—is not desirable. In [9] a specific class of non-prioritized belief revision operators is investigated. A *selective revision* is a two-step revision that consists of 1.) filtering new information using a *transformation function* and 2.) revising the beliefs with the result of the filtering in a prioritized way. In [9], no concrete implementations of the transformation function are given but several results are proven that show how specific properties for the transformation function and the *inner* prioritized revision translate to specific properties for the *outer* non-prioritized revision.

In this paper we propose a specific implementation of a transformation function that makes use of *deductive argumentation* [2]. A deductive argumentation theory is a set of propositional sentences and an argument for some sentence ϕ is a minimal proof for ϕ. If the theory is inconsistent there may also be proofs for the complement of a sentence $\neg\phi$ and in order to decide whether ϕ or $\neg\phi$ is to be believed, an argumentative evaluation is performed that compares arguments with counterarguments. We use the framework of [2] to implement a transformation function for selective revision that decides for each individual piece of information whether to accept it for revision or not, based on its argumentative evaluation. In particular, we consider the case that revision is to be performed based on a set of pieces of information instead of just a single piece of information. By doing so, we allow new information to contain arguments. As a result, an agent decides whether to accept some new information on the basis of its own evaluation of the information and the arguments that may be contained in this information. Consider the following example.

Example 1. Imagine the agent *Anna* wants to spend her holidays on Hawaii. She is aware of the fact that there has been some volcano activity on Hawaii recently but is convinced there is no immediate danger. Anna's boss *Bob* doesn't want Anna to go on vacation at this time of the year and tells her that she has to do some work here and should not go to Hawaii. However, Anna wants to go surfing and to go to Hawaii instead of staying at work. As a consequence she rejects Bob's argument to stay and does not revise her beliefs. Consider now that *Carl*, a good friend of *Anna*, is a vulcanologist and tells Anna that there is actually an immediate danger of an eruption. Anna does not have sufficient arguments to defeat Carls information, thus accepts the new information and revises her beliefs accordingly.   ∎

In the previous example the decisions of the agent Anna resulted in either accepting or rejecting the new information completely. However, it may also be the case that some of the new information is accepted and some is rejected. Consider the following example.

Example 2. Imagine Bob tells Anna that she has to stay for work because all her colleagues are having a vacation at the same time and she has to fill in for them. Suppose Anna knows that there is no work to do during her planned vacation as all clients of her company are on vacation as well. Then Anna would reject the conclusion of Bob's argument that she has to stay, but might very well accept that all her colleagues will be on vacation as well. ∎

In this paper we develop an approach for selective revision that is capable of deciding whether to accept, reject, or partially accept some new information, based on deductive argumentation. In order to do so we also extend the notions of selective revision to the problem of multiple base revision, i.e., the problem of revising a belief base (instead of a belief set) by a set of sentences.

The rest of this paper is organized as follows. In Section 2 we introduce some necessary technical preliminaries. We go on in Section 3 with providing an overview on the notions of belief revision and extending the approach of selective revision to selective multiple base revision. We continue in Section 4 with presenting the framework of deductive argumentation. In Section 5 we propose our implementation of selective multiple base revision via deductive argumentation and investigate its properties. In Section 6 we review some related work and in Section 7 we conclude.

2   Preliminaries

In this paper we suppose that the beliefs of an agent are given in the form of propositional sentences. Let At be a propositional signature, i.e. a set of propositional atoms. Let $\mathcal{L}(\mathsf{At})$ be the corresponding propositional language generated by the atoms in At and the connectives $\wedge$ *(and)*, $\vee$ *(or)*, $\Rightarrow$ *(implication)*, and $\neg$ *(negation)*. As a notational convenience we assume some arbitrary total order $\succ$ on the elements of $\mathcal{L}(\mathsf{At})$ which is used to enumerate elements of each finite $\Phi \subseteq \mathcal{L}(\mathsf{At})$ in a unique way, cf. [2]. For a finite subset $\Phi \subseteq \mathcal{L}(\mathsf{At})$ the *canonical enumeration* of Φ is the vector $\langle \phi_1, \ldots, \phi_n \rangle$ such that $\{\phi_1, \ldots, \phi_n\} = \Phi$ and $\phi_i \succ \phi_j$ for every $i < j$ with $i, j = 1, \ldots, n$. As $\succ$ is total the canonical enumeration of every finite subset $\Phi \subseteq \mathcal{L}(\mathsf{At})$ is uniquely defined.

We use the operator $\vdash$ to denote classical entailment, i.e., for sets of propositional sentences $\Phi_1, \Phi_2 \subseteq \mathcal{L}(\mathsf{At})$ we say that Φ_2 *follows* from Φ_1, denoted by $\Phi_1 \vdash \Phi_2$, if and only if Φ_2 is entailed by Φ_1 in the classical logical sense. For sentences $\phi, \phi' \in \mathcal{L}(\mathsf{At})$ we write $\phi \vdash \phi'$ instead of $\{\phi\} \vdash \{\phi'\}$. We define the deductive closure $Cn(\cdot)$ of a set of sentences Φ as $Cn(\Phi) = \{\phi \in \mathcal{L}(\mathsf{At}) \mid \Phi \vdash \phi\}$. Two sets of sentences $\Phi, \Phi' \subseteq \mathcal{L}(\mathsf{At})$ are *equivalent*, denoted by $\Phi \equiv^p \Phi'$, if and only if it holds that $\Phi \vdash \Phi'$ and $\Phi' \vdash \Phi$. We also use the equivalence relation $\cong^p$ which is defined as $\Phi \cong^p \Phi'$ if and only if there is a bijection $\sigma : \Phi \to \Phi'$ such that for every $\phi \in \Phi$ it holds that $\phi \equiv^p \sigma(\phi)$. This means that $\Phi \cong^p \Phi'$ if Φ and Φ' are *element-wise* equivalent. Note that $\Phi \cong^p \Phi'$ implies $\Phi \equiv^p \Phi'$ but not vice versa. In particular, it holds that e.g $\{a \wedge b\} \equiv^p \{a, b\}$ but $\{a \wedge b\} \not\cong^p \{a, b\}$. For sentences $\phi, \phi' \in \mathcal{L}(\mathsf{At})$ we write $\phi \equiv \phi'$ instead of $\{\phi\} \equiv \{\phi'\}$ if $\equiv \in \{\equiv^p, \cong^p\}$. If $\Phi \vdash \bot$ we say that Φ is *inconsistent*.

For a set S let $\mathfrak{P}(S)$ denote the power set of S, i.e. the set of all subsets of S. For a set S let $\mathfrak{P}\mathfrak{P}(S)$ denote the set of multi-sets of S, i.e. the set of all subsets of S where an element may occur more than once. To distinguish sets from multi-sets we use brackets "$\langle$" and "$\rangle$" for the latter.

3   Selective Multiple Base Revision

The field of belief revision is concerned with the change of beliefs when more recent or more reliable information is at hand. The most important description of properties of *prioritized* belief change operators is given by Alchourrón, Gärdenfors and Makinson in their seminal paper [4]. The usual framework for representing beliefs considered for belief revision is that of *belief sets* which are revised by a single sentence. A *belief set* S is a subset of $\mathcal{L}(\mathsf{At})$ that is deductively closed, i.e., $S = Cn(S)$. Working with belief sets in practice is unmanageable due to their infinite size. The more practical representation form are *belief bases* which are finite sets of sentences. These also come with the advantage of making it possible to differentiate between explicit and inferred beliefs, cf. [12]. In this work we consider the problem of *multiple base revision*. That is, we employ belief bases for knowledge representation and we consider revising a belief base by a set of sentences, cf. the notion of *multiple revision* in [12].

Let $\mathcal{K} \subseteq \mathcal{L}(\mathsf{At})$ be a belief base, $\Phi \subseteq \mathcal{L}(\mathsf{At})$ be some set of sentences, and consider the problem of changing $\mathcal{K}$ in order to entail Φ. If $\mathcal{K} \cup \Phi$ is consistent then there is no need for contracting the existing beliefs and the problem can be solved via *expansion* $\mathcal{K} + \Phi$ which is characterized via $\mathcal{K} \cup \Phi$. If $\mathcal{K} \cup \Phi$ is inconsistent, conflicts arising from the addition of Φ to $\mathcal{K}$ have to be resolved. In general, this means that some of the current beliefs have to be given up in order to come up with a consistent belief base. The AGM framework [4] proposes several basic postulates a revision operator should obey. As we consider belief bases for knowledge representation we start with the corresponding postulates for belief base revision [12] adapted to revision by sets of sentences [8]. Let $*$ be a multiple base revision operator—i.e., if $\mathcal{K}$ and Φ are sets of sentences so is $\mathcal{K} * \Phi$—and consider the following postulates.

Success. $\mathcal{K} * \Phi \vdash \Phi$.
Inclusion. $\mathcal{K} * \Phi \subseteq \mathcal{K} + \Phi$.
Vacuity. If $\mathcal{K} \cup \Phi \nvdash \bot$ then $\mathcal{K} + \Phi \subseteq \mathcal{K} * \Phi$.
Consistency. If Φ is consistent then $\mathcal{K} * \Phi$ is consistent.
Relevance. If $\alpha \in (\mathcal{K} \cup \Phi) \setminus (K * \Phi)$ then there is a set H such that $\mathcal{K} * \Phi \subseteq H \subseteq \mathcal{K} \cup \Phi$ and H is consistent but $H \cup \{\alpha\}$ is inconsistent.

Another important property for the framework of [4] is *extensionality* which can be phrased for multiple base revision as follows.

Extensionality. If $\Phi \equiv^p \Psi$, then $\mathcal{K} * \Phi \equiv^p \mathcal{K} * \Psi$.

The above property is usually not considered for the problem of base revision as base revision is motivated by observing syntax and not (only) semantic contents. In particular, for the problem of multiple base revision, satisfaction of *extensionality* imposes that $\mathcal{K} * \{a, b\} \equiv^p \mathcal{K} * \{a \wedge b\}$ as $\{a, b\} \equiv^p \{a \wedge b\}$. Identifying the "comma"-operator with the logical "AND"-operator is not always a reasonable thing to do, see e.g. [5] for a discussion. However, we consider the following weakened form of *extensionality*.

Weak Extensionality. If $\Phi \cong^p \Phi'$ then $\mathcal{K} * \Phi \equiv^p \mathcal{K} * \Phi'$.

The property *weak extensionality* only demands that the outcomes of the revisions $\mathcal{K} * \Phi$ and $\mathcal{K} * \Phi'$ are equivalent if Φ and Φ' are element-wise equivalent.

Definition 1. *A revision operator* $*$ *is called a* prioritized multiple base revision operator *if* $*$ *satisfies* success, inclusion, vacuity, consistency, relevance, *and* weak extensionality.

For non-prioritized multiple base revision the properties *inclusion, vacuity, consistency, relevance,* and *weak extensionality* can also be regarded as desirable. This is not the case for *success* is general but we can replace *success* by weakened versions, cf. [11]. We denote with $\circ$ a non-prioritized belief revision operator, i.e., $\mathcal{K} \circ \Phi$ is the non-prioritized revision of $\mathcal{K}$ by Φ. Then consider the following properties for $\circ$, cf. [9].

Weak Success. If $\mathcal{K} \cup \Phi \nvdash \bot$ then $\mathcal{K} \circ \Phi \vdash \Phi$.
Consistent Expansion. If $\mathcal{K} \nsubseteq \mathcal{K} \circ \Phi$ then $\mathcal{K} \cup (\mathcal{K} \circ \Phi) \vdash \bot$.

Note that *weak success* follows from *vacuity*, and *consistent expansion* follows from *vacuity* and *success*, cf. [9].

Definition 2. *A revision operator* $\circ$ *is called* non-prioritized multiple base revision operator *if* $\circ$ *satisfies* inclusion, consistency, weak extensionality, weak success, *and* consistent expansion.

We do not require *relevance* to be satisfied by non-prioritized multiple base revisions as it is hardly achievable in the context of selective revision, see below. For the following, bear in mind that the main difference between a prioritized multiple base revision operator $*$ and a non-prioritized multiple base revision operator $\circ$ is that $\mathcal{K} * \Phi \vdash \Phi$ is required but $\mathcal{K} \circ \Phi \vdash \Phi$ is not.

A specific approach to non-prioritized belief revision is *selective revision* [9]. There, the problem of revising a belief set S with a single sentence α is realized by applying a *transformation function* f to α, obtaining a new sentence α', and then revising S by α' in a prioritized way. The transformation function f is supposed to determine whether α should be accepted as a whole or whether it should be somewhat weakened. We adopt the notions of [9] for the problem of *selective multiple belief base revision* and still consider the problem of revising a belief base $\mathcal{K}$ by some set Φ of sentences. Following the ideas of [9] we define the selective multiple base revision $\mathcal{K} \circ \Phi$ via

$$\mathcal{K} \circ \Phi = \mathcal{K} * f_{\mathcal{K}}(\Phi) \tag{1}$$

with a transformation function $f_{\mathcal{K}} : \mathfrak{P}(\mathcal{L}(\mathsf{At})) \to \mathfrak{P}(\mathcal{L}(\mathsf{At}))$ and some (prioritized) multiple base revision $*$. In [9] several properties for transformation functions in the context of belief set revision are discussed. We rephrase some of them here slightly to fit the framework of multiple base revision. Let $\mathcal{K} \subseteq \mathcal{L}(\mathsf{At})$ be consistent and let $\Phi, \Phi' \subseteq \mathcal{L}(\mathsf{At})$.

Inclusion. $f_{\mathcal{K}}(\Phi) \subseteq \Phi$
Weak Inclusion. If $\mathcal{K} \cup \Phi$ is consistent then $f_{\mathcal{K}}(\Phi) \subseteq \Phi$
Extensionality. If $\Phi \equiv^p \Phi'$ then $f_{\mathcal{K}}(\Phi) \equiv^p f_{\mathcal{K}}(\Phi')$
Consistency Preservation. If Φ is consistent then $f_{\mathcal{K}}(\Phi)$ is consistent
Consistency. $f_{\mathcal{K}}(\Phi)$ is consistent
Maximality. $f_{\mathcal{K}}(\Phi) = \Phi$
Weak Maximality. If $\mathcal{K} \cup \Phi$ is consistent then $f_{\mathcal{K}}(\Phi) = \Phi$

We also consider the following novel property.

Weak Extensionality. If $\Phi \cong^p \Phi'$ then $f_{\mathcal{K}}(\Phi) \cong^p f_{\mathcal{K}}(\Phi')$

Not all of the above properties may be desirable for a transformation function that is to be used for selective revision. For example, the property *maximality* states that $f_{\mathcal{K}}$ should not modify the set Φ. Satisfaction of this property makes (1) equivalent to $\mathcal{K} * \Phi$. As $*$ is meant to be a prioritized revision function we lose the possibility for non-prioritized revision.

Note that for *weak extensionality* we demand $f_{\mathcal{K}}(\Phi)$ and $f_{\mathcal{K}}(\Phi')$ to be element-wise equivalent instead of just equivalent (in contrast to the property *weak extensionality* for revision). We do this because $f_{\mathcal{K}}$ is supposed to be applied in the context of base revision which is sensitive to syntactic variants. We introduce the postulate *weak extensionality* for transformation functions with the same motivation as we do for multiple base revision. However, for the case of transformation functions the problem with satisfaction of *extensionality* is more apparent. Consider again $\Phi = \{a, b\}$ and $\Phi' = \{a \wedge b\}$. It follows that $\Phi \equiv^p \Phi'$ and if $f_{\mathcal{K}}$ satisfies *extensionality* this results in $f_{\mathcal{K}}(\{a, b\}) \equiv^p f_{\mathcal{K}}(\{a \wedge b\})$. If $f_{\mathcal{K}}$ also satisfies *inclusion* it follows that $f_{\mathcal{K}}(\{a \wedge b\}) \in \{\emptyset, \{a \wedge b\}\}$ and therefore $f_{\mathcal{K}}(\{a, b\}) \in \{\emptyset, \{a, b\}\}$. In general, if $f_{\mathcal{K}}$ satisfies both *inclusion* and *extensionality* it follows that either $f_{\mathcal{K}}(\Phi) = \emptyset$ or $f_{\mathcal{K}}(\Phi) = \Phi$ for every $\Phi \subseteq \mathcal{L}(\mathsf{At})$ (as Φ is equivalent to a Φ' that consists of a single formula that is the conjunction of the formulas in Φ and $f_{\mathcal{K}}(\Phi') = \emptyset$ or $f_{\mathcal{K}}(\Phi') = \Phi'$ due to *inclusion*). As we are interested in a more graded approach to belief revision we want to be able to accept or reject specific pieces of Φ and not just Φ as a whole. Consequently, we consider *weak extensionality* as a desirable property instead of *extensionality*. Note that *extensionality* implies *weak extensionality* as $\Phi \cong^p \Phi'$ implies $\Phi \equiv^p \Phi'$.

In [9] several representation theorems are given that characterize non-prioritized belief revision by selective revision via (1) and specific properties of $*$ and $f_{\mathcal{K}}$. In particular, it is shown that a reasonable non-prioritized belief revision operator $\circ$ can be characterized by an AGM revision $*$ and a transformation function $f_{\mathcal{K}}$ that satisfies *extensionality, consistency preservation*, and *weak maximality*. Note, however, that [9] deals with the problem of revising a belief set by a single sentence. Nonetheless, we can carry over the results of [9] to the problem of multiple base revision and obtain the following result.

Proposition 1. *Let $*$ be a prioritized multiple base revision operator and let $f_{\mathcal{K}}$ satisfy* inclusion, weak extensionality, consistency preservation, *and* weak maximality. *Then $\circ$ defined via (1) is a non-prioritized multiple base revision operator.*

Proof. We have to show that $\circ$ satisfies inclusion, consistency, weak extensionality, weak success, *and* consistent expansion.

  Inclusion. *It holds that $f_{\mathcal{K}}(\Phi) \subseteq \Phi$ as $f_{\mathcal{K}}$ satisfies* inclusion. *Also, $*$ satisfies* inclusion *and it follows $\mathcal{K} * f_{\mathcal{K}}(\Phi) \subseteq \mathcal{K} \cup f_{\mathcal{K}}(\Phi) \subseteq \mathcal{K} \cup \Phi$.*
  Consistency. *If Φ is consistent so is $f_{\mathcal{K}}(\Phi)$ as $f_{\mathcal{K}}$ satisfies* consistency preservation. *As $*$ satisfies* consistency *it follows that $\mathcal{K} * f_{\mathcal{K}}(\Phi)$ is consistent.*
  Weak Extensionality. *If $\Phi \cong^p \Phi'$ then $f_{\mathcal{K}}(\Phi) \cong^p f_{\mathcal{K}}(\Phi')$ as $f_{\mathcal{K}}$ satisfies* weak extensionality. *It follows that $\mathcal{K} * f_{\mathcal{K}}(\Phi) \equiv^p \mathcal{K} * f_{\mathcal{K}}(\Phi')$ as $*$ satisfies* weak extensionality.
  Weak Success. *If $\mathcal{K} \cup \Phi$ is consistent it follows that $f_{\mathcal{K}}(\Phi) = \Phi$ as $f_{\mathcal{K}}$ satisfies* weak maximality. *As $*$ satisfies* vacuity *it follows $\mathcal{K} + \Phi \subseteq \mathcal{K} * f_{\mathcal{K}}(\Phi)$. Hence, $\circ$ satisfies* vacuity *as well and therefore* weak success.
  Consistent Expansion. *Suppose $\mathcal{K} \not\subseteq \mathcal{K} * f_{\mathcal{K}}(\Phi)$. Note that $*$ satisfies* consistent expansion *as $*$ satisfies* vacuity *and* success, *cf. [9]. It follows that $\mathcal{K} \cup \{\mathcal{K} * f_{\mathcal{K}}(\Phi)\}$ is inconsistent.* □

Note that *relevance* does not hold for $\mathcal{K} \circ \Phi$ defined via (1) in general. Consider for example the transformation function $f_{\mathcal{K}}^0$ defined via $f_{\mathcal{K}}^0(\Phi) = \Phi$ if $\mathcal{K} \cup \Phi$ is consistent and $f_{\mathcal{K}}^0(\Phi) = \emptyset$ otherwise. Then $f_{\mathcal{K}}^0$ satisfies all properties for transformation functions except *maximality*. But it is easy to see that $\mathcal{K} \circ \Phi$ defined via (1) using $f_{\mathcal{K}}^0$ and a prioritized multiple base revision operator $*$ fails to satisfy *relevance*. We leave it to future work to investigate further properties for transformation functions that may enable *relevance* to hold in general.

   In the following we aim at implementing a selective multiple base revision using deductive argumentation and go on with introducing the latter.

4   Deductive Argumentation

Argumentation frameworks [1] allow for reasoning with inconsistent information based on the notions of arguments, counterarguments and their relationships. Since the seminal paper [6] interest has grown in research in computational models for argumentation that allow for a coherent procedure for consistent reasoning in the presence of inconsistency. In this paper we use the framework of *deductive argumentation* as proposed by Besnard and Hunter [2]. This framework bases on classical propositional logic and is therefore apt for our aim to use argumentation to realize a transformation function f. The central notion of the framework of deductive argumentation is that of an *argument*.

Definition 3 (Argument). *Let $\Phi \subseteq \mathcal{L}(\text{At})$ be a set of sentences. An argument $\mathcal{A}$ for a sentence $\alpha \in \mathcal{L}(\text{At})$ in Φ is a tuple $\mathcal{A} = \langle \Psi, \alpha \rangle$ with $\Psi \subseteq \Phi$ that satisfies 1.) $\Psi \nvdash \bot$, 2.) $\Psi \vdash \alpha$, and 3.) there is no $\Psi' \subsetneq \Psi$ with $\Psi' \vdash \alpha$. For an argument $\mathcal{A} = \langle \Psi, \alpha \rangle$ we say that α is the* claim *of $\mathcal{A}$ and Ψ is the* support *of $\mathcal{A}$.*

Hence, an argument $\mathcal{A} = \langle \Psi, \alpha \rangle$ for α is a minimal proof for entailing α. Given a set $\Phi \subseteq \mathcal{L}(\text{At})$ of sentences there may be multiple arguments for α. As in [2] we are interested in arguments that are most cautious.

Definition 4 (Conservativeness). *An argument $\mathcal{A} = \langle \Psi, \alpha \rangle$ is more* conservative *than an argument $\mathcal{B} = \langle \Phi, \beta \rangle$ if and only if $\Psi \subseteq \Phi$ and $\beta \vdash \alpha$.*

In other words, an $\mathcal{A}$ is more conservative than an argument $\mathcal{B}$ if $\mathcal{B}$ has a smaller support (with respect to set inclusion) and a more general conclusion. An argument $\mathcal{A}$ is *strictly more conservative* than an argument $\mathcal{B}$ if and only if $\mathcal{A}$ is more conservative than $\mathcal{B}$ but $\mathcal{B}$ is not more conservative than $\mathcal{A}$. If $\Phi \subseteq \mathcal{L}(\text{At})$ is inconsistent there are arguments with contradictory claims.

Definition 5 (Undercut). *An argument $\mathcal{A} = \langle \Psi, \alpha \rangle$ is an* undercut *for an argument $\mathcal{B} = \langle \Phi, \beta \rangle$ if and only if $\alpha = \neg(\phi_1 \wedge \ldots \wedge \phi_n)$ for some $\phi_1, \ldots, \phi_n \subseteq \Phi$.*

If $\mathcal{A}$ is an undercut for $\mathcal{B}$ then we also say that $\mathcal{A}$ *attacks* $\mathcal{B}$. In order to consider only those undercuts for an argument that are most general we restrain the notion of undercut as follows.

Definition 6 (Maximally conservative undercut). *An argument $\mathcal{A} = \langle \Psi, \alpha \rangle$ is a* maximally conservative undercut *for an argument $\mathcal{B} = \langle \Phi, \beta \rangle$ if and only if $\mathcal{A}$ is an undercut of $\mathcal{B}$ and there is no undercut $\mathcal{A}'$ for $\mathcal{B}$ that is strictly more conservative than $\mathcal{A}$.*

Definition 7 (Canonical undercut). *An argument $\mathcal{A} = \langle \Psi, \neg(\phi_1 \wedge \ldots \wedge \phi_n) \rangle$ is a* canonical undercut *for an argument $\mathcal{B} = \langle \Phi, \beta \rangle$ if and only if $\mathcal{A}$ is a maximally conservative undercut for $\mathcal{B}$ and $\langle \phi_1, \ldots, \phi_n \rangle$ is the canonical enumeration of Φ.*

It can be shown that it suffices to consider only the canonical undercuts for an argument in order to come up with a reasonable argumentative evaluation of some claim α [2]. Having an undercut $\mathcal{B}$ for an argument $\mathcal{A}$ there may also be an undercut $\mathcal{C}$ for $\mathcal{B}$ which *defends* $\mathcal{A}$. In order to give a proper evaluation of some argument $\mathcal{A}$ we have to consider all undercuts for its undercuts as well, and so on. This leads to the notion of an *argument tree*.

Definition 8 (Argument tree). *Let $\alpha \in \mathcal{L}(\text{At})$ be some sentence and let $\Phi \subseteq \mathcal{L}(\text{At})$ be a set of sentences. An* argument tree *$\tau_\Phi(\alpha)$ for α in Φ is a tree where the nodes are arguments and that satisfies*

1. *the root is an argument for α in Φ,*
2. *for every path $[\langle \Phi_1, \alpha_1 \rangle, \ldots, \langle \Phi_n, \alpha_n \rangle]$ in $\tau_\Phi(\alpha)$ it holds that $\Phi_n \nsubseteq \Phi_1 \cup \ldots \cup \Phi_{n-1}$, and*

3. *the children* $\mathcal{B}_1, \ldots, \mathcal{B}_m$ *of a node* $\mathcal{A}$ *consist of all canonical undercuts for* $\mathcal{A}$
   that obey 2.).

Let $\mathcal{T}(\mathsf{At})$ *be the set of all argument trees.*

An argument tree is a concise representation of the relationships between different arguments that favor or reject some argument $\mathcal{A}$. In order to evaluate whether a claim α can be justified we have to consider all argument trees for α and all argument trees for $\neg\alpha$. For an argument tree τ let $\mathsf{root}(\tau)$ denote the root node of τ. Furthermore, for a node $\mathcal{A} \in \tau$ let $\mathsf{ch}_\tau(\mathcal{A})$ denote the children of $\mathcal{A}$ in τ and $\mathsf{ch}_\tau^{\mathcal{T}}(\mathcal{A})$ denote the set of sub-trees rooted at a child of $\mathcal{A}$.

Definition 9 (Argument structure). *Let* $\alpha \in \mathcal{L}(\mathsf{At})$ *be some sentence and let* $\Phi \subseteq \mathcal{L}(\mathsf{At})$ *be a set of sentences. The* argument structure $\Gamma_\Phi(\alpha)$ *for* α *with respect to* Φ *is the tuple* $\Gamma_\Phi(\alpha) = (\mathcal{P}, \mathcal{C})$ *such that* $\mathcal{P}$ *is the set of argument trees for* α *in* Φ *and* $\mathcal{C}$ *is the set of arguments trees for* $\neg\alpha$ *in* Φ.

The argument structure $\Gamma_\Phi(\alpha)$ of a $\alpha \in \mathcal{L}(\mathsf{At})$ gives a complete picture of the reasons for and against α. The argument structure has to be evaluated in order to determine the status of sentences. We introduce the powerful evaluation mechanisms from [2] and give examples of how adequate and simple instantiations can be realized.

Definition 10 (Categorizer). *A* categorizer γ *is a function* $\gamma : \mathcal{T}(\mathsf{At}) \to \mathbb{R}$.

A categorizer is meant to assign a value to an argument tree τ depending on how strongly this argument tree favors the root argument. In particular, the larger the value of $\gamma(\tau)$ the better justification of believing in the claim of the root argument. For an argument structure $\Gamma_\Phi(\alpha) = (\{\tau_1^p, \ldots, \tau_n^p\}, \{\tau_1^c, \ldots, \tau_m^c\})$ and a categorizer γ we abbreviate

$$\gamma(\Gamma_\Phi(\alpha)) = (\langle \gamma(\tau_1^p), \ldots, \gamma(\tau_n^p)\rangle, \langle \gamma(\tau_1^c), \ldots, \gamma(\tau_m^c)\rangle) \in \mathfrak{PP}(\mathbb{R}) \times \mathfrak{PP}(\mathbb{R}).$$

Definition 11 (Accumulator). *An* accumulator κ *is a function* $\kappa : \mathfrak{PP}(\mathbb{R}) \times \mathfrak{PP}(\mathbb{R}) \to \mathbb{R}$.

An accumulator is meant to evaluate the categorization of argument trees for or against some sentence α.

Definition 12 (Acceptance). *We say that a set of sentences* $\Phi \subseteq \mathcal{L}(\mathsf{At})$ *accepts a sentence* α *with respect to a categorizer* γ *and an accumulator* κ, *denoted by*

$$\Phi \mathrel{|\!\sim}_{\kappa,\gamma} \alpha \quad \text{if and and only if} \quad \kappa(\gamma(\Gamma_\Phi(\alpha))) > 0$$

A set of sentences $\Phi \subseteq \mathcal{L}(\mathsf{At})$ *rejects a sentence* α *with respect to a categorizer* γ *and an accumulator* κ, *denoted by*

$$\Phi \mathrel{|\!\not\sim}_{\kappa,\gamma} \alpha \quad \text{if and and only if} \quad \kappa(\gamma(\Gamma_\Phi(\alpha))) < 0$$

If Φ *neither accepts nor rejects* α *with respect to* γ *and* κ *we say that* Φ *is* undecided *about* α *with respect to* γ *and* κ.

Some simple instances of categorizers and accumulators are as follows.

Example 3. Let τ be some argument tree. The classical evaluation of an argument tree—as e.g. employed in *Defeasible Logic Programming* [10]—is that each leaf of the tree is considered "undefeated" and an inner node is "undefeated" if all its children are "defeated" and "defeated" if there is at least one child that is "undefeated". This intuition can be formalized by defining the *classical categorizer* γ_0 recursively via

$$\gamma_0(\tau) = \begin{cases} 1 & \text{if } \mathsf{ch}_\tau(\mathrm{root}(\tau)) = \emptyset \\ 1 - \max\{\gamma_0(\tau') \mid \tau' \in \mathsf{ch}_\tau^{\mathcal{T}}(\mathrm{root}(\tau))\} & \text{otherwise} \end{cases}$$

Furthermore, a simple accumulator κ_0 can be defined via

$$\kappa_0(\langle N_1,\ldots,N_n\rangle, \langle M_1,\ldots,M_m\rangle) = N_1 + \ldots + N_n - M_1 - \ldots - M_m.$$

For example, a set of sentences $\Phi \subseteq \mathcal{L}(\mathsf{At})$ accepts a sentence α with respect to γ_0 and κ_0 if and only if there are more argument trees for α where the root argument is undefeated than argument trees for $\neg\alpha$ where the root argument is undefeated. ∎

More examples of categorizers and accumulators can be found in [2]. Using those notions we are able to state for every sentence $\phi \in \Phi$ whether ϕ is accepted in Φ or not, depending on the arguments that favor α and those that reject α.

5   Selective Revision by Deductive Argumentation

Using the deductive argumentation framework presented in the previous section one is able to decide for each sentence $\alpha \in \Phi$ whether α is justifiable with respect to Φ. Note that the framework of deductive argumentation heavily depends on the actual instances of categorizer and accumulator. In the following we only consider categorizer and accumulator that comply with the following minimal requirements.

Definition 13 (Well-behaving categorizer). *A categorizer γ is called well-behaving if $\gamma(\tau) > \gamma(\tau')$ whenever τ consists only of one single node and τ' consists of at least two nodes.*

In other words, a categorizer γ is well-behaving if the argument tree that has no undercuts for its root is considered the best justification for the root.

Definition 14 (Well-behaving accumulator). *An accumulator κ is called well-behaving if and only if $\kappa((\mathcal{P},\mathcal{C})) > 0$ whenever $\mathcal{P} \neq \emptyset$ and $\mathcal{C} = \emptyset$.*

This means, that if there are no arguments against a claim α and at least one argument for α in Φ then α should be accepted in Φ. Note that both γ_0 and κ_0 are well-behaving as well as all categorizers and accumulators considered in [2].

Note further that if Φ is consistent then every sentence $\alpha \in \Phi$ is accepted by Φ with respect to every well-behaving categorizer and well-behaving accumulator.

Let $\mathcal{K} \subseteq \mathcal{L}(\mathsf{At})$ be a consistent set of sentences, and let γ be some well-behaving categorizer and κ be some well-behaving accumulator. We consider again a selective revision $\circ$ of the form (1). In order to determine the outcome of the non-prioritized revision $\mathcal{K} \circ \Phi$ for some $\Phi \subseteq \mathcal{L}(\mathsf{At})$ we implement a transformation function f that checks for every sentence $\alpha \in \Phi$ whether α is accepted in $\mathcal{K} \cup \Phi$. Note that although $\mathcal{K}$ is consistent the union $\mathcal{K} \cup \Phi$ is not necessarily consistent which gives rise to an argumentative evaluation. In the following, we consider *two* different transformation functions based on deductive argumentation.

Definition 15 (Skeptical Transformation Function). *We define the* skeptical transformation function $\mathsf{S}_{\mathcal{K}}^{\gamma,\kappa}$ *via*

$$\mathsf{S}_{\mathcal{K}}^{\gamma,\kappa}(\Phi) = \{\alpha \in \Phi \mid \mathcal{K} \cup \Phi \mathrel{\vdash}_{\kappa,\gamma} \alpha\}$$

for every $\Phi \subseteq \mathcal{L}(\mathsf{At})$.

Definition 16 (Credulous Transformation Function). *We define the* credulous transformation function $\mathsf{C}_{\mathcal{K}}^{\gamma,\kappa}$ *via*

$$\mathsf{C}_{\mathcal{K}}^{\gamma,\kappa}(\Phi) = \{\alpha \in \Phi \mid \mathcal{K} \cup \Phi \mathrel{\not\vdash}_{\kappa,\gamma} \neg\alpha\}$$

for every $\Phi \subseteq \mathcal{L}(\mathsf{At})$.

In other words, the value of $\mathsf{S}_{\mathcal{K}}^{\gamma,\kappa}(\Phi)$ consists of those sentences of Φ that are accepted in $\mathcal{K} \cup \Phi$ and the value of $\mathsf{C}_{\mathcal{K}}^{\gamma,\kappa}(\Phi)$ consists of those sentences of Φ that are not rejected in $\mathcal{K} \cup \Phi$. There is a subtle difference in the behavior of those two transformation functions as the following example shows.

Example 4. Let $\mathcal{K}_1 = \{a\}$ and $\Phi_1 = \{\neg a\}$. Note that there is exactly one argument tree τ_1 for $\neg a$ and one argument tree τ_2 for a in $\mathcal{K}_1 \cup \Phi$. In τ_1 the root is the argument $\mathcal{A} = \langle\{\neg a\}, \neg a\rangle$ which has the single canonical undercut $\mathcal{B} = \langle\{a\}, a\rangle$. In τ_2 the situation is reversed and the root of τ_2 is the argument $\mathcal{B}$ which has the single canonical undercut $\mathcal{A}$. Therefore, the argument structure for $\neg a$ is given via $\Gamma_{\mathcal{K} \cup \Phi}(\neg a) = (\{\tau_1\}, \{\tau_2\})$. It follows that $\gamma_0(\tau_1) = \gamma_0(\tau_2) = 0$ and $\kappa_0(\gamma_0(\Gamma_{\mathcal{K} \cup \Phi}(a))) = \kappa_0(\langle 0, 0\rangle) = 0$. It follows that $\mathcal{K} \cup \Phi$ is undecided about both $\neg a$ and a. Consequently, it follows that

$$\mathsf{S}_{\mathcal{K}_1}^{\gamma_0,\kappa_0}(\Phi_1) = \emptyset \qquad\qquad \mathsf{C}_{\mathcal{K}_1}^{\gamma_0,\kappa_0}(\Phi_1) = \{\neg a\}.$$

∎

Let $*$ be some (prioritized) multiple base revision operator, γ some categorizer, and κ some accumulator. Using the skeptical transformation function we can define the *skeptical argumentative revision* $\circ_S^{\gamma,\kappa}$ following (1) via

$$\mathcal{K} \circ_S^{\gamma,\kappa} \Phi = \mathcal{K} * \mathsf{S}_{\mathcal{K}}^{\gamma,\kappa}(\Phi) \tag{2}$$

for every $\Phi \subseteq \mathcal{L}(\mathsf{At})$ and using the credulous transformation function we can define the *credulous argumentative revision* $\circ_C^{\gamma,\kappa}$ via

$$\mathcal{K} \circ_C^{\gamma,\kappa} \Phi = \mathcal{K} * C_\mathcal{K}^{\gamma,\kappa}(\Phi) \qquad (3)$$

for every $\Phi \subseteq \mathcal{L}(\mathsf{At})$.

Example 5. We continue Example 4. Let $*$ be some prioritized multiple base revision. Then it follows that $\mathcal{K}_1 \circ_S^{\gamma_0,\kappa_0} \Phi_1 = \{a\}$ and $\mathcal{K}_1 \circ_C^{\gamma_0,\kappa_0} \Phi_1 = \{\neg a\}$. ∎

We now investigate the formal properties of the transformation functions $S_\mathcal{K}^{\gamma,\kappa}$ and $C_\mathcal{K}^{\gamma,\kappa}$ and the resulting revision operators $\circ_S^{\gamma,\kappa}$ and $\circ_C^{\gamma,\kappa}$.

Proposition 2. *Let γ be a well-behaving categorizer and κ be a well-behaving accumulator. Then the transformation functions $S_\mathcal{K}^{\gamma,\kappa}$ and $C_\mathcal{K}^{\gamma,\kappa}$ satisfy inclusion, weak inclusion, weak extensionality, consistency preservation and weak maximality.*

Proof.

Inclusion. *This is satisfied by definition as for $\alpha \in S_\mathcal{K}^{\gamma,\kappa}(\Phi)$ and each $\alpha \in C_\mathcal{K}^{\gamma,\kappa}(\Phi)$ it follows $\alpha \in \Phi$.*

Weak Inclusion. *This follows directly from the satisfaction of* inclusion.

Weak Extensionality. *Let $\Phi \cong^p \Phi'$ and let $\sigma : \Phi \to \Phi'$ be a bijection such that for every $\phi \in \Phi$ it holds that $\phi \equiv^p \sigma(\phi)$. We extend σ to $\mathcal{K}$ via $\sigma(\psi) = \psi$ for every $\psi \in \mathcal{K}$. If $\Psi \subseteq \mathcal{K} \cup \Phi$ we abbreviate*

$$\sigma(\Psi) = \bigcup_{\psi \in \Psi} \{\sigma(\psi)\}.$$

Let $\langle \Psi, \phi \rangle$ be an argument for some $\phi \in \Phi$ with respect to $\mathcal{K} \cup \Phi$. Then $\langle \sigma(\Psi), \sigma(\phi) \rangle$ is an argument for $\sigma(\phi)$ in $\mathcal{K} \cup \Phi'$. It follows that if τ is an argument tree for $\langle \Psi, \phi \rangle$ in $\mathcal{K} \cup \Phi$ then τ' is an argument tree for $\langle \sigma(\Psi), \sigma(\phi) \rangle$ in $\mathcal{K} \cup \Phi'$ where τ' is obtained from τ by replacing each sentence ϕ with $\sigma(\phi)$. This generalizes also to argument structures and it follows that

$$\kappa(\gamma(\Gamma_{\mathcal{K} \cup \Phi}(\phi))) = \kappa(\gamma(\Gamma_{\mathcal{K} \cup \Phi'}(\sigma(\phi)))).$$

Hence, $\phi \in S_\mathcal{K}^{\gamma,\kappa}(\Phi)$ if and only if $\sigma(\phi) \in S_\mathcal{K}^{\gamma,\kappa}(\Phi')$ for every $\phi \in \Phi$. It follows that $S_\mathcal{K}^{\gamma,\kappa}(\Phi) \cong^p S_\mathcal{K}^{\gamma,\kappa}(\Phi')$. The same is true for $C_\mathcal{K}^{\gamma,\kappa}$.

Consistency Preservation. *Every subset of a consistent set of sentences is consistent and, due to* inclusion, *it holds that $S_\mathcal{K}^{\gamma,\kappa}(\Phi), C_\mathcal{K}^{\gamma,\kappa}(\Phi) \subseteq \Phi$ with consistent Φ.*

Weak Maximality. *If $\mathcal{K} \cup \Phi$ is consistent then for all arguments for a sentence $\alpha \in \Phi$ there do not exist any undercuts as these would have to entail the negation of some sentence of the argument for α which implies inconsistency of $\mathcal{K} \cup \Phi$. The argument structure $\Gamma_\Phi(\alpha) = (\mathcal{P}, \mathcal{C})$ consists of one or more single node trees $\mathcal{P}$ and $\mathcal{C} = \emptyset$. As both γ and κ are well-behaving it follows that $\kappa(\gamma(\Gamma_\Phi(\alpha))) > 0$ for each $\alpha \in \Phi$ and therefore $S_\mathcal{K}^{\gamma,\kappa}(\Phi) = \Phi$ and $C_\mathcal{K}^{\gamma,\kappa}(\Phi) = \Phi$.* □

In particular, note that both $S_\mathcal{K}^{\gamma,\kappa}$ and $C_\mathcal{K}^{\gamma,\kappa}$ do not satisfy either *consistency* or *maximality* in general.

Corollary 1. *Let γ be a well-behaving categorizer and κ be a well-behaving accumulator. Then both $\circ_S^{\gamma,\kappa}$ and $\circ_C^{\gamma,\kappa}$ are non-prioritized multiple base revision operators.*

Proof. This follows directly from Propositions 1 and 2.   □

Example 6. We continue Examples 1 and 2 and consider $\mathsf{At} = \{c, a, q, b, r, s\}$ with the following informal interprations.

$$c: \quad \text{Anna has financial problems}$$
$$a: \quad \text{Anna travels to Hawaii}$$
$$q: \quad \text{There is volcano activity on Hawaii}$$
$$b: \quad \text{Anna has a lot of money}$$
$$r: \quad \text{Anna is a surf fanatic}$$
$$s: \quad \text{Anna takes a loan}$$

Now consider Anna's belief base $\mathcal{K}_2$ given via

$$\mathcal{K}_2 = \{r, \quad r \Rightarrow a, \quad s, \quad s \Rightarrow b, \quad b \Rightarrow a, \quad b \Rightarrow \neg c\}.$$

This means that Anna believes that she is a surf fanatic (r), that a surf fanatic should travel to Hawaii ($r \Rightarrow a$), that she takes a loan (s), that taking a loan means that she has a lot of money ($s \Rightarrow b$), that having a lot of money implies she should travel to Hawaii ($b \Rightarrow a$), and that having a lot of money she does not have financial problems. Note that $\mathcal{K} \vdash a$, i. e. Anna intends to go to Hawaii. Now consider the new information $\Phi_2 = \{c, \quad c \Rightarrow \neg a, \quad q, \quad q \Rightarrow \neg a\}$ stemming from communication with Anna's mother. In Φ_2 the mother of Anna tells her not to travel to Hawaii. In particular, Φ_2 states that Anna has financial problems (c), that having financial problems Anna should not travel to Hawaii ($c \Rightarrow \neg a$), that there is also volcano activity on Hawaii (q), and that given volcano activity Anna should not travel to Hawaii ($q \Rightarrow \neg a$).

As one can see there a several arguments for and against a in $\mathcal{K}_2 \cup \Phi_2$, e. g., $\langle r, r \Rightarrow a, a \rangle$, $\langle c, c \Rightarrow \neg a, \neg a \rangle$. We do not go into details regarding the argumentative evaluation of the sentences in Φ_2. We only note that $\mathcal{K}_2 \cup \Phi_2$ is undecided about c but accepts $c \Rightarrow \neg a$, q, and $q \Rightarrow \neg a$ with respect to γ_0 and κ_0. Consequently, the values of $\mathsf{S}_{\mathcal{K}_2}^{\gamma_0,\kappa_0}(\Phi_2)$ and $\mathsf{C}_{\mathcal{K}_2}^{\gamma_0,\kappa_0}(\Phi_2)$ are given via

$$\mathsf{S}_{\mathcal{K}_2}^{\gamma_0,\kappa_0}(\Phi_2) = \Phi_2 \setminus \{c\} \quad \text{and} \quad \mathsf{C}_{\mathcal{K}_2}^{\gamma_0,\kappa_0}(\Phi_2) = \Phi_2.$$

Let $*$ be some prioritized multiple base revision operator and define $\circ_S^{\gamma_0,\kappa_0}$ and $\circ_C^{\gamma_0,\kappa_0}$ via (2) and (3), respectively. Then some possible revisions of $\mathcal{K}_2$ with Φ_2 are given via

$$\mathcal{K}_2 \circ_S^{\gamma_0,\kappa_0} \Phi_2 = \{r, \quad s \Rightarrow b, \quad b \Rightarrow a, \quad b \Rightarrow \neg c, \quad c \Rightarrow \neg a, \quad q \Rightarrow \neg a, \quad q\}$$
$$\mathcal{K}_2 \circ_C^{\gamma_0,\kappa_0} \Phi_2 = \{r, \quad s \Rightarrow b, \quad b \Rightarrow a, \quad b \Rightarrow \neg c, \quad c \Rightarrow \neg a, \quad c, \quad q \Rightarrow \neg a, \quad q\}.$$

Note that it holds $\mathcal{K}_2 \circ_S^{\gamma_0,\kappa_0} \Phi_2 \vdash \neg a$ and $\mathcal{K}_2 \circ_C^{\gamma_0,\kappa_0} \Phi_2 \vdash \neg a$. Hence, Anna accepts the conclusion of her mother's arguments not to travel Hawaii. However, if she revises her beliefs in a skeptical way she does not accept that she has financial problems. ■

6   Related Work

In terms of related work there are mainly two areas that are related to the work presented here. On the one hand, non-prioritized belief revision and on the other hand belief revision by argumentation.

In the former area we instantiate and extended the non-prioritized revision operator of selective revision presented in [9] towards multiple revision and to revision of belief bases. Selective revision is one of the most general non-prioritized revision operator of the type *decision+revision* [11]. Moreover it allows for partial acceptance of the input, in contrast to most other approaches. Apart from *decision+revision* approaches there are *expansion+consolidation* approaches to non-prioritized belief revision. These perform a simple expansion by the new information, i.e. $\mathcal{K} \cup \varPhi$, and then apply a consolidation operator ! that restores consistency, i.e. $\mathcal{K} * \varPhi = (\mathcal{K} \cup \varPhi)!$. This approach is limited to belief bases since all inconsistent belief sets are equal, i.e. $Cn(\bot) = \mathcal{L}(\mathsf{At})$. An instantiation of such an operator that is similar to the setup used in this work has been presented in [8]. The considered input to the revision consists of a set of sentences that form an explanation of some claim in the same form as the argument definition used here. However, as with all approaches of the type *expansion+consolidation*, new and old information are completely equal to the consolidation operator. In contrast, the approach presented here which makes use of two different mechanisms to first decide about if, and which part, of the input shall be accepted just considering the new information, and then performing prioritized belief revision of the old information. Also, there are *integrated choice* approaches that do not feature a two step process but a single step process applying the same technique for the selection and revision process. Mostly these approaches need some meta information, e. g. an epistemic entrenchment relation, and thus differ on the basic process as well as on the information needed.

While there has been some work on the revision of argumentation systems, very little work on the application of argumentation techniques for the revision process has been done so far, cf. [7]. In fact, the work most related to the work presented here makes use of negotiation techniques for belief revision [3,13], without argumentation. In the general setup of [3] a symmetric merging of information from two sources is performed by means of a negotiation procedure that determines which source has to reduce its information in each round. The information to be given up is determined by another function. The negotiation ends when a consistent union of information is reached. While this can be seen as a one step process of merging or consolidation in general, the formalism also allows to differentiate between the information given up from the first source and the second source. In [3], this setting is then successively biased towards prioritizing the second source which leads to representation theorems for operations equivalent to selective revision satisfying *consistent expansion* and for classic AGM operators. While those results are interesting, the negotiation framework used in [3] is very different from the argumentation formalism used here and also very different from the setup of selective revision. Moreover, the functions for the negotiation and concession are left abstract.

In [13] mutual belief revision is considered where two agents revise their respective belief state by information of the other agent. Both agents agree in a negotiation on the information that is accepted by each agent. The revisions of the agents are split into a selection function and two iterated revision functions which leads to operators satisfying *consistent expansion*. The selection function is then a negotiation function on two sets of beliefs that represent the sets of belief that each agent is willing to accept from the other agent that might obey game theoretic principles. This setting has a very different focus as ours and also does not specify the selection function.

7 Conclusion

In this paper we combined the research strains of selective revision and deductive argumentation in order to implement non-prioritized multiple base revision operators that only revise by those portions of the new information that are justified. We only took some first steps in investigating the properties of those revision operators but were able to show that those comply with many desirable properties for non-prioritized revision. We discussed the performance of our operators by examples and briefly compared our approach to related work.

Future work includes a more in depth analysis of the revisions $\circ_S^{\gamma,\kappa}$ and $\circ_C^{\gamma,\kappa}$ and a more thorough comparison with related work, in particular we want to evaluate in more depth the synergetic effects of combining argumentation and belief revision with respect to the results obtained from either framework. Moreover introducing notions of preference seems to be natural in this setting and we plan to extend our framework to preferences on the argumentation as well as on the belief revision side and investigate the connections to epistemic entrenchment.

References

1. Bench-Capon, T.J.M., Dunne, P.E.: Argumentation in artificial intelligence. Artificial Intelligence 171, 619–641 (2007)
2. Besnard, P., Hunter, A.: A logic-based theory of deductive arguments. Artificial Intelligence 128(1-2), 203–235 (2001)
3. Booth, R.: A negotiation-style framework for non-prioritised revision. In: Proceedings of TARK 2001, pp. 137–150 (2001)
4. Makinson, D., Alchourron, C.E., Gärdenfors, P.: On the logic of theory change: Partial meet contraction and revision functions. The Journal of Symbolic Logic 50(2), 510–530 (1985)
5. Delgrande, J.P., Jin, Y.: Parallel belief revision. In: Proceedings of the 23rd National Conference on Artificial Intelligence, AAAI 2008 (2008)
6. Dung, P.M.: On the Acceptability of Arguments and its Fundamental Role in Nonmonotonic Reasoning, Logic Programming and n-Person Games. Artificial Intelligence 77(2), 321–358 (1995)
7. Falappa, M.A., Kern-Isberner, G., Simari, G.R.: Belief revision and argumentation theory. In: Argumentation in Artificial Intelligence, pp. 341–360. Springer, Heidelberg (2009)

8. Falappa, M.A., Kern-Isberner, G., Simari, G.R.: Explanations, belief revision and defeasible reasoning. Artificial Intelligence 141(1), 1–28 (2002)
9. Fermé, E., Hansson, S.O.: Selective revision. Studia Logica 63, 331–342 (1999)
10. García, A.J., Simari, G.R.: Defeasible Logic Programming: An Argumentative Approach. Theory and Practice of Logic Programming 4(1–2), 95–138 (2004)
11. Hansson, S.O.: A survey of non-prioritized belief revision. Erkenntnis 50(2-3), 413–427 (1999)
12. Hansson, S.O.: A Textbook of Belief Dynamics. Kluwer Academic Publishers, Norwell (2001)
13. Zhang, D., Foo, N., Meyer, T., Kwok, R.: Negotiation as mutual belief revision. In: Proceedings of AAAI 2004, pp. 317–322 (2004)

A Three-Layer Argumentation Framework

Paulo Maio and Nuno Silva

GECAD – School of Engineering – Polytechnic of Porto
Rua Dr. Bernardino de Almeida 431, 4200-072 Porto, Portugal
{pam,nps}@isep.ipp.pt

Abstract. Argumentation frameworks which are abstract are suitable for the study of independent properties of any specific aspect (e.g. arguments sceptical and credulous admissible) that are relevant for any argumentation context. However, its direct adoption on specific application contexts requires dealing with questions such as the argument structure, the argument categories, the conditions under which an attack/support is established between arguments, etc. This paper presents a generic argumentation framework which comprehends a conceptualization layer to capture the expressivity and semantics of the argumentation data employed in a specific context and simplifies its adoption by applications. The conceptualization layer together with the defined argument structure is exploited to automatically derive the attack and support relationships between arguments.

Keywords: Argumentation Frameworks, Argument Instantiation, Argument Schemes, Bipolar Argumentation, Agents, MAS.

1 Introduction

A crucial problem on BDI agents as described by Wooldridge [1] concerns what should be the agent beliefs and how those beliefs are used (i) to form new intentions, or (ii) to redraw/revise current intentions. On this matter, contributions of the argumentation research field may be exploited internally by BDI agents since argumentation can be used either for reasoning about what to believe (i.e. theoretical reasoning) and/or for deciding what to do (i.e. practical reasoning). Despite existing differences between both, according to [2], from a standpoint of first-personal reflection, a set of considerations for and against a particular conclusion are drawn on both. Yet, agents in multi-agent systems (MAS) may apply argumentation externally during interactions between agents, i.e. agents' dialogues (cf. [3] for details). Within this context, argumentation is seen as an activity where each participant tries to increase (or decrease) the acceptability of a given standpoint for the other participants by presenting arguments. Therefore, argumentation is foreseen as an adequate modeling formalism to reduce the gap between models governing the internal and external agent behavior.

In which concerns to argumentation, there is an abundance of relevant literature in argumentation and argumentation systems. With regards to argumentation modeling

S. Modgil, N. Oren, and F. Toni (Eds.): TAFA 2011, LNAI 7132, pp. 163–180, 2012.

formalisms, the abstract argumentation frameworks such as the AF [4], the BAF [5] and the VAF [6] are suitable to represent many different situations without being committed to any domain of application. Due to their abstract nature they are also suitable for the study of independent properties of any specific aspect (e.g. arguments sceptical and credulous admissible) that are relevant for any argumentation context that can be captured and formalized accordingly. On the other hand, this abstract nature represents an expressiveness limitation to the direct adoption of specific application contexts [7, 8]. To overcome this limitation, argumentation systems usually adopt an abstract argumentation framework and extend it in order to get a less abstract formalism, dealing in particular with (i) the construction of arguments and their structure, (ii) the conditions under which argument-relations (i.e. attack and/or support) are established, (iii) categories of arguments, etc. Nevertheless, abstract argumentation frameworks do not provide any machinery facilitating and governing how applications should extend or instantiate the framework. As a result, a significant gap between abstract argumentation frameworks and applications exist.

Regarding arguments acceptability, argumentation systems (e.g. the Prakken version of ASPIC [8]) use the abstract level as an abstraction of the overall system to make logical inferences. That is, systems start with a knowledge base, which is used to instantiate the adopted argumentation framework and then apply a given abstract argumentation semantics such as the ones described in [7] to select the conclusions of the associated sets of arguments. However, as studied in [8] and [9], in light of the arguments' content it is still possible that sets of arguments selected by an abstract argumentation criterion yield to inconsistent conclusions.

This paper proposes a less abstract argumentation framework whose purpose is to reduce existing gaps between abstract argumentation frameworks and applications, namely which concerns with the arguments' instantiation. For that, the proposed framework (i) adopts a general and intuitive argument structure, (ii) includes a conceptual layer for the specification of the semantics of argumentation data applied in a specific domain of application (e.g. e-commerce, legal reasoning and decision making) and (iii) defines a novel conceptual relation between argument-schemes called arguments affectation. In addition, the proposed framework exploits the conceptual information and the defined argument structure to automatically derive the attack and support relationships between arguments. Despite the arguments' acceptability issue is not directly addressed in this paper, applications still profiting from the inherent suitability of abstract argumentation frameworks on the study of independent properties, since information represented according to the proposed argumentation framework is easily transformed (or converted) to BAF [5]. Despite having these new features, the proposed argumentation framework remains general, but less abstract than AF [4], BAF [5] and VAF [6].

The rest of the paper is organized as follows. The next section introduces background concepts about abstract argumentation frameworks. Section 3 presents the proposed argumentation framework. Next, in section 4, an example is provided to illustrate the application of the proposed argumentation framework. Section 5 complements the proposed argumentation framework with a process to automatically derive the attack and support relationships between arguments. Section 6 compares and discusses the proposed framework with the related work. Finally, Section 7 draws conclusions and discusses future work.

2   Abstract Argumentation Frameworks

This section briefly describes the main concepts of the most referenced abstract argumentation frameworks found in the literature: the Argumentation Framework proposed by Dung (AF) [4], the Value Argumentation Framework (VAF) [6] and the Bipolar Argumentation Framework (BAF) [5].

As proposed by Dung [4], the AF core entities are *Argument*, and a binary relation between arguments (R_{att}) as depicted in Fig. 1a. The R_{att} relation is known as the attack relation. An AF can be defined as a tuple $AF = (A, R_{att})$ where A is a set of arguments and R_{att} is a relation on A such that $R_{att} \subseteq A \times A$.

An AF instance may be represented by a directed graph whose nodes are arguments and edges represent the attack relation. For any two arguments, say a_1 and a_2, such that $a_1, a_2 \in A$, one says that a_1 attacks a_2 iif $(a_1, a_2) \in R_{att}$.

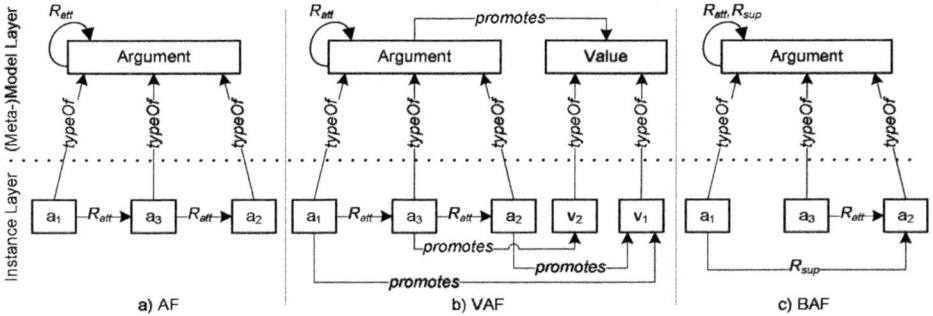

Fig. 1. The main concepts of abstract argumentation frameworks

In Dung's work attacks always succeed (i.e. it defeats the attacked arguments). Yet, one says that an argument y is attacked by a set of arguments S such that $S \subseteq A$ if S contains at least one argument attacking y. Grounded on that, the following notions were defined:

- An argument $a \in A$ is *acceptable* with respect to a set of arguments S, i.e. $acceptable(a, S)$, iif $\forall x: x \in A \land (x, a) \in R_{att} \to \exists y: y \in S \land (y, x) \in R_{att}$;
- A set of arguments S if *conflict-free* iif $\nexists x, y: x, y \in S \land (x, y) \in R_{att}$;
- A *conflict-free* set of arguments S is admissible iif $\forall x: x \in S \to acceptable(x, S)$;
- A set of arguments S is a *preferred extension* iif it is maximal (with respect to set inclusion) *admissible* set of A.

A *preferred extension* represents a consistent position within an AF instance, which is defensible against all attacks and cannot be further extended without introducing a conflict. Yet, multiple *preferred extensions* can exist in an AF instance due to the presence of cycles of *even* length in the graph. Given that, one considers that (i) an argument is *sceptical admissible* if it belongs to any *preferred extension* and (ii) an argument is *credulous admissible* if it belongs to at least one *preferred extension*.

While it is reasonable that attacks always succeed when dealing with deductive arguments, in domains where arguments lack this coercive force, arguments provide reasons which may be more or less persuasive and their persuasiveness may vary according to their audience. Accordingly, it is necessary to distinguish between attacks and successful attacks (i.e. defeats) prescribing different strengths to arguments on the basis of the values they promote and/or their motivation in order to accommodate the different interests and preferences of an audience. With that purpose, the VAF [6] extended the AF [4] with (i) the concept of *Value* and (ii) the function *promotes* relating an *Argument* with a single *Value* (depicted in Fig. 1b). Therefore, a VAF can be defined as 4-uple $VAF = (A, R_{att}, V, promotes)$ where A and R_{att} means the same as in the AF, a non-empty set of values V and the function $promotes: A \rightarrow V$ to map elements from A to elements of V. Consequently, an *audience* for a VAF instance corresponds to a binary preference relation $P \subseteq V \times V$ which is transitive, irreflexive and asymmetric. If a pair $(v_1, v_2) \in P$ means that value v_1 is preferred to v_2 in the audience P. An attack between two arguments (i.e. $(a_1, a_2) \in R_{att}$) where a_1 promotes a value v_1 and a_2 promotes a value v_2 succeeds (i.e. a_1 defeats a_2) iif the adopted audience prefers v_1 to v_2 otherwise the attack fails. As a result, previous notions (i.e. *acceptable, admissible, conflict-free* and *preferred extension*) were redefined accordingly (cf. [6] for details). Notice that for the same audience multiple *preferred extensions* are possible and different audiences may also lead to a unique *preferred extension*. In this way, different agents (each one represented by one audience) can have different perspectives (i.e. *preferred extensions*) over the same arguments.

The AF and the VAF assume that an argument a_1 supports an argument a_2 if a_1 attacks and therefore defeats an argument a_3 that attacks argument a_2. Thus, these frameworks only explicitly represent the negative interaction (i.e. attack), while the positive interaction (i.e. defense/support) of an argument a_1 to another argument a_2 is implicitly represented by the attack of a_1 to a_3. Since support and attack are related notions, this modeling approach adopts a parsimonious strategy, which is neither a complete nor a correct modeling of argumentation [10]. Conversely, the BAF [5] assumes the attack relation is independent of the support relation and both have a diametrically opposed nature and represent repellent forces. As a result, BAF [5] extended the AF [4] with the support relation (R_{sup}) in order to be explicitly represented (depicted in Fig. 1c). Thus, a BAF can be defined as a 3-uple $BAF = (A, R_{att}, R_{sup})$ where A and R_{att} means the same as in the AF and R_{sup} is a binary relation on A such that $R_{sup} \subseteq A \times A$. Given that, for any two arguments, say a_1 and a_2, such that $a_1, a_2 \in A$, one says that a_1 supports a_2 iif $(a_1, a_2) \in R_{sup}$. Consequently, the notions of *acceptable* and *conflict-free* arguments as well as the notion of a *preferred extension* were redefined accordingly (cf. [5] for details).

For all of these frameworks, an argument is anything that may attack/support or be attacked/supported by another argument. The absence of an argument structure and semantics enables the study of independent properties of any specific aspect that are relevant for any argumentation context that can be captured and formalized accordingly. On the other hand, this emphasizes the limited semantics for direct adoption in specific application contexts [7, 8]. Indeed, a given application context requires a less abstract formalism to deal with (i) the construction of arguments and their structure, (ii) the conditions for an argument attack/support another, (iii) categories of arguments, etc.

3   Three-Layer Argumentation Framework

This section presents the proposed argumentation framework, which is denominated as Three-Layer Argumentation Framework (TLAF). First, we give an informal overview of the framework main concepts and their relations. Further, the framework is formally defined.

3.1   Informal Overview

Unlike the abstract argumentation frameworks described, the TLAF features three modeling layers as depicted in Fig. 2 (the line ending with a hollow triangle means specialization/generalization).

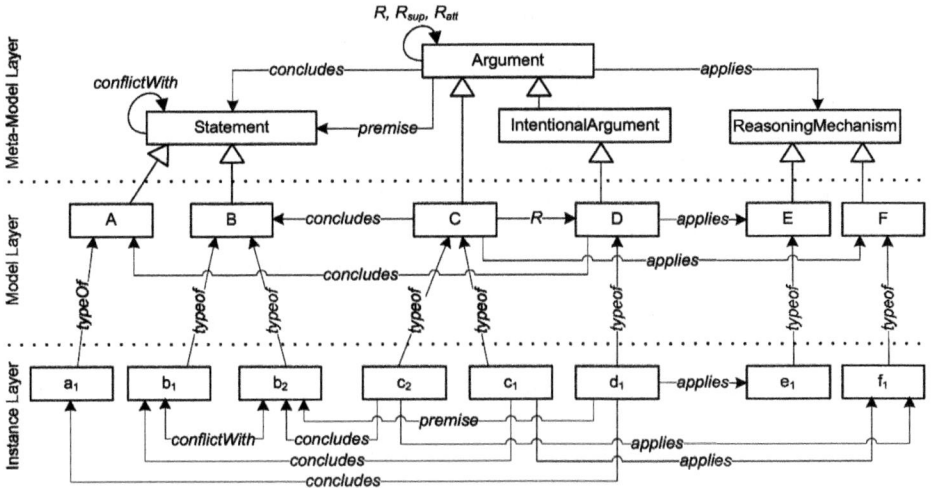

Fig. 2. The three modeling layers of the proposed argumentation framework

Despite existing differences, the TLAF Meta-Model Layer and the TLAF Instance Layer have the same purpose as those of AF [4], BAF [5] and VAF [6] layers with the same name. The TLAF Model Layer intends to capture the semantics of argumentation data (e.g. argument types/schemes) applied in a specific domain of application (e.g. e-commerce, legal reasoning and decision making) and the relations existing between them. In that sense, the model layer is important for the purpose of enabling knowledge sharing and reuse between agents. In this context, a model is a specification used for making model commitments. Practically, a model commitment is an agreement to use a vocabulary in a way that is consistent (but not complete) with respect to the theory specified by a model [11, 12]. Agents then commit to models and models are designed so that the knowledge can be shared among these agents. Accordingly, the content of this layer directly depends on (i) the domain of application to be captured and (ii) the perception one (e.g. a community of agents) has about that domain. Due to this, we adopt the vocabulary of (i) argument (or statement)-instance as an instance of an (ii) argument (or statement)-type defined at the Model Layer. Similarly, we adopt the vocabulary of (i) relation between types, and (ii) relationship between instances.

In TLAF, the meta-model layer defines an argument which is made of three parts: (i) a set of premise-statements, (ii) a conclusion-statement and (iii) an inference from premises to the conclusion enabled by a reasoning mechanism. This argument structure is very intuitive and corresponds to the minimal definition presented by Walton in [13]. For that, the meta-model layer defines the notion of *Argument*, *Statement* and *Reasoning Mechanism*, and a set of relations between these concepts. Following the notion of the BDI model [14, 15], an *IntentionalArgument* is the type of argument whose content corresponds to an intention. Domain data and its meaning are captured by the notion of *Statement*. This mandatorily includes the domain intentions, but also the desires and beliefs. The distinction between arguments and statements allows the application of the same domain data (i.e. statement) in and by different means to arguments. Also the same statement can be concluded by different arguments, and serve as the premise of several arguments. The notion of *Reasoning Mechanism* captures the rules, methods, or processes applied by arguments.

At the model layer, an argument-type (or argument scheme) is characterized by the statement-type it concludes, the applied class of reasoning mechanism (e.g. Deductive, Inductive, Heuristic) and the set of affectation relations (i.e. R) it has. The R relation is a conceptual abstraction of the attack (i.e. R_{att}) and support (i.e. R_{sup}) relationships. The purpose of R is to define at the conceptual level that argument-instances of an argument-type may affect (either positively or negatively) instances of another argument-type. For example, according to the model layer of Fig. 2, $(C, D) \in R$ means instances of argument-type C may attack or may support instances of argument-type D depending on the instances content. On the other hand, if $(X, Y) \notin R$ it means that instances of argument-type X cannot (in any circumstance) attack/support instances of argument-type Y. Yet, the R relation is also used to determine the types of statements that are admissible as premises of an argument-instance. So, an argument-instance of type X can only have as premises statements of type S iif S is concluded by an argument-type Y and Y affects X (i.e. $(Y, X) \in R$). For example, considering again the model layer of Fig. 2, instances of argument-type D can only have as premises statements of type B because D is affected by argument-type C only.

At the instance layer, an argument-instance *applies* a concrete reasoning mechanism to *conclude* a conclusion-statement-instance from a set of *premise*-statement-instances. The relation *conflictWith* is established between two statement-instances only. A statement-instance b_1 is said to be in conflict with another statement-instance b_2 when b_1 states something that implies or suggests that b_2 is not true or do not holds. The *conflictWith* relation is asymmetric (in Fig. 2 b_2 conflicts with b_1 too). In this case, for example, b_1 may represent the statement "Peter is an expert on PCs." and b_2 may represent the statement "Peter is not an expert on PCs". While the R_{att} and R_{sup} relations are established between argument-instances as in BAF [5], these relationships are automatically inferred in TLAF exploiting (i) the argument statements (i.e. conclusion and premises), (ii) the existing conflicts between statement-instances and (iii) based on the R relations defined at the model layer (cf. section 5 for details). It is worth noticing that all instances existing in the instance layer must have an existing type in the model layer and according to the type characterization.

3.2   Formal Definition

The TLAF is formally described as follows.

Definition 1 (TLAF). A TLAF structure is a singleton $TLAF = (E)$, where E is the set of entities of a TLAF.

A TLAF represents a self-contained unit of structured information. Elements in a TLAF are called argumentation entities.

Definition 2 (TLAF Model Layer). A model layer associated with a TLAF is a 6-tuple $ML(TLAF) = (A, IA, S, M, R, \sigma)$ where:

— $A \subseteq E$ is a set of argument-types;
— $IA \subseteq A$ is the sub-set of argument-types whose instances claim corresponds to an *intention* ([14, 15]);
— $S \subseteq E$ is the a set of statement-types;
— $M \subseteq E$ is the set of reasoning mechanisms;
— $R \subseteq A \times A$ establishes a reflexive relation between two argument-types called arguments' affectation. If a pair $(a_1, a_2) \in R$ then argument-instances of type a_1 may affect (positively or negatively) argument-instances of type a_2;
— σ is a function that assigns to every argument-type (i) the concluded statement-type and (ii) the reasoning mechanism applied, such as $\sigma: A \to S \times M$ where:

— function $concl: A \to S$;
— function $reason: A \to M$.

Each TLAF has a model layer associated with it. Information captured within the model layer plays an important role by conducting and governing the instantiation process of the framework by an application, namely which concerns the construction and semantics of instances and existing relations between them. In that sense, the model layer can also be used to validate the TLAF Instance Layer.

Notice that argument-types do not define their statement-types used as premises. Instead, these are derived from the R relation established between arguments.

Definition 3 (TLAF Instance Layer). An instance layer associated with a TLAF is a 6-tuple $IP(TLAF) = (I, instA, instS, instM, \Sigma, sconflict)$ where:

— $I \subseteq E$, is a set of instances;
— function $instA: A \to 2^I$ relates an argument-type with a set of instances. Consequently, the set of all argument instances AI is defined according to equation 1 (see below). Furthermore, we define the inverse function as $instA^-: AI \to A$;
— function $instS: S \to 2^I$ relates a statement-type with a set of instances. Consequently, the set of all statement instances SI is defined according to equation 1. Furthermore, we define the inverse function as $instS^-: SI \to S$;

- function $instM: M \to 2^I$ relates a reasoning mechanism with a set of instances. Consequently, the set of all reasoning mechanism instances MI is defined according to equation 1. Furthermore, we define the inverse function as $instM^-: MI \to M$;
- function $\Sigma: AI \to SI \times MI \times 2^{SI}$, defines for every argument-instance (i) the statement-instance concluded, (ii) the reasoning mechanism instance used to infer the conclusion and (iii) the set of statement-instances used as premises, where:

  - function $iconcl: AI \to SI$, defines the statement-instance that plays the role of conclusion on an argument-instance. Indeed, an argument-instance has only one statement-instance as conclusion while a statement-instance is concluded by at least one argument-instance;
  - function $ireason: AI \to MI$, defines the reasoning mechanism instance that is used by an argument-instance.
  - function $ipremise: AI \to 2^{SI}$, defines the statement-instances used as premises on an argument-instance. Moreover, statement-instances used as premises are also concluded by other arguments;

- function $sconflict: SI \to 2^{SI}$, defines the statement-instances that are in conflict with a statement-instance.

$$AI = \bigcup_{\forall x: x \in A} instA(x), \qquad SI = \bigcup_{\forall x: x \in S} instS(x), \qquad MI = \bigcup_{\forall x: x \in M} instM(x) \qquad (1)$$

As the reader might have noticed, the instance layer definition is concerned with the generation of argument-instances, statement-instances and their inter-relationships (Σ and $sconflict$). Despite the fact that this is a domain dependent process, it profits from the subjacent TLAF model, namely due to the rules complementing the $iconcl$, $ipremise$ (see next definition) and $sconflict$ (see section 5), that have the ability to conduct and simplify the process.

Definition 4 (TLAF Interpretation). An interpretation of a TLAF is a structure $\mathfrak{T} = (\Delta^{\mathfrak{T}}, A^{\mathfrak{T}}, S^{\mathfrak{T}}, M^{\mathfrak{T}}, I^{\mathfrak{T}})$ where:

- $\Delta^{\mathfrak{T}}$ is the domain set;
- $A^{\mathfrak{T}}: A \to 2^{\Delta^{\mathfrak{T}}}$ is an argument interpretation function that maps each argument-type to a subset of the domain set;
- $S^{\mathfrak{T}}: S \to 2^{\Delta^{\mathfrak{T}}}$ is a statement interpretation function that maps each statement-type to a subset of the domain set;
- $M^{\mathfrak{T}}: M \to 2^{\Delta^{\mathfrak{T}}}$ is a reasoning mechanism interpretation function that maps each reasoning mechanism to a subset of the domain set;
- $I^{\mathfrak{T}}: I \to \Delta^{\mathfrak{T}}$ is an instance interpretation function that maps each instance to a single element in the domain set;

An interpretation is a model of TLAF if it satisfies the following properties:

- $\forall a, i: a \in A \land i \in instA(a) \Rightarrow I^{\mathfrak{T}}(i) \in A^{\mathfrak{T}}(a)$;
- $\forall s, i: s \in S \land i \in instS(s) \Rightarrow I^{\mathfrak{T}}(i) \in S^{\mathfrak{T}}(s)$;
- $\forall m, i: m \in M \land i \in instM(m) \Rightarrow I^{\mathfrak{T}}(i) \in M^{\mathfrak{T}}(m)$;

— $\forall a: a \in IA \Rightarrow A^{\mathfrak{T}}(a) \text{ are intentions}$
— $\forall a, i: a \in A \wedge i \in instA(a) \Rightarrow I^{\mathfrak{T}}(iconcl(i)) \in S^{\mathfrak{T}}(concl(a)) \wedge$
$$I^{\mathfrak{T}}(ireason(i)) \in M^{\mathfrak{T}}(reason(a));$$
— $\forall a, i, p: a \in A \wedge i \in instA(a) \wedge p \in ipremise(i) \Rightarrow$
$$\exists x, y: I^{\mathfrak{T}}(y) \in A^{\mathfrak{T}}(x) \wedge p = iconcl(y) \wedge (x, a) \in R$$
— $\forall a, s: a \in A \wedge s \in S \Rightarrow A^{\mathfrak{T}}(a) \cap S^{\mathfrak{T}}(s) = \emptyset;$
— $\forall a, m: a \in A \wedge m \in M \Rightarrow A^{\mathfrak{T}}(a) \cap M^{\mathfrak{T}}(m) = \emptyset;$
— $\forall s, m: s \in S \wedge m \in M \Rightarrow S^{\mathfrak{T}}(s) \cap M^{\mathfrak{T}}(m) = \emptyset.$

Definition 5 (Argument Properties). An argument-type $a \in A$ and all its argument-instances (i.e. $\forall a_i: a_i \in AI \wedge a_i \in instA(a)$) are said to be:

— *intentional* if $a \in IA$;
— *non-intentional* if $a \notin IA$;
— *defeasible* if $\exists x: x \in A \wedge x \neq a \wedge (x, a) \in R$;
— *indefeasible* if $\forall x: x \in A \wedge x \neq a \wedge (x, a) \notin R$.

Arguments may be used with two purposes: (i) to represent and communicate *intentions* (i.e. intentional arguments) and (ii) to provide considerations (i.e. *beliefs*, *desires*) for and against those *intentions* (i.e. non-intentional arguments). Thus, an intentional argument may be affected by several non-intentional arguments. Additionally, to capture dependency between *intentions*, intentional arguments may be also affected (directly or indirectly) by other intentional arguments. A defeasible argument is affected by other (sub-) arguments (i.e. the ones concluding its premises or the ones undermining those premises) while an indefeasible argument can only be affected by its negation since it cannot have premises. Given that, in a TLAF Model Layer, intentional arguments should be always defeasible. On the contrary, non-intentional arguments can be both defeasible and indefeasible.

4   A Walk-through Example

This section provides an example whose purpose is to show the application of TLAF. For that, we decide on a common and simple scenario such as buying digital cameras. First, for the scenario in hands a possible TLAF model is introduced and discussed. Next, a short and somewhat contrived dialogue is used to demonstrate how the TLAF model guides the instantiation process of TLAF.

4.1   A TLAF Model

Consider the partial TLAF model layer graphically depicted in Fig. 3[1], where the rectangles denote non-intentional argument types, the rectangles with rounded corners denote intentional argument-types and the oriented arrows denote an R-relation between two argument types.

---

[1] Instead of a formal definition, we present a partial graphical view of the model layer because we consider it to be more informative to the reader.

The intention of buying a camera is captured by the argument-type *BuyCamera* which is affected by considerations about (i) the *Requirement* to buy a camera, (ii) the general trend of received *Reviews*, (iii) the general perspective about the cameras' *Features* and (iv) the *PriceRelation* (i.e. expensive vs. cheap). The *PriceRelation* grounds on considerations about the *CurrentPrice* and the *PastPrice*. The *Requirement* is affected by two types of considerations: (i) *HobbyReq* (i.e. a hobby requirement) or (ii) a *JobReq* (i.e. job requirement). *Reviews* are affected by each individual opinion (i) of friends (*FriendsReview*) and (ii) of experts (*ExpertReview*). The latter requires that the reviewer is considered an expert (*PersonExpert*). The *Features* are affected by considerations about the *Zoom* which is made based on the *DigitalZoom* and *OpticalZoom*. Additionally, for the sake of brevity, consider that each of these arguments concludes a statement-type with a similar name (e.g. argument *OpticalZoom* concludes *OpticalZoomStmt*) and applies a heuristic or presumptive reasoning mechanism. Notice that the provided conceptualization do not intends to be neither complete nor the most accurate approach for the scenario in hands.

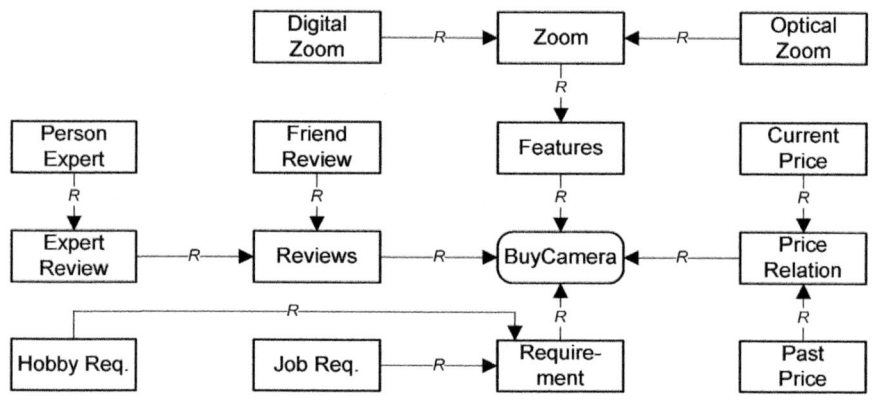

Fig. 3. A partial view over a TLAF model layer for buying cameras

This TLAF model has several indefeasible argument-types (e.g. *PersonExpert*, *CurrentPrice*, *PastPrice*) and several defeasible argument-types (e.g. *Reviews*, *Requirement*, *PriceRelation*). Regarding the former ones, agents are only able to agree or disagree with the conclusions of those argument-instances. For example, an agent can agree or disagree with other agent on the fact that someone is expert on digital cameras but it cannot argue about the information behind such position (i.e. belief). On the contrary, agents are able to argue about the information behind the conclusions of defeasible arguments. For example, an agent that does not agree about the general trend of reviews about a given digital camera presented by another agent is able to present a set of individual reviews (provided by friends and/or experts) supporting its position, which may lead the other agent to change its initial position.

Since a TLAF model captures the perception, the understanding and the rationality that someone (e.g. an agent or a community of agents) has on a given moment about a domain of application, it may evolve over time. For example, this model may evolve in order to allow agents to argue about the fact of someone to be or not to be an expert.

The information used for that purpose (e.g. the person's skills) should be conceptual analyzed and captured on the TLAF model. The resulting statement and argument types must be connected with the already existing argument types through R-relations.

4.2   Instantiating a TLAF Model

Consider the following dialogue takes place between husband (H) and wife (W). In the light of previous TLAF model, the relevant statements (i.e. domain data) uttered by both are marked as st_i (with $i > 0$).

H. I am looking forward to buy camera X (st_1).
W. Why? We don't need it (st_2).
H. That is not true (st_3). I need a camera to perform the task that Sam assigned to me (st_4). Besides that, the camera received several good reviews on a website (st_5).
W. Susan and Mary bought that camera and they told me that they regret their option (st_6 and st_7).
H. Oh, come on Honey. Peter Noble is an expert on the matter (st_8) and he says great things about the camera (st_9).
W. How much it costs? Is it expensive?
H. No! Currently, there is a great opportunity in the city mall (st_{10}). It only costs 100€ (st_{11}). Last week, the price was 150€ (st_{12}).
W. That camera is a discontinued product.
H. I don't care about that.
W. I am reading in this magazine that it lacks some minimal features (st_{13}) such as zoom (st_{14}).
H. Nonsense! Camera X has a digital zoom of "80x" (st_{15});
W. Yeah! But, the optical zoom is only of "4x" (st_{16}).

It is worth noting that: (i) the information stating camera X is a discontinued product did not give raise any statement because it was not envisioned in the TLAF model layer being used; and (ii) despite Susan and Mary have the same opinion, two statements (i.e. st_6 and st_7) were identified such that each statement corresponds to the opinion of a single person (i.e. Susan and Mary respectively), which is consistent with the semantics of the underlying TLAF model.

   Even though this is a short dialogue, it already may be difficult to keep track of all information used and how it is inter-related in the form of argument-instances. As the result of an instantiation process, consider the arguments, the statements and the relationships between arguments and statements presented in Table 1.

   To make evident how the information captured in a TLAF model can be exploit during the instantiation process, let us roughly describe the one adopted here. It consists of three distinct and complementary steps. First step, each statement identified in the dialogue gives raise to one argument-instance concluding that statement. Second step, because the premises of argument-instances are not always explicit in the dialogue, the instantiation process infers the premises through the information captured in the model layer. Thus, it sets as premise-statements of an argument-instance t_1 of type t the statement-instances concluded by argument-instances whose types affect t and that show to support the idea concluded by t_1. For example, st_5 is set as premise of argument a_1 of type *BuyCamera* because st_5 is concluded by a_5 of type *Reviews* and $(Reviews, BuyCamera) \in R$ and the idea underlying st_5 somehow contributes for the idea expressed by the conclusion of a_1 which is st_1. Third, conflicts between statement-instances are established based on two conditions:

— two statement instances are in mutual conflict if both statement-instances are of the same type but they express contradictory ideas (e.g. st_2 and st_3); or
— a statement-instance s_1 is in conflict with a statement-instance s_2 if both are concluded by two distinct argument-types (say t_1 and t_2 respectively) and t_1 affects t_2 (i.e. $(t_1, t_2) \in R$) and the idea expressed by s_1 suggest that s_2 is not true or do not holds (e.g. st_2 and st_1).

Table 1. Instances of arguments and statements constructed and their relationships

ID	Argument Type	Premise Statements	Conclusion-Statement Statement	conflictWith
a_1	BuyCamera	st_3, st_5, st_{10}	st_1	
a_2	Requirement		st_2	st_1, st_3
a_3	Requirement	st_4	st_3	st_2
a_4	JobReq		st_4	
a_5	Reviews	st_9	st_5	
a_6	FriendReview		st_6	st_5
a_7	FriendReview		st_7	st_5
a_8	PersonExpert		st_8	
a_9	ExpertReview	st_8	st_9	
a_{10}	PriceRelation	st_{11}, st_{12}	st_{10}	
a_{11}	CurrentPrice		st_{11}	
a_{12}	PastPrice		st_{12}	
a_{13}	Features	st_{14}	st_{13}	st_1
a_{14}	Zoom	st_{16}	st_{14}	
a_{15}	DigitalZoom		st_{15}	st_{14}
a_{16}	OpticalZoom		st_{16}	

It is envisaged that each scenario of application may require an instantiation process able to deal with its own particularities. However, it is our conviction that most of those processes may take advantage of the R-relations in a very similar way to the described process.

Once the instantiation process ends, support and attack relationships between argument-instances are inferred automatically. This is the subject of next section.

5   Deriving Arguments Relationships

According to the formal definitions introduced in section 3.2, the R_{att} and R_{sup} relationships between argument-instances of an $IP(TLAF)$ are not explicitly defined. Instead, these relationships are derived based on two distinct kinds of information:

— extensional information (existing at the instance layer):
    — the premises and conclusions of the argument-instances;
    — the conflicts between statement-instances, and;

— conceptual information (existing at the model layer), namely the R relations defined between argument-types.

5.1   Deriving Support Relationships

A support relationship between two argument-instances (say x and y) is established (i.e. $(x, y) \in R_{sup}$) when the argument-type of x (say a) affects the argument-type of y (say b), i.e. $(a, b) \in R$, and either (i) the conclusion of x is a premise of y or (ii) both argument-instances have the same conclusion. The following rules (graphically depicted in Fig. 4) capture the conditions required to establish support relationships between argument-instances:

R1. $\forall a, b, x, y: a, b \in A \wedge (a, b) \in R \wedge x \in instA(a) \wedge y \in instA(b) \wedge x \neq y \wedge$
$\quad iconcl(x) \in ipremise(y) \Rightarrow (x, y) \in R_{sup}$ (Fig. 4a);

R2. $\forall a, b, x, y: a, b \in A \wedge (a, b) \in R \wedge x \in instA(a) \wedge y \in instA(b) \wedge x \neq y \wedge$
$\quad iconcl(x) = iconcl(y) \Rightarrow (x, y) \in R_{sup}$ (Fig. 4b).

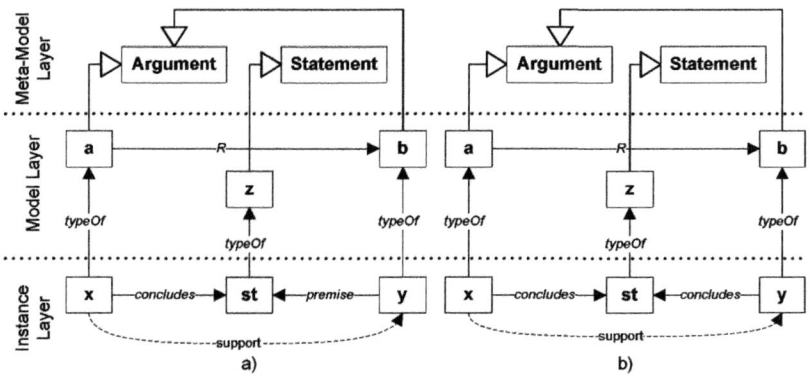

Fig. 4. Conditions to derive a support relationship between two argument-instances

Notice that two argument-instances might achieve the same conclusion starting from a different set of premises and/or reasoning mechanisms. In those circumstances, a support relation between argument-instances exists if there is a R relation between both (depicted in Fig. 4b). For a mutual support, two R relationships are required: one from a to b (i.e. $(a, b) \in R$) and another one from b to a (i.e. $(b, a) \in R$).

5.2   Deriving Attack Relationships

An attack relationship between two argument-instances (say x and y) is established (i.e. $(x, y) \in R_{att}$) when the argument-type of x (say a) affects the argument-type of y (say b), i.e. $(a, b) \in R$, and either (i) the conclusion of x is in conflict with any premise of y or (ii) the conclusion of x is in conflict with the conclusion of y. The following rules (graphically depicted in Fig. 5) capture the conditions required to establish attack relationships between argument-instances:

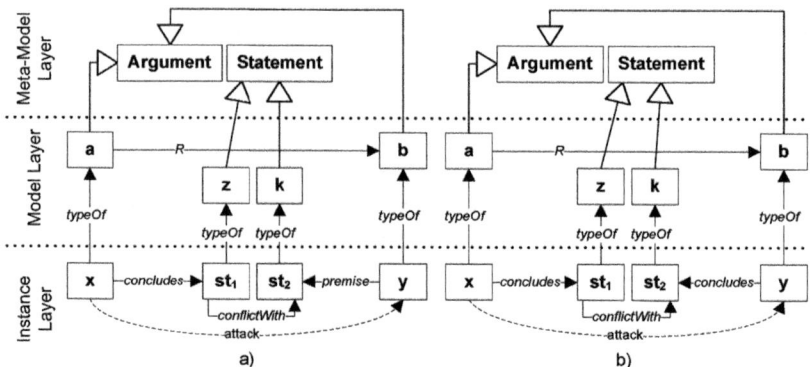

Fig. 5. Conditions to derive an attack relationship between two argument-instances

R3. $\forall a, b, x, y, s: a, b \in A \wedge (a, b) \in R \wedge x \in instA(a) \wedge y \in instA(b) \wedge x \neq y \wedge$
$s \in ipremise(y) \wedge s \in sconflict(iconcl(x)) \Rightarrow (x, y) \in R_{att}$ (Fig. 5a);

R4. $\forall a, b, x, y: a, b \in A \wedge (a, b) \in R \wedge x \in instA(a) \wedge y \in instA(b) \wedge x \neq y \wedge$
$iconcl(y) \in sconflict(iconcl(x)) \Rightarrow (x, y) \in R_{att}$ (Fig. 5b).

According to the rule/scenario depicted in Fig. 5b, one cannot say that argument y also attacks argument x because the conflict relation between statements is asymmetric. However, that would happen iif statement st_2 is also in conflict with statement st_1 (i.e. $st_1 \in sconflict(st_2)$) and a R relationship between b and a (i.e. $(b, a) \in R$) exists too.

5.3   Exploiting the Derivation Process

The application process used to identify and establish conflicts between statement-instances may exploit the knowledge embedded in rules R3 and R4 to reduce and drive the search/combination space between statements. Indeed, it is worth establishing a conflict relationship between two statement-instances (say st_1 and st_2) iif their statement-types (say z and k respectively) satisfy at least one of the following conditions:

— There is an argument-type (say a) concluding z that affects any other argument-type (say b), i.e. $(a, b) \in R$, where statement-instances of type k can be used as premises of argument-instances of type b;

— There is an argument-type (say a) concluding z that affects any other argument-type (say b), i.e. $(a, b) \in R$, where k is concluded by b.

Notice that, these conditions can be verified using the information captured at the model layer only. On the other hand, if a conflict relationship is established between two statement-instances and none of these conditions apply then it has no impact on derived attack relationships between arguments.

Regarding the argument-instances of the example introduced in section 4, these four rules would derive the support and attack relationships graphically depicted in Fig. 6.

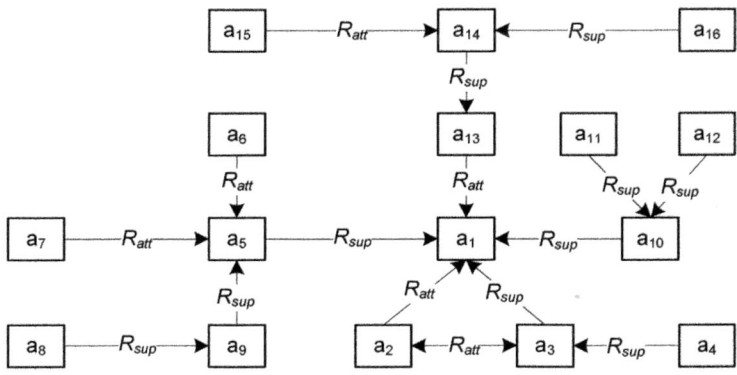

Fig. 6. Derived support and attack relationships between argument-instances of the example

6   Related Work

In this paper the advantages of having a conceptual model layer and the consequent adoption of a structured argumentation are exploited to reduce the existing gap between the most referenced abstract argumentation frameworks and its adoption by applications, namely which concerns to the instantiation process. Regarding the conceptual model only, to the best of our knowledge the most similar work existing in literature is the Description Logic formalizations of the Argument Interchange Format (AIF) [16] proposed by Iyad Rahwan in [17, 18]. In common to the AIF-based work, the TLAF has mainly two aspects:

— the adopted argument structure suggested by Walton [13]; and
— the possibility of the TLAF model layer being represented by means of an OWL ontology as the reader may confirm on [19].

However, although both works adopt the same argument structure they diverge on their purpose and consequently on the modeling approach taken. While the main purpose of the AIF-based work is to take advantage of the powerful reasoning capabilities of OWL to automatically classify argument types (or argument schemes) and argument instances, the TLAF purpose is to show the advantages that applications have with respect to the argument instantiation process by adopting an argumentation framework which comprehends a model layer to specify the types of arguments used and how they affect each other. Consequently, the modeling approach taken by both works diverge on several issues too. The most evident is that TLAF explicitly distinguishes between argument-types and the reasoning mechanisms, while in the AIF-based work the reasoning mechanisms are implicit in the name of the argument-scheme. However, the most relevant difference concerns the way premises of argument-types are defined. In the AIF-based work each argument-type defines explicitly the set of statement-types it has as premises. On the contrary, in the TLAF the set of admissible statement-types that an argument-type has as premises is inferred through the R-relations established between argument-types. This lets you constrain that an argument-type only accepts a given statement-type as a premise when it is concluded by a specific reasoning mechanism. Moreover, similarly to the Carneades framework [20], in TLAF an argument has zero or more statements as premise.

On the contrary, in the AIF-based work an argument has at least one statement as premise. Another difference between the AIF-based work and TLAF is the fact that in the former an argument-instance can be classified into several types (one or more) while in the latter an argument-instance is classified into one type only, which must be the most specific/representative one of that instance. While the multi-classification of argument-instances is useful for several tasks (e.g. querying of arguments), it raises acceptability problems that are not completely understood yet.

In the general abstract framework for rule-based argumentation described by Prakken [8] arguments apply either a *strict* or a *defeasible* rule over a set of axioms (i.e. premises) to conclude another axiom, such that axioms are defined in a logical language. In TLAF, these two kinds of rules may correspond to two kinds of reasoning mechanisms and the concrete rules may correspond to instances of those reasoning mechanisms. However, TLAF does not constraint rules to be classified only in two types. Yet, the three types of attack relationship between argument-instances: (i) rebutting, (ii) undercutting and (iii) undermining described in [8] are captured by the TLAF rules to derive such relationship. Prakken work also describes arguments as trees of inference rules such that an argument contains other sub-arguments concluding intermediate conclusions and so on. TLAF comprehends such trees of arguments at the model layer (through the R-relation) and also at the instance layer such that the root of the trees are intentional arguments. Still, since TLAF allows capturing mutual dependency between two intentional arguments, one can think on arguments as graphs rather than trees. Contrary to the Carneades framework [20], where it is assumed that argument graphs contain no cycles, the argument graphs of TLAF may contain cycles since no restriction exists at the model layer level.

Finally, as claimed by Prakken, other relevant work on structured argumentation, such as DefLog [21], is a special case of its general framework [8]. In that sense, no additional discussion with such work is provided.

7   Conclusions and Future Work

This paper describes the Three-Layer Argumentation Framework (TLAF) that reduces the existing gap between the most referenced abstract argumentation frameworks and its adoption by applications. The main novelty of the proposed argumentation framework relies on its conceptualization layer (i.e. model layer), namely the R relation. This layer captures the structure and semantics of the argumentation data employed in a specific context constraining and conducting the modeling process of the argumentation specific scenario. Even though, for the same scenario very different modeling approaches are possible.

Despite being generic, TLAF is mainly targeted to be adopted by autonomous agents. In relation to that, the TLAF adopts and follows some terminology from the BDI model, namely by distinguishing between intentional arguments and non-intentional arguments. Based on the conceptual relations captured by the framework and the defined argument structure, a clear and minimal set of conditions was established for an argument-instance to attack/support another one. Given that, the support and attack relations between argument-instances are automatically derived according to the subjacent TLAF model. Despite the fact that the argument-instances generation process, and the Σ and *sconflict* functions are fully domain dependent, their definition profits from the established TLAF model.

While not directly addressed in this paper, the TLAF has the following advantages: (i) when generating statements it constrains the scope in which it is valuable to establish a conflict relationship between statements (i.e. *sconflict*), and therefore simplifies the automation of the process that discovers or instantiates the *sconflict* relation, by reducing and driving the search/combination space between statements; (ii) when generating arguments upon existing statements, it constraints the type of conclusion and premises, and the reasoning mechanism associated with an argument-instance, therefore simplifying the automation of the process that instantiates arguments, that establishes the premises and conclusion relationships with statements and establishes the R_{att} and R_{sup} relationships between arguments.

Besides the new features provided by TLAF, it is generic enough to be adopted by different domain applications. Moreover, a TLAF instance can be easily represented in a more abstract formalism such BAF [5], where the *AI* set corresponds to the set of arguments of BAF and the derived argument-instances relationships, i.e. R_{sup} and R_{att}, correspond to the BAF binary relations with the same name respectively. This is especially relevant because TLAF does not impose any particular argument evaluation process. Therefore, one can use this feature to apply an argument evaluation process such as the ones proposed in [10, 22-24]. However, because none of these processes is able to take advantage of the TLAF Model Layer we are working to propose one as well. For that, we need to take into consideration the argumentation abstract semantics described in literature as well the rationality postulates introduced by Caminada and Amgoud [9] and Prakken [8].

The authors consider that no experiences would be relevant for the evaluation of the proposed framework, as its application depends on the modeling approaches of the domain, and less of the framework. This suggests the need for further development of methods and methodologies for argument modeling.

In order to simplify the modeling process and profit from experience, for example, in the software engineering and ontology development fields, the authors envisage the need to provide modularity and extensibility modeling features to TLAF. These new features potentially promote TLAF in the scope of heterogeneous, ill-specified, emergent multi-agent systems as it provides the mechanisms to model private argumentation models in respect (specializing) to other argumentation models, thus inheriting a common model.

Acknowledgments. This work is partially supported by the Portuguese projects: COALESCE (PTDC/EIA/74417/2006) of MCTES-FCT and World Search (QREN11495) of FEDER. The authors would like to acknowledge Jorge Santos, Maria João Viamonte, Jorge Coelho and Besik Dundua for their useful counsels and Owen Gilson for his revisions of the document.

References

1. Wooldridge, M.: Reasoning about rational agents. The MIT press, Cambridge (2000)
2. Moran, R.: Authority and Estrangement: An Essay on Self-Knowledge. Princeton University Press (2001)
3. Walton, D.N., Krabbe, E.C.W.: Commitment in dialogue. Suny Press (1995)

4. Dung, P.M.: On the acceptability of arguments and its fundamental role in nonmonotonic reasoning, logic programming and n-person games. Artificial Intelligence 77, 321–357 (1995)
5. Cayrol, C., Lagasquie-Schiex, M.C.: On the Acceptability of Arguments in Bipolar Argumentation Frameworks. In: Godo, L. (ed.) ECSQARU 2005. LNCS (LNAI), vol. 3571, pp. 378–389. Springer, Heidelberg (2005)
6. Bench-Capon, T.J.M.: Persuasion in Practical Argument Using Value-based Argumentation Frameworks. J. Logic Computation 13, 429–448 (2003)
7. Baroni, P., Giacomin, M.: Semantics of Abstract Argument Systems. In: Argumentation in Artificial Intelligence, pp. 25–44 (2009)
8. Prakken, H.: An abstract framework for argumentation with structured arguments. Argument & Computation 1, 93 (2010)
9. Caminada, M., Amgoud, L.: On the evaluation of argumentation formalisms. Artificial Intelligence 171, 286–310 (2007)
10. Cayrol, C., Lagasquie-Schiex, M.C.: Gradual Valuation for Bipolar Argumentation Frameworks. In: Godo, L. (ed.) ECSQARU 2005. LNCS (LNAI), vol. 3571, pp. 366–377. Springer, Heidelberg (2005)
11. Gruber, T.R.: A translation approach to portable ontology specifications. Journal of Knowledge Acquisition 5, 199–220 (1993)
12. Gruber, T.: What is an Ontology?, http://www-ksl.stanford.edu/kst/what-is-an-ontology.html
13. Walton, D.N.: Fundamentals of critical argumentation. Cambridge Univ. Pr. (2006)
14. Bratman, M.: Intention, Plans and Practical Reason. Harvard University Press, Cambridge (1987)
15. Wooldridge, M.: An Introduction to MultiAgent Systems. Wiley (2009)
16. Chesñevar, C., McGinnis, J., Modgil, S., Rahwan, I., Reed, C., Simari, G., South, M., Vreeswijk, G., Willmott, S.: Towards an Argument Interchange Format. The Knowledge Engineering Review 21, 293–316 (2006)
17. Rahwan, I., Banihashemi, B.: Arguments in OWL: A Progress Report. In: Proceeding of the 2008 Conference on Computational Models of Argument: Proceedings of COMMA 2008, pp. 297–310. IOS Press, Amsterdam (2008)
18. Rahwan, I., Banihashemi, B., Reed, C., Walton, D., Abdallah, S.: Representing and classifying arguments on the semantic web. The Knowledge Engineering Review (2011)
19. Maio, P., Silva, N.: TLAF Meta-Model Layer as an Ontology, http://www.dei.isep.ipp.pt/~pmaio/TLAF/Ontology/TLAF_Ontology.owl
20. Gordon, T.F., Prakken, H., Walton, D.: The Carneades model of argument and burden of proof. Artif. Intell. 171, 875–896 (2007)
21. Verheij, B.: DefLog: on the Logical Interpretation of Prima Facie Justified Assumptions. Journal of Logic and Computation 13, 319–346 (2003)
22. Amgoud, L., Cayrol, C., Lagasquie-Schiex, M.C., Livet, P.: On bipolarity in argumentation frameworks. Int. J. Intell. Syst. 23, 1062–1093 (2008)
23. Karacapilidis, N., Papadias, D.: Computer Supported Argumentation And Collaborative Decision Making: The Hermes System. Information Systems 26, 259–277 (2001)
24. Verheij, B.: On the existence and multiplicity of extensions in dialectical argumentation, cs/0207067 (2002)

Stable Extensions in Timed Argumentation Frameworks

Maria Laura Cobo, Diego C. Martinez, and Guillermo R. Simari

Artificial Intelligence Research and Development Laboratory (LIDIA)
Department of Computer Science and Engineering, Universidad Nacional del Sur
Av. Alem 1253 - (8000) Bahía Blanca - Bs. As. - Argentina
{mlc,dcm,grs}@cs.uns.edu.ar
http://www.cs.uns.edu.ar/lidia

Abstract. A Timed Abstract Argumentation Framework is a novel formalism where arguments are only valid for consideration in a given period of time, which is defined for every individual argument. Thus, the attainability of attacks and defenses is related to time, and the outcome of the framework may vary accordingly. In this work we study the notion of stable extensions applied to timed-arguments. The framework is extended to include intermittent arguments, which are available with some repeated interruptions in time.

1 Introduction

One of the main concerns in Argumentation Theory is the search for rationally based positions of acceptance in a given scenario of arguments and their relationships. This task requires some level of abstraction in order to study pure semantic notions. Abstract argumentation systems [10,15,3,2] are formalisms for argumentation where some components remain unspecified, being the structure of an argument the main abstraction. In this kind of system, the emphasis is put on the semantic notion of finding the set of accepted arguments. Most of these systems are based on the single abstract concept of *attack* represented as an abstract relation, and extensions are defined as sets of possibly accepted arguments. For two arguments $\mathcal{A}$ and $\mathcal{B}$, if $(\mathcal{A}, \mathcal{B})$ is in the attack relation, then the acceptance of $\mathcal{B}$ is conditioned by the acceptance of $\mathcal{A}$, but not the other way around. It is said that argument $\mathcal{A}$ *attacks* $\mathcal{B}$, and it implies a priority between conflicting arguments.

The simplest abstract framework is defined by Dung in [10]. It only includes a set of abstract arguments and a binary relation of attack between arguments. Several semantics notions are defined and the Dung's argument extensions became the foundation of further research. Other proposals extends Dung's framework by the addition of new elements, such as preferences between arguments [3,6] or subarguments [13]. Other authors use the original framework to elaborate new extensions [11,5]. All of these proposals are based on varied abstract formalizations of arguments and attacks.

In this scenario, the combination of time and argumentation is a novel research line. In [12] a calculus for representing temporal knowledge is proposed, and defined in terms of propositional logic. This calculus is then considered with respect to argumentation, where an argument is defined in the standard way: an argument is a pair constituted by a minimally consistent subset of a database entailing its conclusion. This work is thus related to [4].

S. Modgil, N. Oren, and F. Toni (Eds.): TAFA 2011, LNAI 7132, pp. 181–196, 2012.

In [7,8] a novel framework is proposed, called *Timed Abstract Framework* (TAF), combining arguments and temporal notions. In this formalism, arguments are relevant only in a period of time, called its *availability interval*. This framework maintains a high abstract level in an effort to capture intuitions related with the dynamic interplay of arguments as they become available and cease to be so. The notion of *availability interval* refers to an interval of time in which the argument can be legally used for the particular purpose of an argumentation process. Thus, this kind of timed-argument has a limited influence in the system, given by the temporal context in which these arguments are taken into account. For example, consider the following argument:

My client committed no crime, since he was drafted for the ongoing war

In order to argue about an alleged crime, this argument can only be used when there is an actual war in which the defended is involved. The same argument cannot be used as a defense for crimes committed after the soldier was discharged. Thus, this argument has a temporal relevance. Timed abstract frameworks capture the previous argument model by assigning arguments to an availability interval of time. In [8] a skeptical, timed interval-based semantics is proposed, using admissibility notions. As arguments may get attacked during a certain period of time, defense is also time-dependent, requiring a proper adaptation of classical acceptability. In [7], algorithms for the characterization of defenses between timed arguments are presented.

In this work we formalize a natural expansion of timed argumentation frameworks by considering arguments with more than one availability interval. These are called *intermittent arguments*. These arguments are available with (possibly) some repeated interruptions in time. For instance, a traveling salesman may have a set of arguments to be used in negotiations, but maybe some of them are irrelevant (or politically incorrect) in different cultures, and then these arguments can only be used while staying in certain cities. In some legal procedures, a lawyer can use some arguments depending on the actual stage of the process (initial disclosure of evidence, the trial, the appeal). Some countries apply sets of specific legal and economic rules during certain periods of time, like financial crises, natural disasters or wartime. This has an impact on, for instance, political argumentation. In all of these scenarios arguments may become relevant, or cease to be so, depending on time-related factors. Using this extended timed argumen-tation framework, we analyze the notion of *stable extension*. A stable extension is a set of arguments attacking every other argument not in the same set. Since in a timed context arguments are attacked sporadically, the characterization of a stable set of argu-ments requires a deeper analysis. For instance, consider two laws about environmental protection, both of them enacted in different years. Law A was created in 1960 and states that the minimal budget for environmental protection is determined by procedure X. Law B was created in 2000 and states that minimal budget for rainforest protection is determined by procedure Y. Since 2000, there is a conflict between both laws when funds for rainforest protection must be established. Hence, there are two valid positions regarding this issue. The first one privileges Law A from 1960 until present. The second one privileges Law B since 2000. Both positions are clearly in conflict. However, when discussing about rainforest protection in the nation, this last position should recognizes that there is a law that was applied between 1980 an 2000. Thus, the two positions are

– Position 1: Law A, from 1980 to 2011
– Position 2: Law A, from 1980 to 2000 and Law B from 2000 to 2011.

A stable extension captures knowledge that attacks every piece of knowledge not in the extension. When time is relevant, it is possible for a stable extension to include conflictive knowledge (as in position 2 in the previous example) although this knowledge is considered in different periods of time. This timed notion of stable extension is the main contribution of this work.

The paper is organized as follows. In the next section we recall classic argumentation semantic notions. Thereafter, time-intervals and the terminology used in this work are defined, towards the presentation of our Timed Abstract Argumentation Framework with intermittent arguments in Section 4. The notion of stable extension is presented in Section 5. Finally, conclusions and future work are discussed.

2   Classic Abstract Argumentation

Dung defines several argument extensions that are used as a reference for many authors. The formal definition of the classic argumentation framework follows.

Definition 1. [10] *An argumentation framework is a pair $AF = \langle AR, attacks \rangle$ where AR is a set of arguments, and attacks is a binary relation on AR, i.e. attacks $\subseteq AR \times AR$.*

Arguments are denoted by labels starting with an uppercase letter, leaving the underlying logic unspecified. A set of accepted arguments is characterized in [10] using the concept of *acceptability*, which is a central notion in argumentation, formalized by Dung in the following definition.

Definition 2. [10] *An argument $A \in AR$ is acceptable with respect to a set of arguments S if and only if every argument B attacking A is attacked by an argument in S.*

If an argument A is acceptable with respect to a set of arguments S then it is also said that S *defends* A. Also, the attackers of the attackers of A are called *defenders* of A. We will use these terms throughout this paper.

Acceptability is the main property of Dung's semantic notions, which are summarized in the following definition.

Definition 3. *A set of arguments S is said to be*
– *conflict-free if there are no arguments A, B in S such that A attacks B.*
– *admissible if it is conflict-free and defends all its elements.*
– *a preferred extension if S is a maximal (for set inclusion) admissible set.*
– *a complete extension if S is admissible and it includes every acceptable argument w.r.t. S.*
– *a grounded extension if and only if it is the least (for set inclusion) complete extension.*
– *a stable extension if S is conflict-free and it attacks each argument not belonging to S.*

The grounded extension is also the least fixpoint of a simple monotonic *characteristic* function:

$$F_{AF}(S) = \{\mathcal{A} : \mathcal{A} \text{ is acceptable with respect to } S\}.$$

In [10], theorems stating conditions of existence and equivalence between these extensions are also introduced.

Example 1. *Consider the argumentation framework* $AF_1 = \langle AR, attacks \rangle$, *where* $AR = \{\mathcal{A}, \mathcal{B}, \mathcal{C}, \mathcal{D}, \mathcal{E}, \mathcal{F}, \mathcal{G}, \mathcal{H}\}$ *and* $attacks = \{(\mathcal{B}, \mathcal{A}), (\mathcal{C}, \mathcal{B}), (\mathcal{D}, \mathcal{A}), (\mathcal{E}, \mathcal{D}), (\mathcal{G}, \mathcal{H}), (\mathcal{H}, \mathcal{G})\}$. *Then*
- $\{\mathcal{A}, \mathcal{C}, \mathcal{E}\}$ *is an admissible set of arguments.*
- $\{\mathcal{A}, \mathcal{C}, \mathcal{E}, \mathcal{F}, \mathcal{G}\}$ *is a preferred extension. It is also a complete extension.*
- $\{\mathcal{A}, \mathcal{C}, \mathcal{E}, \mathcal{F}\}$ *is the grounded extension.*

Dung's abstract formalism is sufficient to define some basic extensions on arguments. In this work we study the formalization of intermittent timed-arguments in an abstract framework, and we present an argument extension inspired by the stable semantics. In the following section we prepare the road to timed argumentation by introducing several time-related concepts.

3   Time Representation

In order to capture a time-based model of argumentation, we enrich the classical abstract frameworks with temporal information regarding arguments. The problem of representing temporal knowledge and temporal reasoning arises in a lot of disciplines, including Artificial Intelligence. There are many ways of representing temporal knowledge. A usual way to do this is to determine a *primitive* to represent time, and its corresponding *metric relations* [1,9,14]. In this work we will use *temporal intervals of discrete time* as primitives for time representation, and thus only metric relations for intervals are applied.

Definition 4. *[Temporal Interval] An interval is a pair build from* $a, b \in \mathbb{Z} \cup \{-\infty, \infty\}$, *in one of the following ways:*
- $[a, a]$ *denotes a set of time moments formed only by moment* a.
- $[a, \infty)$ *denotes a set of moments formed by all the numbers in* $\mathbb{Z}$ *since a (including* a).
- $(-\infty, b]$ *denotes a set of moments formed by all the numbers in* $\mathbb{Z}$ *until moment* i *(including* b).
- $[a, b]$ *denotes a set of moments formed by all the numbers in* $\mathbb{Z}$ *moment* i *until moment* j *(including both* a *and* b).
- $(-\infty, \infty)$ *a set of moments formed by all the numbers in* $\mathbb{Z}$.

The moments a, b *are called endpoints.*

The set of all the intervals defined over $\mathbb{Z} \cup \{-\infty, \infty\}$ *is denoted* Υ.

For example, $[5, 12]$ and $[1, 200]$ are intervals. If X is an interval then X^-, X^+ are the corresponding endpoints (*i.e.*, $X = [X^-, X^+]$). And endpoint may be a point of discrete time, identified by an integer number, or infinite.

There are thirteen possible relations between intervals [1]. Seven of them are the basics while the remaining six are defined as their inverses. In the context of this work it is unnecessary to keep track of the difference between some of Allen relations, particularly *starts*, *during* and *finishes*, so we redefine the *during*. On Table 1 we present the relation we are going to use in the paper. The 'x's and 'y's represents the interval X and Y respectively. The table shows the relation between *endpoints*.

Table 1. Qualitative relations among arguments, based on [1]

Relation	Symb	Inverse	e.g.	Relation on Endpoints
X Before Y	ⓑ	ⓑⓘ		$X^+ < Y^-$
X Meets Y	ⓜ	ⓜⓘ		$X^+ = Y^-$
X Overlaps Y	ⓞ	ⓞⓘ		$X^- < Y^-, X^+ > Y^-$
X During Y	ⓓ	ⓓⓘ		$X^- \geq Y^-, X^+ < Y^+$ or $X^- > Y^-, X^+ \leq Y^+$
X Equal Y	ⓔ			$X^- = Y^-, X^+ = Y^+$

We will usually work with sets of intervals (as they will be somehow related to arguments). Thus, we introduce several definitions and properties needed for semantic elaborations.

Definition 5. *Let S be a set of intervals and let i be a moment of time. The exclusion of i from S, denoted $S \ominus i$, is defined as follows:*

$$S \ominus i = \{I : I \in S \land i \notin I\} \qquad \cup$$
$$\{[I^-, i-1] : I \in S \land i \in I, i \neq I^-\} \cup$$
$$\{[i+1, I^+] : I \in S \land i \in I, i \neq I^+\}$$

The exclusion of the interval I from S, noted as $S①i$, can be recursively defined as follows:

$$i. \ S①I = S \ominus I^- \qquad\qquad if \ I^- = I^+$$
$$ii. \ S①I = (S \ominus I^-)①[I^- + 1, I^+] \ if \ I^- \neq I^+$$

The needed operation is the difference among set of intervals, $i.e.S_1 - S_2$ being S_1 and S_2 sets of intervals. This operation can be defined recursively using ①.

Intersection is another relevant operation on intervals. The intersection of two intervals is the interval formed by all the common points of both of them. Its endpoints are the minimal and maximal time points in common.

Definition 6. *Let I_1 and I_2 be two intervals. The intersection is defined as: $I_1 \cap I_2 = [x, y]$ with $x, y \in I_1$ and $x, y \in I_2$ such that there are no $w, z : w, z \in I_1$ and $w, z \in I_2$ with $w < x$ or $y < z$.*

Definition 7. *Let S_1 and S_2 be two sets of intervals. The intersection of these sets, noted as $S_1 \cap\!\!\!\!\!\sqcap S_2$, is: $S_1 \cap\!\!\!\!\!\sqcap S_2 = \{I : I = I_1 \cap I_2 \neq [\,], \forall I_1 \in S_1, I_2 \in S_2\}$*

Definition 8. *Let S be a set of intervals. The partition of S, denoted $\mathsf{Part}(S)$ is defined as:*

- $\mathsf{Part}(S) = S$ *if* $\forall I_1, I_2 \in S, I_1 \cap I_2 = \emptyset$.
- $\mathsf{Part}(S) = \mathsf{Part}(S - \{l_1, l_2\} \cup \{l_1 - (l_1 \cap l_2), l_2 - (l_1 \cap l_2), l_1 \cap l_2\})$, *with* $I_1, I_2 \in S$ *and* $I_1 \cap I_2 \neq \emptyset$

The partition of a set of argument's availability *breaks* overlapped intervals in smaller intervals. This notion simplifies semantic elaborations, since it discretizes the evolution of the framework according to moments where arguments start or cease to be available.

Any set of intervals, either fragmented or not, has several non-fragmented subsets. In fact, any singleton subset is trivially non-fragmented.

Although the previous definition can be given in terms of interval calculus relations (Table 1), we use a different notation to improve their readability.

In the following section we present Timed Abstract Argumentation Frameworks with intermittent arguments.

4   Timed Argumentation Framework

As remarked before, in Timed Argumentation Frameworks the consideration of time restrictions for arguments is formalized through an *availability function*, which defines a temporal interval for each argument in the framework. This interval states the period of time in which an argument is available for consideration in the argumentation scenario. The formal definition of our timed abstract argumentation framework follows.

Definition 9. *A timed abstract argumentation framework (TAF) is a 3-tuple $\langle Args,$ Atts, $Av \rangle$ where Args is a set of arguments, Atts is a binary relation defined over Args and Av is the availability function for timed arguments, defined as $Av : Args \to \wp(\Upsilon)$.*

Example 2. *The triplet $\langle Args, Atts, Av \rangle$, where $Args = \{A, B, C, D, E\}$, $Atts = \{(B, A), (C, B), (D, A), (E, D)\}$ and the availability function is defined as*

$Args$	Av	$Args$	Av
A	$\{[10, 40], [60, 75]\}$	B	$\{[30, 50]\}$
C	$\{[20, 40], [45, 55], [60, 70]\}$	D	$\{[47, 65]\}$
E	$\{(-\infty, 44]\}$		

is a timed abstract argumentation framework.

The framework of Example 2 can be depicted as in Figure 1, using a digraph where nodes are arguments and arcs are attack relations. An arc from argument $\mathcal{X}$ to argument $\mathcal{Y}$ exists if $(\mathcal{X}, \mathcal{Y}) \in Atts$. Figure 1 also shows the time availability of every argument, as a graphical reference of the $\mathcal{A}v$ function. It is basically the framework's evolution in time. Endpoints are marked with a vertical line, except for $-\infty$ and ∞. For space reasons, only some relevant time points are numbered in the figure. As stated before, the availability of arguments is tied to a temporal restriction. Thus, an attack to an argument may actually occur only if both the attacker and the attacked argument are available. In other words, an attack between two arguments may be *attainable*, under certain conditions. *Attainable attacks* are attacks that will eventually occur in some period of time. In order to formalize this, we need to compare time intervals, using the previously defined metric relations.

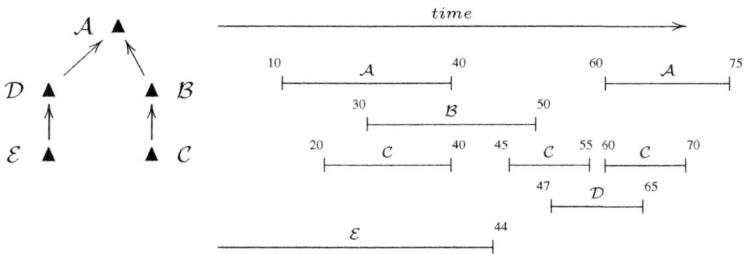

Fig. 1. Framework of Example 2

Definition 10. *Let $\Phi = \langle Args, Atts, \mathcal{A}v \rangle$ be a TAF, and let $\{\mathcal{A}, \mathcal{B}\} \subseteq Args$ such that $(\mathcal{B}, \mathcal{A}) \in Atts$. The attack $(\mathcal{B}, \mathcal{A})$ is said to be attainable if the following conditions holds: $I_{\mathcal{A}} \; R \; I_{\mathcal{B}}$, where $R \in \{ \text{ⓓ}, \text{ⓓⓘ}, \text{ⓞ}, \text{ⓞⓘ} \; \text{ⓔ} \}$ for some $I_{\mathcal{A}} \in \mathcal{A}v(\mathcal{A})$ and $I_{\mathcal{B}} \in \mathcal{A}v(\mathcal{B})$. The attack is said to be attainable in $\mathcal{A}v(\mathcal{A}) \cap \mathcal{A}v(\mathcal{B})$. The set of intervals where an attack $(\mathcal{B}, \mathcal{A})$ is attainable will be noted as $IntSet((\mathcal{B}, \mathcal{A}))$*

Note that an attack is attainable if the availability of both the attacker and the attacked argument eventually overlaps.

Example 3. *Consider the timed argumentation framework of example 2. The attacks $(\mathcal{D}, \mathcal{A})$ and $(\mathcal{B}, \mathcal{A})$ are both attainable in the framework. Attack $(\mathcal{D}, \mathcal{A})$ is attainable since $[47, 65] \; \text{ⓞ} \; [60, 75]$ with $[47, 65] \in \mathcal{A}v(\mathcal{D})$ and $[60, 75] \in \mathcal{A}v(\mathcal{A})$. Attack $(\mathcal{B}, \mathcal{A})$ is attainable since $[30, 50] \; \text{ⓞⓘ} \; [10, 40]$, in $[30, 40]$. Recall that $[30, 50] \in \mathcal{A}v(\mathcal{B})$, $[10, 40] \in \mathcal{A}v(\mathcal{A})$. The attack $(\mathcal{C}, \mathcal{B})$ is also attainable. Since $\mathcal{A}v(\mathcal{C}) = \{[20, 40], [30, 50]\}$ and $\mathcal{A}v(\mathcal{B}) = \{[30, 50]\}$ then we can assure the attainability of the attack by one of the following relations:$[20, 40] \; \text{ⓞ} \; [30, 50]$, $[45, 55] \; \text{ⓞⓘ} \; [30, 50]$. The attack is then attainable at $\{[30, 40], [45, 50]\}$, i.e. $\mathcal{A}v(\mathcal{C}) \cap \mathcal{A}v(\mathcal{B})$. The attack $(\mathcal{E}, \mathcal{D})$ is not attainable, since $(-\infty, 45] \; \text{ⓑ} \; [47, 65]$. The arguments involved in this attack are never available at the same time.*

The set of all attainable attacks in the framework Φ is denoted $AttAtts_{\Phi}$. It is also possible to define the attainability of attacks at a particular timed interval, as shown next.

Definition 11. *Let* $\Phi = \langle Args, Atts, Av \rangle$ *be a TAF, and let* $\{A, B\} \subseteq Args$ *such that* $(B, A) \in Atts$. *The attack* (B, A) *is said to be attainable at* I *if:* $I \cap Av(A) \neq [\ ]$ *and the following condition holds:* $I \cap I_A$ R I_B, *where* $R \in \{\text{ⓓ}, \text{ⓓⓘ}, \text{ⓞ}, \text{ⓞⓘ}, \text{ⓔ}\}$, *for some* $I_A \in Av(A)$ *and* $I_B \in Av(B)$.

The set of attainable attacks of Φ at interval I is denoted $AttAtts_\Phi^I$.

Example 4. *Consider the timed argumentation framework of Example 2. The set* $AttAtts_\Phi$ *is:* $\{(D, A), (B, A), (C, B)\}$. *The set* $AttAtts_\Phi^{[35,40]}$ *is* $\{(B, A), (C, B)\}$. *The attack* (D, A) *is in* $AttAtts_\Phi$ *but it is not in* $AttAtts_\Phi^{[35,40]}$, *since* $[35, 40] \cap [47, 65]$ *is the empty set. The attack* (B, A) *is in* $AttAtts_\Phi$ *and is also in* $AttAtts_\Phi^{[35,40]}$, *since* $[35, 40] \cap [10, 40] = [35, 40]$ *and* $[35, 40]\text{ⓓ}[30, 50]$. *Note that* $[10, 40] \in Av(A)$ *and* $[30, 50] \in Av(B)$.

The definition of attainability of attacks can be attached to particular time points too. The set of attainable attacks of Φ at moment i is denoted $AttAtts_\Phi(i)$ and is defined as $AttAtts_\Phi(i) = AttAtts_\Phi^{[i,i]}$.

5   Semantics for Timed Argumentation

As attacks may occur only on a period of time (that in which the participants are available), argument defense is also occasional. In [8] a skeptical, timed interval-based semantics is proposed, using admissibility notions. The classical definition of acceptability is adapted to a timed context. The complexity of this adaptation lies on the fact that defenses may occur sporadically and hence the focus is put on finding *when* the defense takes place. For example, an argument A may be defended by X in the first half of its time interval, and later by an argument Y in the second half. Although X is not capable of providing a full defense, argument A is defended while A is available. In other words, defenders *take turns* to provide a defense.

In this paper we are mainly interested in the notion of stable set of arguments. In the classical sense, a stable extension of an argumentation framework is a set of arguments S that attacks each argument not belonging to S. A set of arguments S is said to attack an argument A if at least one argument in S is an attacker of A.

In timed argumentation frameworks, an attack may be attainable only in a restricted period of time, and so the existance of an attacker in the set is not enough now.

Definition 12. *Let* $\Phi = \langle Args, Atts, Av \rangle$ *be a TAF, and let* $A \in Args$. *Argument* A *is a threatened argument if there is at least one argument* B, *such that* $(B, A) \in AttAtts_\Phi$.

A threat interval is a period of time in which an argument attacks another. Naturally, it is possible for an argument to have more than one threat interval. Consider again the framework of Example 2. Argument A has two threat intervals since there are two attainable attacks, (B, A) and (D, A). The set of intervals $\tau_B(A)$ is $\{[30, 40]\}$ while $\tau_D(A)$ is $\{[60, 65]\}$. Notice that $\tau_X(Y)$ is a set in general since it is possible that X threats Y in more than one interval, that is the case of the attack (C, B), $\tau_C(B) = \{[30, 40], [54, 50]\}$.

Since attacks are sporadic we need to know in which subintervals of its availability interval an argument is threatened. We need a general structure associating arguments and intervals. This is captured by the notion of *t-profile* as defined next.

Definition 13. *Let* $\Phi = \langle Args, Atts, Av \rangle$ *be a TAF. A t-profile is general structure* $\langle arg, set \rangle$ *where* $arg \in Args$ *and* $set \in \wp(\Upsilon)$ *and for each* $I \in set$, $I \subseteq I_{arg}$ *with* $I_{arg} \in Av(arg)$. *The set of all the t-profiles definable from* Φ *will be noted as* $\mathfrak{P}$.

A t-profile is a record of an argument $\mathcal{A}$ associated with a set of intervals with only one restriction: every interval in the set is a subinterval of an availability interval of $\mathcal{A}$. In other words, t-profiles reflect a particular view of the availability of an argument. The associated set of intervals may denote moments of attacks, or moments of defense, or any other special consideration of an argument regarding a set of intervals. In particular, a set of t-profiles is the consideration of a set of arguments in several moments of time. For instance, the set $S = \{\langle \mathcal{A}, \{I_1\} \rangle, \langle \mathcal{B}, \{I_2\} \rangle\}$ includes two arguments $\mathcal{A}$ and $\mathcal{B}$, but it is possible that these arguments never co-exist if I_1 and I_2 are not overlapping intervals. The set S may denote a special status for arguments, and *when* this status is assigned.

Remember that a stable extension attacks every single external argument. Since we are dealing with arguments restricted to intervals, it is necessary to define what it means for a t-profile to be attacked by a set of t-profiles. This is substantially different than in classic frameworks. In order to figure out why, consider an argument $\mathcal{A}$ attacked by two arguments $\mathcal{B}$ and $\mathcal{C}$. Suppose that $\mathcal{A}$ is attacked by $\mathcal{B}$ and $\mathcal{C}$ in two different, non-overlapping intervals I_1 and I_2, such that $\mathcal{B}$ attacks $\mathcal{A}$ during I_1 and $\mathcal{C}$ attacks $\mathcal{A}$ during I_2. There is no moment of time in which $\mathcal{A}$ is free of an attainable attack, either from $\mathcal{B}$ or $\mathcal{C}$ since both attackers take turns to attack $\mathcal{A}$. Is the set $S_1 = \{\mathcal{B}\}$ attacking $\mathcal{A}$? In the classical sense of stable semantics it does, since an attack finally occurs. However, when time dimension is considered, S_1 does not attack $\mathcal{A}$ in every moment in which $\mathcal{A}$ is considered: in I_2 argument $\mathcal{A}$ is not attacked by any argument in S_1. On the other hand, $\mathcal{A}$ *always* has an attacker in the set $S_2 = \{\mathcal{B}, \mathcal{C}\}$.

Clearly, any further analysis of attacks between timed arguments must take intervals into account. Thus, following definition formalizes an attack between a set of t-profiled and a single t-profile.

Definition 14. *Let* $\Phi = \langle Args, Atts, Av \rangle$ *be a TAF, let* $S \subseteq \mathfrak{P}$ *be a set of t-profiles and let* $t_x = \langle \mathcal{X}, \mathfrak{D}_{\mathcal{X}} \rangle$ *be a t-profile not in* S. *The set of all subintervals of intervals in* t_x *attacked by a t-profile in* S *is denoted as* $\mathfrak{T}(S, t_x)$ *and defined as:*

$$\mathfrak{T}(S, t_x) = \{\tau_{\mathcal{Y}}(\mathcal{X}) \cap \mathfrak{D}_{\mathcal{Y}} : \langle \mathcal{Y}, \mathfrak{D}_{\mathcal{Y}} \rangle \in S \text{ and } \tau_{\mathcal{Y}}(\mathcal{X}) \cap \mathfrak{D}_{\mathcal{Y}} \neq \emptyset\}$$

The set $\mathfrak{T}(S, t_x)$ is formed by all the intersections between an interval of a t-profile in S and an interval in t_x. It captures all of the moments in which argument $\mathcal{X}$ is attacked by S in $\mathfrak{D}_{\mathcal{X}}$. Consider the framework of Figure 1, if $S = \{\langle \mathcal{B}, \{[35, 45]\} \rangle, \langle \mathcal{D}, \{[60, 65]\} \rangle, \langle \mathcal{E}, \{[20, 44]\} \rangle\}$ and $t_x = \langle \mathcal{A}, \{[22, 40], [60, 65]\} \rangle$ then the set $\mathfrak{T}(S, t_x) = \{[35, 40], [60, 65]\}$. The set is the result of the union of:

$$\tau_{\mathcal{B}}(\mathcal{A}) \cap \mathfrak{D}_{\mathcal{B}} = \{[30, 40]\} \cap \{[35, 40]\} = \{[35, 40]\}$$
$$\tau_{\mathcal{D}}(\mathcal{A}) \cap \mathfrak{D}_{\mathcal{D}} = \{[60, 65]\} \cap \{[60, 65]\} = \{[60, 65]\}$$
$$\tau_{\mathcal{E}}(\mathcal{A}) \cap \mathfrak{D}_{\mathcal{E}} = \{\} \cap \{[20, 44]\} = \{\}$$

Since S is a collection of t-profiles, it is important to find whether this collection represents a full attack on a given t-profile t_x, i.e., S leaves no unattacked periods of time in t_x. This is captured in the following definition.

Definition 15. Let $\Phi = \langle Args, Atts, Av \rangle$ be a TAF, let $S \subseteq \mathfrak{P}$ be a set of t-profiles and let $\mathcal{X}$ be an argument in $Args$. The set S attacks a t-profile $t_x = \langle \mathcal{X}, \mathfrak{D}_{\mathcal{X}} \rangle$ if $\mathfrak{D}_{\mathcal{X}} \subseteq \mathfrak{T}(S, t_x)$.

In this sense $t_x = \langle \mathcal{A}, \{[22, 40], [60, 65]\} \rangle$ is not attacked by S since $\{[22, 40], [60, 65]\}$ $\not\subseteq \mathfrak{T}(S, t_x)$, being $S = \{\langle \mathcal{B}, \{[35, 45]\} \rangle, \langle \mathcal{D}, \{[60, 65]\} \rangle, \langle \mathcal{E}, \{[20, 44]\} \rangle \}$.
     A t-profile $t = \langle \mathcal{X}, \mathfrak{D}_{\mathcal{X}} \rangle$ is attacked by a set S of t-profiles if every interval of t overlaps an interval of an attacker in S. In order to define a stable extension, a formal definition of conflict-freeness for sets of t-profile is needed.

Definition 16. Let $\langle Args, Atts, Av \rangle$ be a TAF, and let $S \subseteq \mathfrak{P}$. The set S is said to be conflict-free if there are not two profiles $\langle \mathcal{A}, \mathfrak{D}_{\mathcal{A}} \rangle$ and $\langle \mathcal{B}, \mathfrak{D}_{\mathcal{B}} \rangle$ such that

$$\exists i \in \mathfrak{D}_{\mathcal{B}}, i \in \mathfrak{D}_{\mathcal{B}} : (\mathcal{A}, \mathcal{B}) \in AttAtts_{\Phi}^{[i,i]}$$

Now we are in conditions to define the stable semantics. In order for a set S of t-profiles to be considered a stable extension, two main situations must be considered. First, the set S must attack an argument $\mathcal{X}$ not considered in any t-profile of S. This is consistent with the classical notion of stable extension: if it is not in S, then it must be attacked. Second, if an argument $\mathcal{X}$ has a t-profile in S, then it must be attacked by S in any moment of time in which $\mathcal{X}$ is not considered in S. The formal definition follows.

Definition 17. Let $\Phi = \langle Args, Atts, Av \rangle$ be a TAF and let $S \subseteq \mathfrak{P}$ a set of t-profiles. The set S is an stable extension if and only if:

1. S is conflict-free.
2. $\forall \mathcal{X} \in Args$ such that $\langle \mathcal{X}, \mathfrak{D}_{\mathcal{X}} \rangle \notin S$, S attacks $\langle \mathcal{X}, Av(\mathcal{X}) \rangle$.
3. $\forall \mathcal{X} \in Args$ such that $\langle \mathcal{X}, \mathfrak{D}_{\mathcal{X}} \rangle \in S$, S attacks $\langle \mathcal{X}, Av(\mathcal{X}) - \mathfrak{D}_{\mathcal{X}} \rangle$.

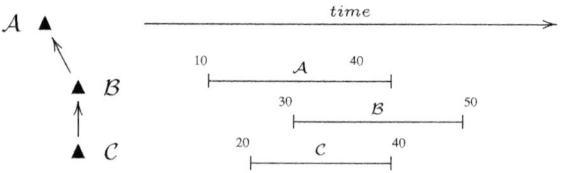

Fig. 2. Framework $\Phi_{\mathcal{E}}$

Example 5. Let $\Phi_{\mathcal{E}} = \langle Args, Atts, Av \rangle$, be the TAF depicted on Figure 2. The set $S = \{\langle \mathcal{A}, [10, 40] \rangle, \langle \mathcal{B}, (40, 50] \rangle, \langle \mathcal{C}, [20, 40] \rangle\}$ is a stable extension of $\Phi_{\mathcal{E}}$. The set S is clearly conflict-free and there are no arguments in $Args$ not considered in S. Thus, only condition (3) of Definition 17 is relevant. In this particular case, the only argument in such condition is $\mathcal{B}$ for which the availabiliy in $\Phi_{\mathcal{E}}$ is $[30, 50]$. However the t-profile of

B in S considers only the sub-interval $(40, 50]$. The rest of the original availability interval is $[30, 40]$, in which B is attacked by C. Argument C is in the set S, and according to the corresponding profile it is associated with all the availability interval $\{[20, 40]\}$.

The set $S_2 = \{\langle A, [10, 20]\rangle, \langle C, [20, 40]\rangle\}$ is not a stable extension, since there is argument B, that is not included in a t-profile in S_2 and it is not attacked by an t-profile in the set. The set S_2 should attack $\langle B, Av(B)\rangle$ in order to be a stable extension, but this is not the case. It can be observed that there is a period of time, $[41, 50]$, in which B is not attacked by any other argument.

Finally the set $S_3 = \{\langle A, [10, 30]\rangle, \langle B, [30, 50]\rangle, \langle C, [20, 30]\rangle\}$ is not stable extension. In this case the problem is related to $\mathfrak{D}_C = \{[20, 30)\}$, the argument C has $\{[20, 40]\}$ as its availability period in the framework and has no attackers there. Then the third condition fails, the t-profile $\langle C, Av(C) - \{[10, 30]\}\rangle = \langle C, Av(C) - \{[10, 19)\}\rangle$ is not attacked by S_3.

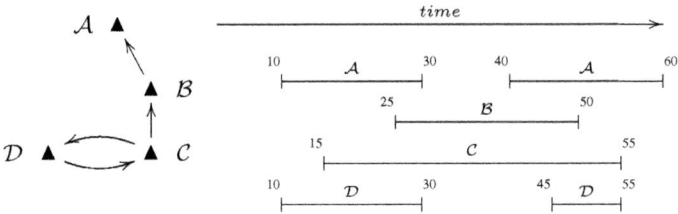

Fig. 3. Framework Φ_S

Example 6. Let $\Phi_S = \langle Args, Atts, Av\rangle$, be the TAF depicted on Figure 3. In this framework we have two stable extensions:

$$S_1 = \{ \langle A, \{[10, 24], [51, 50]\}\rangle, \qquad S_2 = \{ \langle A, Av(A)\rangle,$$
$$\langle B, \{[25, 30], [45, 50]\}\rangle, \qquad \langle C, Av(C)\rangle,$$
$$\langle C, \{[31, 44]\}\rangle, \qquad \langle D, \{[10, 14]\}\rangle \}$$
$$\langle D, Av(D)\rangle \qquad \}$$

This example shows shows that the presence of cycles requires a proper consideration of time. In non-temporal, classical frameworks whenever cycles are presente like the one in Φ_S, only one of the participant appears in the stable extension. In timed argumentation frameworks, the analysis requires an examination of the associated time intervals. In this case both arguments C and D appear in the extension, but with different time restrictions. Note that the set $S_3 = \{\langle A, Av(A)\rangle, \langle C, Av(C)\rangle\}$ is not a stable extension since it is not true that $\forall X \in Args$ such that $\langle X, \mathfrak{D}_X\rangle \notin S$, S attacks $\langle X, Av(X)\rangle$, since S_3 does not attack $\langle D, Av(D)\rangle$. The condition that $Av(D) = \{[10, 30], [45, 55]\}$ must be a subset of $\mathfrak{T}$ does not hold. Since C is the only attacker of D, the set $\mathfrak{T}$ is $\tau_C(D) \cap Av(C)$ i.e.$\{[15, 30], [45, 55]\} \cap \{15, 55\} = \{[15, 30], [45, 55]\}$. It is clear that $\{[10, 30], [45, 55]\} \nsubseteq \{[15, 30], [45, 55]\}$.

A set of t-profiles is a set of arguments that are put in specific contextual time restrictions. Example 6 shows that a set of t-profiles may have arguments with potential conflicts, and this is legal as far as conflictive arguments are considered in different intervals of time.

As shown in previous section, Dung defines the notion of admissibility. This notion can be naturally extended to timed argumentation frameworks. Again, we use t-profiles to associate arguments with sets of intervals. An addmissible set is a set of t-profiles that is conlict-free and defends all of its elements.

Definition 18. *Let* $\Phi = \langle Args, Atts, Av \rangle$ *be a TAF and* $S \subseteq \mathfrak{P}$ *a set of t-profiles. The set of defense intervals for* $\mathcal{A}$ *against* $\mathcal{B}$, *denoted* $\delta_{\mathcal{A}}^{\mathcal{B}}(S)$, *is defined as:*

$$\bigcup \{IntSet((\mathcal{X}, \mathcal{B})) \cap IntSet((\mathcal{B}, \mathcal{A})) \cap \mathfrak{D}_{\mathcal{X}} : \langle \mathcal{X}, \mathfrak{D}_{\mathcal{X}} \rangle \in S\}$$

The set of defense intervals is the set of all the intervals in which an attacker is attacked by another argument. This is obtained by intersection of availability intervals.

Example 7. *Consider the timed argumentation framework of Example 2 and the set* $S = \{\langle \mathcal{C}, \{[35, 40], [48, 52]\} \rangle\}$. *The set of defense intervals for* $\mathcal{A}$ *against* $\mathcal{B}$ *is*

$$\begin{aligned} \delta_{\mathcal{A}}^{\mathcal{B}}(S) &= \{IntSet((\mathcal{C}, \mathcal{B})) \cap IntSet((\mathcal{B}, \mathcal{A})) \cap \mathfrak{D}_{\mathcal{X}} \\ &= \{[30, 40], [45, 50]\} \cap \{[30, 40]\} \cap \{[35, 40], [48, 52]\} \\ &= \{[35, 40]\} \end{aligned}$$

The following definition considers every attainable attacks at a certain period I, in order to grant defense. If there are multiple attackers, then the defense takes place in those moments where the argument has defenders for all of the attainable attacks, that is, no attack succeeds.

Definition 19. *Let* $\Phi = \langle Args, Atts, Av \rangle$ *be a TAF and* $S \subseteq \mathfrak{P}$ *a set of t-profiles. Let* $I \in \mathsf{Part}(\tau_{\Phi}(\mathcal{A}))$ *The set of defended intervals for* $\mathcal{A}$ *at* I, *denoted* $\Delta_{\mathcal{A}}^{S}(I)$, *is defined as:*

$$\Delta_{\mathcal{A}}^{S}(I) = \cap \delta_{\mathcal{A}}^{\mathcal{X}}(S) \cap \{I\} : \mathcal{X} \in AttAtts_{\Phi}^{I}$$

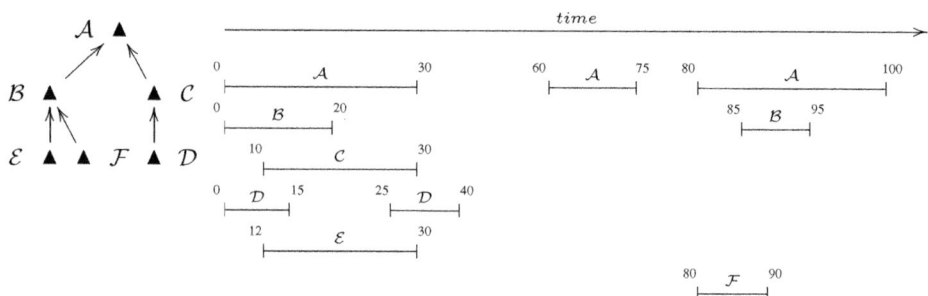

Fig. 4. Timed Framework to ilustrate defenses

Example 8. *Consider the timed argumentation framework of Figure 4. If we consider the argument $\mathcal{A}$ is clear that it is threatened by $\mathcal{B}$ in $\{[0, 20], [85, 95]\}$ and in $\{[10, 30]\}$. The partition of the union of this sets is $\{[0, 9], [10, 20], [21, 30]\}$. This is important because in these small intervals the argument $\mathcal{A}$ needs defense against different attackers, in particular in $[10, 20]$ it requires defense against $\mathcal{B}$ and $\mathcal{C}$.*

Consider the set $S = \{\langle \mathcal{D}, \{[10, 15], [25, 30]\}\rangle, \langle \mathcal{E}, \{[16, 30]\}\rangle\}$. The set $\Delta_{\mathcal{A}}^S([0, 9])$ is the empty set since $\delta_{\mathcal{A}}^{\mathcal{B}}(S) = \{[16, 20]\}$ but $\delta_{\mathcal{A}}^{\mathcal{B}}(S) \Cap \{[0, 9]\} = \emptyset$. The set $\Delta_{\mathcal{A}}^S([21, 30])$ is $\{[25, 30]\}$ since $\delta_{\mathcal{A}}^{\mathcal{C}}(S) = \{[10, 15], [25, 30]\}$ so $\delta_{\mathcal{A}}^{\mathcal{C}}(S) \Cap \{[21, 30]\} = \{[25, 30]\}$. Finally $\Delta_{\mathcal{A}}^S([10, 20]) = \emptyset$ since S fails in providing defense for both attacker at the same time, so the intersection is empty. You can see that

$$\delta_{\mathcal{A}}^{\mathcal{C}}(S) \Cap \{[10, 20]\} = \{[10, 15], [25, 30]\} \Cap \{[10, 20]\} = \{[10, 15]\} \ and$$
$$\delta_{\mathcal{A}}^{\mathcal{B}}(S) \Cap \{[10, 20]\} = \{[16, 20]\} \Cap \{[10, 20]\} = \{[16, 20]\}.$$

The set $\Delta_{\mathcal{A}}^S([10, 20])$ is defined as the intersection of this last sets, i.e. $\{[10, 15]\} \Cap \{[16, 20]\}$, which is clearly the empty set.

Finally the argument is defended in the union of the moments defined in every interval of the partition of its threat-intervals.

Definition 20. *Let $\Phi = \langle Args, Atts, Av \rangle$ be a TAF and $S \subseteq \mathfrak{P}$ a set of t-profiles. The set of defended intervals for $\mathcal{A}$, denoted $\Delta_{\mathcal{A}}^S$, is:*

$$\Delta_{\mathcal{A}}^{\emptyset} = Av(\mathcal{A}) \stackrel{\bot}{-} \tau_{\Phi}(\mathcal{A})$$
$$\Delta_{\mathcal{A}}^S = \Delta_{\mathcal{A}}^{\emptyset} \cup \bigcup_{I \in Part(\tau_{\Phi}(\mathcal{A}))} \Delta_{\mathcal{A}}^S(I) \ when \ S \neq \emptyset.$$

Following the analysis made in Example 8 we can determine $\Delta_{\mathcal{A}}^S$ which in this case is $\{[60, 75], [80, 100], [25, 30]\}$. The set $\{[60, 75], [80, 100]\}$ represents the periods in which $\mathcal{A}$ is not attacked ($\Delta_{\mathcal{A}}^{\emptyset}$), and the set $\{[25, 30]\}$ is the union of $\Delta_{\mathcal{A}}^S(I)$, for all $I \in Part(\tau_{\Phi}(\mathcal{A}))$.

Definition 21. *Let $\Phi = \langle Args, Atts, Av \rangle$ be a TAF and $S \subseteq \mathfrak{P}$ be a set of t-profiles. An argument $\mathcal{A} \in Args$ is acceptable with respect to S if $\Delta_{\mathcal{A}}^S \neq \emptyset$. If $\mathcal{A}$ is acceptable, then it is acceptable at $\Delta_{\mathcal{A}}^S$. Its t-profile of acceptability is $\langle \mathcal{A}, \Delta_{\mathcal{A}}^S \rangle$*

Once this point is reached we can determine if some t-profile is defended or not.

Definition 22. *Let $\Phi = \langle Args, Atts, Av \rangle$ be a TAF, $S \subseteq \mathfrak{P}$ be a set of t-profiles and a t-profile $\langle \mathcal{X}, \mathfrak{D}_{\mathcal{X}} \rangle$. The t-profile $\langle \mathcal{X}, \mathfrak{D}_{\mathcal{X}} \rangle$ is defended by S if $\mathfrak{D}_{\mathcal{X}}$ is included in $\Delta_{\mathcal{X}}^S$.*

Consider the timed argumentation framework of Figure 4 and the set

$$S = \{\langle \mathcal{D}, \{[10, 15], [25, 30]\}\rangle, \langle \mathcal{E}, \{[16, 30]\}\rangle\}.$$

Argument $\mathcal{A}$ is acceptable with respect to S and its t-profile of acceptability is

$$\langle \mathcal{A}, \{[60, 75], [80, 100], [25, 30]\}\rangle.$$

As a consequence the t-profile $\langle \mathcal{A}, \{[80, 90], [28, 30]\}\rangle$ is defended by S while the t-profile $\langle \mathcal{A}, Av\mathcal{A} \rangle$ is not.

Proposition 1. *If an argument $\mathcal{A}$ belongs to a stable extension S with t-profile $\langle \mathcal{C}, \mathfrak{D}_\mathcal{X} \rangle$, then $\langle \mathcal{C}, \mathfrak{D}_\mathcal{X} \rangle$ is defended by S.*

Proof: If $\langle \mathcal{X}, \mathfrak{D}_\mathcal{X} \rangle \in S$ then either

- *$\mathcal{X}$ has no attackers in $\mathfrak{D}_\mathcal{X}$, and then $\langle \mathcal{X}, \mathfrak{D}_\mathcal{X} \rangle$ is acceptable with respect to S.*
- *$\mathcal{X}$ has an attacker $\mathcal{Y}$ in $\mathfrak{D}_\mathcal{X}$. Clearly, since S is conflict free, then no t-profile $\langle \mathcal{Y}, \mathfrak{D}_\mathcal{Y} \rangle$ is included in S such that $\mathfrak{D}_\mathcal{X}$ and $\mathfrak{D}_\mathcal{Y}$ have moments in common. Then, there is a t-profile $t = \langle \mathcal{Y}, \mathfrak{D}_\mathcal{Y} \rangle$ outside S that attacks $\mathcal{X}$, i.e. $\mathfrak{D}_\mathcal{X}$ and $\mathfrak{D}_\mathcal{Y}$ have moments in common. Since t is not in S, then S attacks t. This means that every interval in $\mathfrak{D}_\mathcal{Y}$ has moments in common (i.e. the intersection is not empty) t-profiles in S. But then, the same time $\mathcal{X}$ is attacked by $\mathcal{Y}$, it is defended by another argument in S. Thus, S defends $\mathcal{X}$ in $\mathfrak{D}_\mathcal{X}$.* □

Since arguments in a stable extension are defended by the extension, the following proposition is induced.

Definition 23. *A t-profile $\langle \mathcal{X}, \mathfrak{D}_\mathcal{X} \rangle$ is said to attack another t-profile $\langle \mathcal{Y}, \mathfrak{D}_\mathcal{Y} \rangle$ if $(\mathcal{X}, \mathcal{Y}) \in Atts$ and at least one interval in $\mathfrak{D}_\mathcal{X}$ overlaps an interval in $\mathfrak{D}_\mathcal{Y}$.*

The attainability of attacks is a necessary condition for attacking t-profiles. Note, however, that since t-profiles are restricted versions of the intervals of availability of an argument, a t-profile t_1 may not attack a t-profile t_2 even when both arguments have an attainable attack in the framework.

Proposition 2. *A stable extension S of a timed argumentation framework is admissible.*

Proof: Suppose there is a t-profile $t_1 = \langle \mathcal{A}, \mathfrak{D}_\mathcal{A} \rangle \in S$ such that there exists a t-profile $t_2 = \langle \mathcal{X}, \mathfrak{D}_\mathcal{X} \rangle \notin S$ attacking t_1. Then there exists an interval $I = I_A \cap I_X$ with $I_A \in \mathfrak{D}_\mathcal{A}$ and $I_X \in \mathfrak{D}_\mathcal{X}$. Consider the set $\mathfrak{T}(S, t_2)$ of all intervals of t_2 attacked by S. Since S is a stable set and then S attacks t_2. This means that $I \subseteq \mathfrak{T}(S, t_2)$ and then t_1 is defended by S during $I \subseteq I_A$. □

It is important to preserve the rationality behind classical semantics for argumentation frameworks and the new semantics for timed argumentation. This means that whenever a relation between classical semantic notions is established, the same relation is expected to be found in a timed context. This is perhaps the most difficult aspect of the task of elaborating semantic notions for new timed argumentation frameworks, and it is an active part of this line of research. In the following section conclusions and future work are discussed.

6   Conclusions and Future Work

In this work we presented an extension of previously defined Timed Argumentation Frameworks in which arguments with more than one availability interval are considered. These arguments are called *intermittent arguments*, and are temporarily available with some repeated interruptions in time. Using this extended timed argumentation framework, we studied the notion of *stable extension*, which requires the consideration of time as a new dimension, leading to the definition of *t-profiles* of timed arguments. Since arguments are related to time, a t-profile of an argument $\mathcal{X}$ is the formal reference of $\mathcal{X}$ within several frames of time, which are subintervals of the availability intervals of $\mathcal{X}$. A t-profile attacks another t-profile if an attack is formally defined between its arguments and at least an interval of time of each profile is overlapping. A set of t-profiles is a collection of arguments which are considered within different intervals of time. We defined the timed notion of *stable extension* as a set S of t-profiles such that every t-profile outside the set is attacked by a t-profile in the set according to the availability intervals in S. Notoriously, an argument $\mathcal{X}$ may appear in t-profiles inside and outside a stable set simultaneously, but with different intervals of time since an argument may become attacked or not as time evolves. Thus, an argument may gets in and out a stable set depending on time.

Future work has several directions. The relation between different timed semantics needs to be addressed. In classical argumentation there are conditions for which several semantics coincide. For instance, in well-formed argumentation frameworks [10] there is only one extension that is grounded, preferred and stable. The notion of *well-formed* applied to timed argumentation frameworks is being analyzed. We are also interested in the evolution of the framework through time. For a given semantic notion S, such as stable as presented in this paper, there may be intervals of time in which the extensions induced by S do not change, even when some arguments become or cease to be available during these intervals. These are called *steady periods* of the framework and are also an interesting topic. It may be used to model *eras of thinking* for a rational agent or a society, and the impact of including new arguments. New semantics elaborations based in this notion are being studied.

References

1. Allen, J.: Maintaining knowledge about temporal intervals. Communications of the ACM (26), 832–843 (1983)
2. Amgoud, L., Cayrol, C.: A reasoning model based on the production of acceptable arguments. Annals of Mathematics and Artificial Intelligence
3. Amgoud, L., Cayrol, C.: On the acceptability of arguments in preference-based argumentation. In: 14th Conference on Uncertainty in Artificial Intelligence (UAI 1998), pp. 1–7. Morgan Kaufmann (1998)
4. Augusto, J.C., Simari, G.R.: Temporal defeasible reasoning. Knowl. Inf. Syst. 3(3), 287–318 (2001)
5. Baroni, P., Giacomin, M.: Resolution-based argumentation semantics. In: Proc. of 2nd International Conf. on Computational Models of Argument (COMMA 2008), pp. 25–36 (2008)
6. Bench-Capon, T.: Value-based argumentation frameworks. In: Proc. of Nonmonotonic Reasoning, pp. 444–453 (2002)
7. Cobo, M., Martinez, D., Simari, G.: An approach to timed abstract argumentation. In: Proc. of Int. Workshop of Non-monotonic Reasoning 2010 (2010)
8. Cobo, M.L., Martinez, D.C., Simari, G.R.: On admissibility in timed abstract argumentation frameworks. In: Coelho, H., Studer, R., Wooldridge, M. (eds.) ECAI. Frontiers in Artificial Intelligence and Applications, vol. 215, pp. 1007–1008. IOS Press (2010)

9. Dechter, R., Meiri, I., Pearl, J.: Temporal constaints networks. In: Proceedings KR 1989, pp. 83–93 (1989)
10. Dung, P.M.: On the acceptability of arguments and its fundamental role in nonmonotonic reasoning, logic programming and n-person games. Artificial Intelligence 77(2), 321–358
11. Jakobovits, H.: Robust semantics for argumentation frameworks. Journal of Logic and Computation 9(2), 215–261 (1999)
12. Mann, N., Hunter, A.: Argumentation using temporal knowledge. In: Proc. of 2nd International Conf. on Computational Models of Argument (COMMA 2008), pp. 204–215 (2008)
13. Martínez, D.C., García, A.J., Simari, G.R.: Modelling well-structured argumentation lines. In: Proc. of XX IJCAI 2007, pp. 465–470 (2007)
14. Meiri, I.: Combining qualitative and quantitative contraints in temporal reasoning. In: Proceedings of AAAI 1992, pp. 260–267 (1992)
15. Vreeswijk, G.A.W.: Abstract argumentation systems. Artificial Intelligence 90(1–2), 225–279 (1997)

Computing with Infinite Argumentation Frameworks: The Case of AFRAs

Pietro Baroni[1], Federico Cerutti[1],
Paul E. Dunne[2], and Massimiliano Giacomin[1]

[1] Dipartimento di Ingegneria dell'Informazione, University of Brescia, via Branze, 38,
25123, Brescia, Italy
{pietro.baroni,federico.cerutti,massimiliano.giacomin}@ing.unibs.it
[2] Department of Computer Science, Ashton Building, University of Liverpool,
Liverpool, L69 7ZF, United Kingdom
P.E.Dunne@liverpool.ac.uk

Abstract. In recent years a large corpus of studies has arisen from Dung's seminal abstract model of argumentation, including several extensions aimed at increasing its expressiveness. Most of these works focus on the case of finite argumentation frameworks, leaving the potential practical applications of infinite frameworks largely unexplored. In the context of a recently proposed extension of Dung's framework called AFRA (Argumentation Framework with Recursive Attacks), this paper makes a first step to fill this gap. It is shown that, under some reasonable restrictions, infinite frameworks admit a compact finite specification and that, on this basis, computational problems which are tractable for finite frameworks may preserve the same property in the infinite case. In particular we provide a polynomial-time algorithm to compute the finite representation of the (possibly infinite) grounded extension of an AFRA with infinite attacks. An example concerning the representation of a moral dilemma is introduced to illustrate and instantiate the proposal and gives a preliminary idea of its potential applicability.

1 Introduction

Infinite argumentation frameworks, though encompassed by Dung's theory of abstract argumentation [6], have received relatively limited attention in the literature so that their use as a modelling tool and the relevant computational issues are largely unexplored.

This paper provides a first step towards filling this gap, by considering the case of existence of infinite attacks in a recently proposed extension of Dung's framework called AFRA (*Argumentation Framework with Recursive Attacks*) [2] where "attacks" may themselves be attacked by arguments. The idea of encompassing attacks to attacks in abstract argumentation framework has been first considered in [4], and subsequently investigated and developed, for instance, in [2,11,13]. Computational issues in this kind of extended frameworks have been first addressed in [8] for the finite case of EAF [13]. In this paper, we show that,

S. Modgil, N. Oren, and F. Toni (Eds.): TAFA 2011, LNAI 7132, pp. 197–214, 2012.
© Springer-Verlag Berlin Heidelberg 2012

under some mild restrictions, an AFRA with infinite attacks can be represented through a deterministic finite automaton (DFA), which provides the basis for the efficient solution of semantics-related computational problems. To demonstrate this, we show in particular that a DFA representing the (possibly infinite) grounded extension of an AFRA with infinite attacks can be derived in polynomial time from the DFA representing the AFRA itself.

From a general perspective, the ultimate aim of this paper is to provide an enabling technique for practical applications of infinite argumentation frameworks. While this is a largely open issue, we illustrate the theoretical concepts developed throughout the paper using a preliminary example concerning moral dilemma representation. Of course, the value of the methodology goes beyond both the simple example at hand and the use of the AFRA framework. Indeed the main contribution of this paper is twofold: on one hand, we address the topic of representing an Argumentation Framework through a formal language; and, secondly, we show that this kind of representation can be useful to compute semantics extensions also in the case of infinite Argumentation Frameworks.

The paper is organized as follows. After recalling the preliminary background concepts in Sect. 2, we provide an example encompassing infinite attacks in Sect. 3 and discuss specification mechanisms for AFRAs with infinite attacks in Sect. 4. Section 5 describes the actual specification mechanism adopted in the paper, called DFA$^+$, and Sect. 6 provides a polynomial time algorithm to compute the (representation of) the (possibly infinite) grounded extension of an AFRA starting from its DFA$^+$ specification. Finally Sect. 7 concludes the paper.

2   Preliminary Background

In this section we define the abstract argumentation models which are the core focus of this article: the AF model [6] with a finite set of arguments and the AFRA model [2].

Definition 1. *A finite argumentation framework (AF) is a pair $\langle \mathcal{X}, \mathcal{A} \rangle$, in which $\mathcal{X}$ is a finite set of arguments and $\mathcal{A} \subseteq \mathcal{X} \times \mathcal{X}$ is the attack relationship. A pair $\langle x, y \rangle \in \mathcal{A}$ is referred to as 'y is attacked by x' or 'x attacks y'; $x \in \mathcal{X}$ is acceptable with respect to $S \subseteq \mathcal{X}$ if for every $y \in \mathcal{X}$ that attacks x there is some $z \in S$ that attacks y. The characteristic function, $\mathcal{F} : 2^{\mathcal{X}} \to 2^{\mathcal{X}}$ is the mapping which, given $S \subseteq \mathcal{X}$, returns the set of $y \in \mathcal{X}$ for which y is acceptable w.r.t. S. For any set S we define $\mathcal{F}^0(S) = \emptyset$ and for $k \geq 1$ $\mathcal{F}^k(S) = \mathcal{F}(\mathcal{F}^{k-1}(S))$. The grounded extension is the (unique) least fixed point of $\mathcal{F}$. We denote by $GE(\langle \mathcal{X}, \mathcal{A} \rangle) \subseteq \mathcal{X}$ the grounded extension of $\langle \mathcal{X}, \mathcal{A} \rangle$.*

Definition 2. *An Argumentation Framework with Recursive Attacks (AFRA) is described by a pair $\langle \mathcal{X}, \mathcal{R} \rangle$ where $\mathcal{X}$ is a (finite) set of arguments and $\mathcal{R}$ consists of pairs of the form $\langle x, \alpha \rangle$ where $x \in \mathcal{X}$ and $\alpha \in \mathcal{X} \cup \mathcal{R}$. For $\alpha = \langle x, \beta \rangle \in \mathcal{R}$, the source (src) and target (trg) of α are defined by $src(\alpha) = x$ and $trg(\alpha) = \beta$. In order to avoid a surfeit of brackets, we describe elements of $\mathcal{R}$ as finite length sequences of arguments, so that $x_k\ x_{k-1}\ x_{k-2}\ \cdots\ x_2\ x_1 \in \mathcal{R}$*

if $\{x_1, \ldots, x_k\} \subseteq \mathcal{X}$ (note that an argument may occur more than once in this sequence), $\langle x_2, x_1 \rangle \in \mathcal{R}$ (i.e. $x_2 x_1 \in \mathcal{R}$) and $\langle x_j \langle x_{j-1} \langle \cdots x_1 \rangle \rangle \rangle \in \mathcal{R}$, with $2 < j \leq k$. Letting $\mathcal{C} = \mathcal{R} \cup \mathcal{X}$, for $\alpha \in \mathcal{R}$ and $\beta \in \mathcal{C}$, α is said to defeat β $(\alpha \to \beta)$ whenever any of the following hold:

1. $trg(\alpha) = \beta$
2. $trg(\alpha) = src(\beta)$ i.e. $\beta \in \mathcal{R}$, $\alpha = xy$ and $\beta = y\gamma$ $(y \in \mathcal{X})$.

Definition 3. Let $\langle \mathcal{X}, \mathcal{R} \rangle$ be an AFRA, $\alpha, \beta \in \mathcal{R}$, $V, W \in \mathcal{X} \cup \mathcal{R}$, $\mathcal{S} \subseteq \mathcal{X} \cup \mathcal{R}$; then:

- W is acceptable w.r.t. $\mathcal{S}$ (or, equivalently is defended by $\mathcal{S}$) iff $\forall \alpha \in \mathcal{R}$ s.t. $\alpha \to W \exists \beta \in \mathcal{S}$ s.t. $\beta \to \alpha$;
- the characteristic function $\mathbb{F}_{\langle \mathcal{X}, \mathcal{R} \rangle}$ is defined as follows: $\mathbb{F}_{\langle \mathcal{X}, \mathcal{R} \rangle} : 2^{\mathcal{X} \cup \mathcal{R}} \mapsto 2^{\mathcal{X} \cup \mathcal{R}}$; $\mathbb{F}_{\langle \mathcal{X}, \mathcal{R} \rangle}(\mathcal{S}) = \{V | V \text{ is acceptable w.r.t. } \mathcal{S}\}$;
- the grounded extension (denoted as $GE^{\text{AFRA}}(\langle \mathcal{X}, \mathcal{R} \rangle)$ is the least fixed point of $\mathbb{F}_{\langle \mathcal{X}, \mathcal{R} \rangle}$.

By considering the (Dung) style AF, $\langle \tilde{\mathcal{X}}, \tilde{\mathcal{R}} \rangle$ constructed from an AFRA $\langle \mathcal{X}, \mathcal{R} \rangle$ by $\tilde{\mathcal{X}} = \mathcal{X} \cup \mathcal{R}$ and $\tilde{\mathcal{R}} = \{\langle \alpha, \beta \rangle : \alpha \to \beta\}$ (for further details see [2]), a correspondence between semantics structures (e.g. the basic notions of conflict-free and admissible sets and the extensions of various semantics) in an AFRA $\langle \mathcal{X}, \mathcal{R} \rangle$ and the analogous (Dung style) structures within $\langle \tilde{\mathcal{X}}, \tilde{\mathcal{R}} \rangle$ is obtained. In particular we will exploit the fact that the grounded extension of an AFRA coincides with the (Dung style) grounded extension of the corresponding AF, i.e. with $GE(\langle \tilde{\mathcal{X}}, \tilde{\mathcal{R}} \rangle)$.

3 An Example: Moral Dilemmas

The recursive form of $\mathcal{R}$ in an AFRA, $\langle \mathcal{X}, \mathcal{R} \rangle$, in principle, admits the capability of describing infinite attack structures even though $\mathcal{X}$ is a finite set. To exemplify the potential utility of this kind of structures as a modelling tool we consider a case of moral dilemma.

Fred is the network administrator of a large company and among his duties he has to release emails, addressed to staff members, that have been accidentally blocked by the security filters. One day he gets a helpdesk request from Eve, a staff member and his best friend's wife, requesting the release of an email. As part of the procedure he has to ensure that the email is safe by scanning its contents. He finds out that it's actually an email addressed to Eve from her lover. He releases the email, and his initial reaction is to call his friend up and tell him about the affair. However, the law forbids him to reveal the information. This is a case of conflict of obligations, and, following [5,1], we can model this situation with abstract argumentation[1].

---

[1] A detailed comparison of alternative argumentation-based approaches to practical reasoning is beyond the scope of this paper. The interested reader may refer to [2] for a comparison between AFRA and Modgil's EAF, or to [1] for an illustration of the modelling approach adopted in the example.

First of all, the reasons for the alternative actions can be represented as practical arguments [14]. Indeed, since Fred wants to be a good friend, then he should tell his friend what he knows (**T**), but since Fred wants to be a good citizen, then he should not (**D**). These two arguments are obviously attacking each other. Moreover, both **D** and **T** are related to values [5], respectively Legality and Friendship. These values can be represented as arguments (**L** and **F**) [14,1] which affect the evaluation of the two practical arguments. For instance, in the case at hand, the value of Friendship would resolve the dilemma by making **T** prevail over **D**. From an argumentation point of view, this means that **F** would allow **T** to defeat **D**. This can be modelled by making ineffective the attack from **D** to **T** (α in Fig. 1) by attacking it through an attack (β) whose source is the value of Friendship **F**. This can be read as: even if the attack from **D** to **T** (α) holds because **D** and **T** support conflicting actions, nevertheless, in the case at hand, α is undermined by the moral commitment of **F**. Obviously, **L** states a similar moral commitment, namely making **D** prevail over **T**. This requires **L** to undermine both the attack (γ) from **T** to **D** and the attack (β) from **F** against α. In turn **F** should make ineffective the latter attacks (δ and η) and **L** and **F** will continue attacking each other's attacks forever. This infinite construction reveals an unresolved dilemma.

Finally, let us suppose that Fred chooses to pursue Legality rather than Friendship. This can be represented by another argument (**M**) (a "must" argument in the terminology of [1]). The argument **M** represents a choice between values in the case at hand. Therefore, **M** will undermine any moral commitment of **F** over the two actions **T** and **D** by attacking the (infinite number of) attacks whose source is **F**.

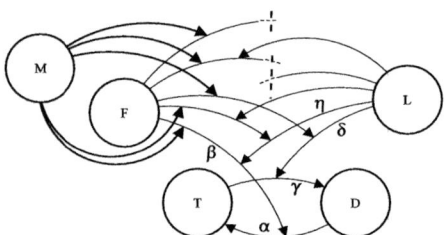

Fig. 1. Fred's dilemma

An AFRA representing Fred's dilemma is shown in Fig. 1. It consists of a finite set of arguments $\mathcal{X}_F = \{\mathbf{D}, \mathbf{T}, \mathbf{L}, \mathbf{F}, \mathbf{M}\}$ and of an infinite set of attacks $\mathcal{R}_F = \{\mathbf{DT}, \mathbf{TD}, \mathbf{LTD}, \mathbf{FDT}, \mathbf{FLTD}, \mathbf{LFDT}, \mathbf{LFLTD}, \mathbf{FLFDT}, \dots, \mathbf{MFDT}, \mathbf{MFLTD}, \mathbf{MFLFDT}, \dots\}$.

4   Representing $\mathcal{R}$ in AFRAS

Given the potential practical interest in AFRAs with infinite attacks, the following question arises.

When $\mathcal{R}$ is infinite what characterises suitable specification mechanisms for describing $\mathcal{R}$?

In order to pursue this question, we need some terminology.

Definition 4. *For $\mathcal{X}$ a finite set of arguments, we denote by $\mathcal{X}^*$ the set of all finite length sequences (or words) that can be formed using arguments in $\mathcal{X}$ (noting this includes ε the so-called empty sequence comprising no arguments). Given $w \in \mathcal{X}^*$, $|w|$ denotes its length, i.e. the number of arguments occurring in its definition. Note that repetitions of the same arguments contribute to $|w|$ so that, e.g. $|x_1 x_2 x_1| = 3$ (and not 2). Given $w \in \mathcal{X}^*$ we will denote as $\bar{w}$ the sequence obtained by reversing the order of the symbols in w, namely, given $w = x_1 x_2 \ldots x_n$, $\bar{w} = x_n \ldots x_2 x_1$.*

Given $u = u_1 u_2 \ldots u_r$ and $v = v_1 v_2 \ldots v_k \in \Sigma^*$ we denote by $u \cdot v$ (or simply uv) the word w of length $k + r$ defined by $u_1 u_2 \ldots u_r v_1 v_2 \ldots v_k$. We note that $w \cdot \varepsilon = \varepsilon \cdot w = w$. We say that $\mathcal{L} \subseteq \mathcal{X}^*$ is an attack language *over $\mathcal{X}$ if $\mathcal{L}$ satisfies $\forall\, w \in \mathcal{L}\ \ w = xu$ with $x \in \mathcal{X}$ and either $|u| = 1$ or $u \in \mathcal{L}$.*

If $\mathcal{L}$ is an attack language over $\mathcal{X}$, then the pair $\langle \mathcal{X}, \mathcal{L} \rangle$ certainly describes an AFRA. Classical formal language and computability theory, see e.g. [12], provides a means of capturing the vague concept of "specification mechanism" via *Formal Grammars* and their associated machine models. As well known, given a set of symbols Σ a *formal grammar* G specifies the derivation of a language $L(G) \subseteq \Sigma^*$ called *language generated by* G. A language, $L \subseteq \Sigma^*$, is *recognisable* if there is a formal grammar G for which $w \in L$ if and only if $w \in L(G)$.

As a starting point for "specification mechanisms" for attack languages we can consider descriptions which are formal grammars (so that $\Sigma = \mathcal{X}$ in such cases).

Unsurprisingly, arbitrary attack languages have unhelpful computational properties.

Proposition 1. *Let $\mathcal{X} = \{x, y\}$. There are attack languages, L, over $\mathcal{X}$ which are not recognisable, i.e. for which there is no formal grammar G for which $L(G) = L$.*

Proof. In view of the correspondence from the fact that $L \subseteq \Sigma^*$ is *recursively enumerable* if and only if there is an unrestricted grammar, G such that $L(G) = L$, it suffices to show that there are attack languages which fail to be r.e. First recall that any TM program, M, can be associated with a finite length *codeword*, $\beta(M)$, (over the alphabet $\{0,1\}$) in such a way that given $\beta(M)$ the behaviour of M can be reproduced by another TM program. Furthermore, the language corresponding to the set of valid encodings, i.e. $CODE = \{w \in \{0,1\}^* : w = \beta(M)$ for some TM program, $M\}$ is recursive.[2] With such encodings it is known that the language $L^\varepsilon_{\neg HALT} \subset \{0,1\}^*$ given by $\{\beta(M) :$ The TM program, M, **fails to halt** given the empty word as input$\}$ is not r.e.

---

[2] See e.g. [7, Ch. 4] or any standard introductory text on computability, such as [12, Ch. 8.3].

Now since $CODE \subset \{0,1\}^*$ we can *order* the set of all TM programs simply by ordering words[3] within $\{0,1\}^*$, so that the "first" TM program (M_1) is the first word, w_1 in this ordering of $\{0,1\}^*$ for which $w_1 \in CODE$, the "second" program (M_2) the second word, w_2 in the ordering for which $w_2 \in CODE$, and so on.

We are now ready to define a suitable attack language, $\mathcal{R} \subset \{x,y\}^*$ establishing the proposition's claim: $\mathcal{R} = \{ xy^k : k \geq 2 \text{ and } M_k \in L^\varepsilon_{\neg HALT} \} \cup \{y^n : n \geq 2\}$. This is easily seen to be an attack language[4] and, furthermore, cannot be r.e. For suppose, $\mathcal{R}$ is r.e. with AL a TM accepting exactly the words in $\mathcal{R}$ then $L^\varepsilon_{\neg HALT}$ could be shown r.e. as follows: given $\beta(M)$ determine the index k for which M is the k'th TM program. Then $\beta(M) \in L^\varepsilon_{\neg HALT}$ if and only if xy^k is accepted by AL.   □

As a consequence of Propn. 1 there will be attack languages for which it is not possible to present any specification (as a formal grammar). Of course the nature of such languages is unlikely to be of practical concern: Propn. 1 merely establishes a technical limitation affecting attack languages but certainly does not invalidate their use. In practice we would wish to consider only attack languages that are presented in some "verifiable form". What is the notion of "verifiable form" intended to capture here? Presented with a formal grammar G, there are two immediate issues which we would like to ensure can be addressed:

Q1. How easily can it be verified that $L(G)$ *does* describe an attack language?

Q2. Assuming $L(G)$ is verified as describing *some* attack language, $\mathcal{R}$ over $\mathcal{X}$, given $\alpha \in \mathcal{X}^*$ how easily can it be decided whether α *is* an attack in $\langle \mathcal{X}, \mathcal{R} \rangle$, i.e. whether $\alpha \in L(G)$?

It can be easily derived from Rice's Theorem (see, e.g. [12, pp. 185–195]) that unrestricted grammars face problems with respect to Q1.

Proposition 2. *Given an unrestricted grammar G, the problem of determining if $L(G)$ is an attack language is undecidable.*

On the other hand the family of *regular languages* [12] provides the basis for a positive result, using automata as representation mechanism.

Definition 5. *A deterministic finite automaton (DFA) is defined via a 5-tuple, $M = \langle \Sigma, Q, q_0, F, \delta \rangle$ where $\Sigma = \{\sigma_1, \ldots, \sigma_k\}$ is a finite set of input symbols, $Q = \{q_0, q_1, \ldots, q_m\}$ a finite set of states; $q_0 \in Q$ the initial state; $F \subseteq Q$ the set of accepting states; and $\delta : Q \times \Sigma \to Q$ the state transition function. For $q \in Q$ and $w \in \Sigma^*$, the reachable state from q on input w is*

$$\rho(q,w) = \begin{cases} q \text{ if } w = \varepsilon \\ \delta(q,w) \text{ if } |w| = 1 \\ \delta(\rho(q,u),x) \text{ if } w = u \cdot x \end{cases}$$

---

[3] For example using the standard lexicographic ordering under which $0 <_{\text{lex}} 1$ and $u <_{\text{lex}} w$ whenever $|u| < |w|$.

[4] The reader concerned by the fact that this includes a self-attacking argument (y) may note that we may use $xy^k x$ and $y^n x$ $(n \geq 1)$ to achieve the same effect without self-attacking arguments.

A sequence $w = w_1 w_2 \ldots w_n \in \Sigma^*$ is accepted by the DFA $\langle \Sigma, Q, q_0, F, \delta \rangle$ if $\rho(q_0, \bar{w}) = \rho(q_0, w_n w_{n-1} \ldots w_1) \in F$, i.e. the sequence of states (consistent with the state transition function δ) which processes every symbol in w in reverse order ends in an accepting state. For a DFA, $M = \langle \Sigma, Q, q_0, F, \delta \rangle$, $L(M)$ is the subset of Σ^* accepted by M.

Fact 1. The language $L \subseteq \Sigma$ is regular if and only if there is a DFA $M = \langle \Sigma, Q, q_0, F, \delta \rangle$ for which $L(M) = L$.

The following lemma shows that the conditions for an automaton to recognize an attack language are relatively simple.

Lemma 1. Let $M = \langle \mathcal{X}, Q, q_0, F, \delta \rangle$ be a DFA. Then $L(M)$ is an attack language if and only if both the following conditions hold:

C1. $\forall\, w \in \{\varepsilon\} \cup \mathcal{X}, \rho(q_0, w) \notin F$.
C2. $\forall\, q \in (Q \setminus \{q_0\}), \forall x \in \mathcal{X}$ if $q' = \delta(q, x) \notin F$ then $\forall w \in \mathcal{X}^*$ it holds that $\rho(q', \bar{w}) \notin F$.

Proof. Suppose first that $L(M)$ is an attack language. Since every $w \in L(M)$ satisfies $|w| \geq 2$ it is immediate that M satisfies C1. To see that C2 must hold, consider any $q \in (Q \setminus \{q_0\})$ and $x \in \mathcal{X}$ such that $q' = \delta(q, x) \notin F$. Furthermore consider any $u \in \mathcal{X}^*$ such that $q = \rho(q_0, \bar{u})$. Since $q \neq q_0$, $|u| \geq 1$ and, since $q' = \delta(q, x) \notin F$, $xu \notin L(M)$ and $|xu| \geq 2$. Since $L(M)$ is an attack language $\nexists p \in \mathcal{X}^*$ such that $p = vxu \in L(M)$, i.e. it is not possible to reach an accepting state from $q' = \delta(q, x)$.

For the converse direction, we show that if M satisfies both C1 and C2 then $L(M)$ is an attack language, i.e. $\forall\, w = xu \in L(M)$ either $|u| = 1$ or $u \in L(M)$. Since C1 holds, it is immediate that $|w| \geq 2$ for every $w \in L(M)$. Suppose now $w = yu \in L(M)$ with $|u| > 1$. Assume by contradiction $u \notin L(M)$, i.e. letting $q' = \rho(q_0, \bar{u})$ it holds that $q' \notin F$. Since $|u| > 1$ it must be the case that $q' = \delta(q, x)$ for some x with $q \neq q_0$. By C2, this implies that $\forall w \in \mathcal{X}^*$ $\rho(q', \bar{w}) \notin F$ which contradicts $w = yu \in L(M)$, as this would entail $\delta(q', y) \in F$. ∎

The desired result in Theorem 1 follows directly from Fact 1 and Lemma 1.

Theorem 1. Let $M = \langle \mathcal{X}, Q, q_0, F, \delta \rangle$ be a DFA defining the regular language, $L(M) \subseteq \mathcal{X}^*$. The problem of verifying that $L(M)$ is an attack language is polynomial time decidable.

Proof. Given a DFA, $M = \langle \mathcal{X}, Q, q_0, \delta, F \rangle$ from Lemma 1, in order to verify that $L(M)$ is an attack language it suffices to confirm that M satisfies the conditions C1 and C2 and that these can tested in time polynomial in $|Q|$.

To check that C1 holds we need only confirm that $q_0 \notin F$ giving $\varepsilon \notin L(M)$ and for each $x \in \mathcal{X}$ that $\delta(q_0, x) \notin F$, so that $x \notin L(M)$. To test condition C2, for all non-accepting states q' for which there exists a transition from a state different from q_0, it has to be verified that for all $w \in \mathcal{X}^*$ $\rho(q', \bar{w}) \notin F$. This, however, is simply a directed path problem, i.e. verifying that there is no path from q' to any state in F which is easily solved in polynomial time, e.g. by carrying out a breadth-first search of states reachable from q'.   □

Finally, as to question Q2, given M a DFA describing the attack language $L(M)$ and $w \in \mathcal{X}^*$, we can decide if w is an an attack in the AFRA $\langle \mathcal{X}, L(M) \rangle$ in polynomial time simply by confirming that $\rho(q_0, \bar{w}) \in F$.

5   The DFA$^+$ Representation of AFRAs

Expressing $\mathcal{R}$ within an AFRA, $\langle \mathcal{X}, \mathcal{R} \rangle$ via a DFA, M for which $L(M) = \mathcal{R}$ turns out to have some useful computational benefits in addition to verifiability and deciding whether a specified attack is present. We will demonstrate these advantages as far as the problem of computing the grounded extension is concerned. To this purpose we have first to introduce a representation of a whole AFRA (not just the attack relation) as an automaton and analyze its properties. Given an AFRA $\langle \mathcal{X}, \mathcal{R} \rangle$ where $\mathcal{R} \subset \mathcal{X}^*$ is a regular language represented as a DFA $M = \langle \mathcal{X}, Q_M, q_0, F_M, \delta \rangle$, it is easy to obtain a representation of $\langle \mathcal{X}, \mathcal{R} \rangle$ as a single DFA $M^+ = \langle \mathcal{X}, Q_{M+}, q_0, F_{M+}, \delta^+ \rangle$ (indicated for the sake of brevity as DFA$^+$ in the following) such that for any $w \in \mathcal{X}^*$ it holds $w \in L(M^+)$ if and only if $w \in \mathcal{X} \cup \mathcal{R}$. Let us notice that, in general, there are infinite DFA$^+$s representing a single AFRA. This may raise the problem of defining a canonical DFA$^+$ representation for each AFRA. This problem, not considered in the paper, is left for future work. In the following we will provide some general results that hold for any DFA$^+$ representing an AFRA.

Figure 2 shows $\mathcal{M}_F^+$, a DFA$^+$ which accepts all the words of the regular language $\mathcal{R}_F$ describing Fred's dilemma.

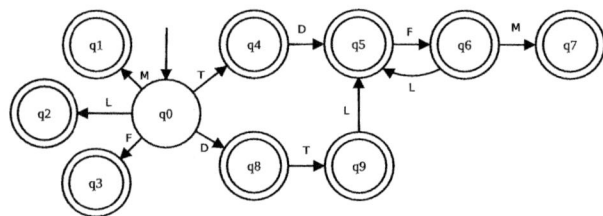

Fig. 2. A DFA$^+$ for Fred's dilemma (double circles represent accepting states)

We introduce also some handy notation concerning neighbor states and "input" symbols for a given state. For $p \in Q_{M+}$ we define $state - out(p) = \{ q \in Q_{M+} : \exists x \in \mathcal{X}$ for which $q = \delta^+(p, x) \}$. For instance, in $\mathcal{M}_F^+$, $state - out(q_0) = \{q_1, q_2, q_3, q_4, q_8\}$. For $p \in F_{M+}$ we define $sym - in(p) = \{x \in \mathcal{X} : \exists q \in Q_{M+}$ for which $p = \delta^+(q, x) \}$ and $state - in(p) = \{ q \in Q_{M+} : \exists x \in \mathcal{X}$ for which $p = \delta^+(q, x) \}$. In $\mathcal{M}_F^+$, $sym - in(q_5) = \{D, L\}$ and $state - in(q_5) = \{q_4, q_9, q_6\}$.

It is now useful to point out several properties of the DFA^+ representation (we will implicitly assume that each accepting state is reachable from q_0, as it should be in order to avoid useless parts in the automaton).

First we can partition the accepting states in $F_{\mathcal{M}+}$ into two sets: *argument states* and *attack states*.

Argument states are in one-to-one correspondence with the elements of $\mathcal{X}$ and are reachable in one step from the initial state q_0: they represent the "additional part" of the DFA^+ w.r.t. the DFA representation. Formally $\forall x \in \mathcal{X}$ $\exists q \in F_{\mathcal{M}+}$ such that $\delta^+(q_0, x) = q$ and $sym - in(q) = \{x\}$. For each $x \in \mathcal{X}$ we will denote the corresponding argument state as $argst(x)$ and, conversely, if $q = argst(x)$ we will say that $x = reparg(q)$. For the whole set of arguments $\mathcal{X}$ in a DFA^+ representation we define $ArgS(\mathcal{M}^+) \triangleq \{argst(x) \mid x \in \mathcal{X}\}$. Hence, $ArgS(\mathcal{M}_F^+) = \{q_1, q_2, q_3, q_4, q_8\}$.

In AFRA an argument can receive only direct defeats from other arguments: an argument x is defeated by an argument y if and only if $\langle x, y \rangle \in \mathcal{R}$ namely if the corresponding two-length string in $\mathcal{X}^*$ is accepted by the DFA^+ (and of course by the original DFA). Formally we can identify the set of direct defeaters of an argument x as $dirdef(x) \triangleq \{y \in \mathcal{X} \mid \delta^+(argst(x), y) \in F_{\mathcal{M}+}\}$. Of course an argument x is *unattacked* in AFRA if and only if $dirdef(x) = \emptyset$. The set of unattacked arguments will be denoted as $unatt - args(\mathcal{M}^+)$. The above definitions can be extended from arguments to argument states in the obvious way.

Attack states are all the accepting states which are not argument states and are defined as $AttS(\mathcal{M}^+) \triangleq F_{\mathcal{M}+} \setminus ArgS(\mathcal{M}^+)$. Hence, $AttS(\mathcal{M}_F^+) = \{q_5, q_6, q_7, q_9\}$. Every attack state q in a DFA^+ (and in the original DFA) corresponds to a (possibly infinite) subset of $\mathcal{R}$, namely to a (nonempty) set of elements of the corresponding attack language, denoted as $AttL(q)$. Formally, for any $q \in AttS(\mathcal{M}^+)$ $AttL(q) \triangleq \{r \in \mathcal{R} \mid \rho(q_0, \bar{r}) = q\}$. Given $r \in AttL(q)$ we will say that q is the *representative state* of r, denoted as $q = repst(r)$. Of course, $\forall r \in \mathcal{R}$ $\exists! q \in AttS(\mathcal{M}^+) \mid q = repst(r)$.

An element r of $\mathcal{R}$ can have both direct defeaters and indirect defeaters (see 1. and 2. in Def. 2). A direct defeater is any argument x which is the source of an attack whose target is r, and then $xr \in \mathcal{R}$. It can then be observed that given an attack state q all elements of $AttL(q)$ have the same direct defeaters. Formally, for any $q \in AttS(\mathcal{M}^+)$ we define $dirdef(q) \triangleq \{x \in \mathcal{X} \mid \delta^+(q, x) \in F_{\mathcal{M}+}\}$ and for any $r \in \mathcal{R}$ $dirdef(r) \triangleq dirdef(repst(r))$.

An indirect defeater is any argument x which is the source of an attack whose target is the source of r: $indirdef(r) \triangleq dirdef(src(r))$.

Given an attack state q it can be noted that the source of any attack represented by q corresponds to one of the elements of $sym-in(q)$: in fact any element of $sym - in(q)$ is the first symbol of some of the elements of the attack language accepted by q. By extension, we can hence define the indirect defeaters of any $q \in AttS(\mathcal{M}^+)$: $indirdef(q) \triangleq \bigcup_{r \in AttL(q)} indirdef(r) = \bigcup_{x \in sym-in(q)} dirdef(x)$.

The whole set of defeaters of an element r of $\mathcal{R}$ will be denoted as $totdef(r) \triangleq dirdef(r) \cup indirdef(r)$. Analogously, for a state q, $totdef(q) \triangleq dirdef(q) \cup indirdef(q)$. We say that an attack state q is unattacked if $totdef(q) = \emptyset$.

For instance in Fig. 2 q_7 is unattacked while $totdef(q_5) = \{\mathbf{F}, \mathbf{T}\}$. In the following we will use the term unattacked states to refer collectively to both unattacked argument states and unattacked attack states. It can be noted that if an attack state q is unattacked then all elements of $AttL(q)$ are unattacked, but it does not hold that if $r \in \mathcal{R}$ is unattacked then $repst(r)$ is unattacked. In fact $totdef(r) = \emptyset$ implies $dirdef(repst(r)) = \emptyset$ but does not imply $indirdef(repst(r)) = \emptyset$ since $repst(r)$ might have indirect defeaters due to other elements of $AttL(q)$. On the other hand it can easily be observed that $totdef(r) = \emptyset$ implies also $indirdef(repst(r)) = \emptyset$ if $|sym - in(repst(r))| = 1$. Under this condition $r \in \mathcal{R}$ is unattacked if and only if $repst(r)$ is unattacked.

Since this is a desirable property, we need to introduce a transformation of DFA$^+$ aimed at ensuring the above condition while leaving unmodified the accepted language. This will be achieved by *splitting* some attack states of the DFA$^+$.

Definition 6. *An attack state p is* splittable *if $|sym - in(p)| > 1$. The set of splittable states of a* DFA$^+$ $\mathcal{M}^+$ *will be denoted as* $split - states(\mathcal{M}^+)$.

In $\mathcal{M}_F^+$, q_5 is splittable since $sym - in(q_5) = \{L, D\}$.

As explained above we need a *complete split* (*csplit* in the following) operator whose goal is transforming a DFA$^+$ (without affecting the language it accepts) so that in the resulting DFA$^+$ there are no splittable states. This is achieved by adding, for each splittable state p, a number $|sym - in(p)| - 1$ new accepting states. Accordingly a *split* operation w.r.t a splittable state can be defined as follows.

Definition 7. *For $\mathcal{M}^+ = \langle \mathcal{X}, Q_{\mathcal{M}^+}, q_0, F_{\mathcal{M}^+}, \delta^+ \rangle$ let p be a splittable state with $sym - in(p) = \{x_1, \ldots, x_n\}$, $(n > 1)$. The* DFA$^+$ *resulting by splitting p, $split(\mathcal{M}^+, p) = \langle \mathcal{X}, Q_{\mathcal{M}^+}^{spl}, q_0, F_{\mathcal{M}^+}^{spl}, \delta^{+spl} \rangle$ is obtained by:*

S1. $Q_{\mathcal{M}^+}^{spl} = Q_{\mathcal{M}^+} \cup \{p_2, \ldots, p_n\}$ *where $p_2, \ldots, p_n$ are new (accepting) states hence included also in $F_{\mathcal{M}^+}^{spl}$.*

S2. *Letting $p_1 = p$ the transition function δ^{+spl} has, for $i = 1 \ldots n$:*
$\delta^{spl}(q', x_i) = p_i$ *if $q' \in state - in(p) \wedge \delta(q', x_i) = p$*
$\delta^{spl}(p_i, y) = \delta(p, y)$
$\delta^{spl}(q, y) = \delta(q, y)$ *if $q \in Q_{\mathcal{M}^+} \setminus state - in(p)$*

In words, a splittable state p is partitioned into several states p_i each with $sym - in(p_i) = \{x_i\}$ and the transitions from p to other states are replicated from each p_i to them. It can be observed that the application of the split operation:

- does not affect the language accepted by the DFA$^+$: for any splittable state p $L(\mathcal{M}^+) = L(split(\mathcal{M}^+, p))$;
- does not affect the cardinality of $sym - in(q)$ for any state $q \neq p$: in fact q may have additional incoming transitions from the elements p_i but they all correspond to elements already present in $sym - in(q)$;
- for each state p_i, letting x_i be the only element of $sym - in(p_i)$, in $split(\mathcal{M}^+, p)$ it holds that $dirdef(p_i) = dirdef(p)$ and $indirdef(p_i) = indirdef(x_i)$.

In virtue of the second point above, it can be noted that it is possible to extend the definition of the split operation to a set of splittable states: given a set P of splittable states of a DFA$^+$ $\mathcal{M}^+$, the result of the operation $split(\mathcal{M}^+, P)$ is the DFA$^+$ resulting from the application of $split(\mathcal{M}^+, p)$ for each $p \in P$ (the order of application of the operations $split(\mathcal{M}^+, p)$ does not matter).

Of course the *csplit* operation is obtained by applying the split operation to all splittable states of a DFA$^+$ $\mathcal{M}^+$: $csplit(\mathcal{M}^+) \triangleq split(\mathcal{M}^+, split - states(\mathcal{M}^+))$. It is easy to observe that the number of states of $split(\mathcal{M}^+, split - states(\mathcal{M}^+))$ is upper bounded by $|Q_{\mathcal{M}^+}| * |\mathcal{X}|$ hence the *csplit* operation can be carried out in polynomial time with respect to the number of states and arguments of $\mathcal{M}^+$.

Figure 3 depicts the result of the application of the *csplit* operator to $\mathcal{M}_F^+$. As we noticed before, q_5 is a splittable state (and it is the only one in $\mathcal{M}_F^+$). $csplit(\mathcal{M}_F^+)$ has an additional state w.r.t. $\mathcal{M}_F^+$, namely q'_5 with $sym - in(q'_5) = \{L\}$ while, after splitting, $sym - in(q_5) = \{D\}$. Moreover, as required by Def. 7, any outgoing transitions from the split state is replicated, giving rise to the transitions from q_5 to q_6 and from q'_5 to q_6, both triggered by F.

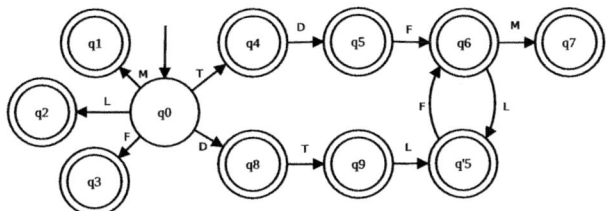

Fig. 3. Graphical representation of $csplit(\mathcal{M}_F^+)$

6   Computing the Grounded Extension with the DFA$^+$ Representation

In this section we show that the grounded extension of AFRAs with DFA$^+$ representation can be computed in polynomial time. Since the grounded extension of an AFRA includes both arguments and attacks, it may be infinite and therefore will, in turn, be expressed through a DFA$^+$, algorithmically derived from the one of the AFRA.

Before illustrating the algorithm we need to consider some properties of AFRAs and of the grounded extension.

First recall a characterization of the grounded extension for finitary argumentation frameworks [6].

Definition 8. *An argumentation framework $\langle \mathcal{X}, \mathcal{A} \rangle$ is finitary iff for each argument x there are only finitely many arguments in $\mathcal{X}$ which attack x.*

Proposition 3. *If an argumentation framework AF is finitary then $GE(AF) = \bigcup_{i=1...\infty} \mathcal{F}^i(\emptyset)$ where $\mathcal{F}$ is the characteristic function of AF (Def. 1).*

It is now easy to see that, for any AFRA, the corresponding AF $\langle \tilde{\mathcal{X}}, \tilde{\mathcal{R}} \rangle$ (see Sect. 2) is finitary:

- the attackers of each element x of $\tilde{\mathcal{X}} \cap \mathcal{X}$ correspond to the direct defeaters of x in AFRA, which are at most $|\mathcal{X}|$;
- the attackers of each element r of $\tilde{\mathcal{X}} \cap \mathcal{R}$ correspond to the direct and indirect defeaters of r in AFRA, which are at most $2 * |\mathcal{X}|$.

On this basis we can now state some relatively straightforward conditions concerning the membership of AFRA arguments and attacks to $GE(\langle \tilde{\mathcal{X}}, \tilde{\mathcal{R}} \rangle) = GE^{\text{AFRA}}(\langle \mathcal{X}, \mathcal{R} \rangle)$, drawing relations between the characteristic function and defeaters in the DFA$^+$ representation.

Proposition 4. *Let $\langle \mathcal{X}, \mathcal{R} \rangle$ be an AFRA with DFA$^+$ representation and $\langle \tilde{\mathcal{X}}, \tilde{\mathcal{R}} \rangle$ be its corresponding AF with characteristic function $\mathcal{F}$, x be an element of $\tilde{\mathcal{X}} \cap \mathcal{X}$, r be an element of $\tilde{\mathcal{X}} \cap \mathcal{R}$. It holds that:*

1. *$x \in \mathcal{F}^1(\emptyset)$ iff $dirdef(x) = \emptyset$*
2. *$r \in \mathcal{F}^1(\emptyset)$ iff $totdef(r) = \emptyset$*
3. *for $i \geq 2, x \in \mathcal{F}^i(\emptyset) \setminus \mathcal{F}^{i-1}(\emptyset)$ iff $\forall y \in dirdef(x)$ $(totdef(yx) \cap \mathcal{F}^{i-1}(\emptyset)) \neq \emptyset \wedge \exists y \in dirdef(x) \mid (totdef(yx) \cap \mathcal{F}^{i-2}(\emptyset)) = \emptyset$*
4. *for $i \geq 2, r \in \mathcal{F}^i(\emptyset) \setminus \mathcal{F}^{i-1}(\emptyset)$ iff $\forall y \in totdef(r)$ $(totdef(yr) \cap \mathcal{F}^{i-1}(\emptyset)) \neq \emptyset \wedge \exists y \in totdef(r) \mid (totdef(yr) \cap \mathcal{F}^{i-2}(\emptyset)) = \emptyset$*

We can now introduce an algorithm (Alg. 1) which builds a DFA accepting the grounded extension of $\langle \mathcal{X}, \mathcal{R} \rangle$. The result of its execution on $\mathcal{M}_F^+$ is illustrated in Fig. 4. After splitting, in the first iteration of the **repeat** cycle the unattacked states q_1, q_2, q_3, q_7 are marked **in**(1) (note that q_7 has no indirect defeaters since q_1 is unattacked). Then, since $state - in(q_7) = \{q_6\}$, q_6 is marked **out** and removed from the set of accepting states. As a consequence, during the second iteration, q'_5 is unattacked and is marked **in**(2). Then, q_9 is marked **out** at line 9 of Alg. 1 and removed from the set of accepting states. Finally, in the third iteration, both q_5 and q_8 are unattacked (note in particular that q_5 is unattacked since $argst(\mathbf{D}) = q_8$ is unattacked). As a consequence they are marked **in**(3) and q_4 is marked **out** at line 9 of Alg. 1. The algorithm will then terminate at the following iteration.

From an argumentation point of view, this result means that the arguments **M**, **L**, **F** and **D** are in the AFRA grounded extension, along with any attack whose source is one of **M**, **L**, and **D**. Therefore, the dilemma's solution is that Fred should not tell his friend what he knows, because in this situation the value of legality prevails over the value of friendship.

Turning back to technical results, correctness of Algorithm 1 follows from the following proposition.

Proposition 5. *Let $\mathcal{M}^+ = \langle \mathcal{X}, Q_{\mathcal{M}^+}, q_0, F_{\mathcal{M}^+}, \delta^+ \rangle$ with $\alpha \in L(\mathcal{M}^+) \Leftrightarrow \alpha \in \mathcal{X} \cup \mathcal{R}$ be a DFA$^+$ describing the AFRA $\langle \mathcal{X}, \mathcal{R} \rangle$, with corresponding AF $\langle \tilde{\mathcal{X}}, \tilde{\mathcal{R}} \rangle$,*

Algorithm 1. Determining $GE(\langle \mathcal{X}, \mathcal{R} \rangle)$ in AFRAS

1: **Input:** DFA$^+$ $\mathcal{M}^+ = \langle \mathcal{X}, Q_{\mathcal{M}+}, q_0, F_{\mathcal{M}+}, \delta^+ \rangle$ with $\alpha \in L(\mathcal{M}^+) \Leftrightarrow \alpha \in \mathcal{X} \cup \mathcal{R}$.

2: **Output:** DFA $\mathcal{M}_G = \langle \mathcal{X}, Q_G, q_0, F_G, \delta_G \rangle$ with $\alpha \in L(\mathcal{M}_G) \Leftrightarrow \alpha \in GE(\langle \tilde{\mathcal{X}}, \tilde{\mathcal{R}} \rangle)$

3: $i := 0$

4: $\mathcal{M}_i := csplit(\mathcal{M}^+)$; with $\mathcal{M}_i = \langle \mathcal{X}, Q_i, q_0, F_i, \delta_i \rangle$

5: **repeat**

6:      $i := i+1$; $\mathcal{M}_i := \mathcal{M}_{i-1}$;

7:      For each (unmarked) unattacked state q of $\mathcal{M}_i$ mark q as **in**(i).

8:      **for** each unattacked state q and every $q' \in state - in(q) \cap F_i$ **do**

9:          Mark q' as **out** and remove q' from F_i.

10:      **end for**

11:      **for** each $x \in \mathcal{X}$ s.t. $argst(x)$ is marked **out do**

12:          For each state $q \in F_i$ with $x \in sym - in(q)$ mark q as **out** and remove q from F_i.

13:      **end for**

14: **until** $\mathcal{M}_i = \mathcal{M}_{i-1}$

15: **for** any $q \in F_i$ which is not marked **in**() **do**

16:      remove q from F_i

17: **end for**

18: **return** $\langle \mathcal{X}, Q_i, q_0, F_i, \delta_i \rangle$

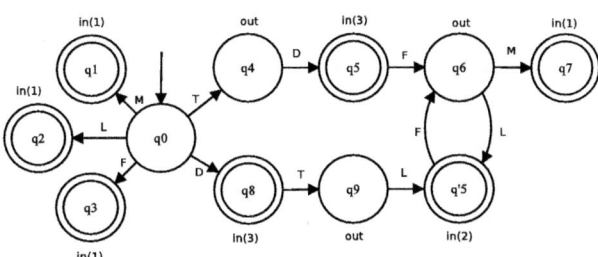

Fig. 4. The DFA$^+$ after the execution of Alg. 1 on $\mathcal{M}_F^+$

and $\mathcal{M}_i = \langle \mathcal{X}, Q_i, q_0, F_i, \delta_i \rangle$ the automaton produced by Algorithm 1 at the i-th iteration of the **repeat** cycle. For $i \geq 0$, let $T_i \subseteq F_i$ be the set of states

$$T_i = \bigcup_{k=1}^{i} \{ q \in F_i : q \text{ is labelled } \textbf{in(k)} \text{ by Algorithm 1} \}$$

and $L_i = \{ \alpha \in \mathcal{X}^* : \rho(q_0, \bar{\alpha}) \in T_i \}$. For every $i \geq 1$ $\alpha \in \mathcal{X} \cup \mathcal{R}$ is in L_i if and only if $\alpha \in \mathcal{F}_i(\emptyset)$, i.e. α is acceptable w.r.t. $\mathcal{F}_{i-1}(\emptyset)$ in $\langle \tilde{\mathcal{X}}, \tilde{\mathcal{R}} \rangle$.

Proof. The proof proceeds by induction on $i \geq 1$. For the inductive base we need $\alpha \in L_1$ if and only if $\alpha \in \mathcal{F}_1(\emptyset)$. Assume first that $\alpha \in L_1$: from l. 7 of Alg. 1 it follows that $q = \rho(q_0, \bar{\alpha})$ is unattacked after the $csplit$ operation has been applied to $\mathcal{M}_0 = \mathcal{M}^+$. If q is an argument state (not affected by the $csplit$ operation) it follows that $dirdef(\alpha) = \emptyset$ both in $\mathcal{M}_0 = \mathcal{M}^+$ and in $\mathcal{M}_1$: hence, by Prop.

4, $\alpha \in \mathcal{F}_1(\emptyset)$. If q is an attack state it either was unattacked in $\mathcal{M}_0 = \mathcal{M}^+$ or it became unattacked in $\mathcal{M}_1$ as a consequence of the splitting of a splittable state in $\mathcal{M}_0$. Taking into account the properties of the split operation discussed in Sec. 5, in both cases it holds that $\alpha \in AttL(q)$ and $totdef(\alpha) = \emptyset$: again by Prop. 4, $\alpha \in \mathcal{F}_1(\emptyset)$.

Assume now that $\alpha \in \mathcal{F}_1(\emptyset)$. From Prop. 4 one of the following two conditions holds: α is an argument with $dirdef(\alpha) = \emptyset$ or α is an attack with $totdef(\alpha) = \emptyset$. In the first case the state $q = argst(\alpha)$ is unattacked in $\mathcal{M}_0 = \mathcal{M}^+$ (and hence also in $\mathcal{M}_1$) and is marked as $\mathbf{in}(1)$ by l. 8, hence $q \in T_1$ and $\alpha \in L_1$. In the second case it follows that the state $q = repst(\alpha)$ is either unattacked or splittable in $\mathcal{M}_0 = \mathcal{M}^+$. In fact q can not have direct defeaters (since α has not), and either has not indirect defeaters (hence being unattacked) or has indirect defeaters (due to other elements of $AttL(q)$) hence being splittable. As a consequence, in both cases after the $csplit$ operation on $\mathcal{M}_0$, in $\mathcal{M}_1$ $q = repst(\alpha)$ is unattacked and is marked as $\mathbf{in}(1)$ by l. 7, hence $q \in T_1$ and $\alpha \in L_1$.

Now inductively assume, for some $k \geq 1$, that for all $i \leq k$ $\alpha \in L_i$ if and only if $\alpha \in \mathcal{F}_i(\emptyset)$. We show $\alpha \in L_{k+1}$ if and only if $\alpha \in \mathcal{F}_{k+1}(\emptyset)$.

Consider any $\alpha \in L_{k+1}$: without loss of generality we may assume $\alpha \in L_{k+1} \setminus L_k$ (since $\mathcal{F}_k(\emptyset) \subseteq \mathcal{F}_{k+1}(\emptyset)$ and, via induction, we have $\alpha \in L_k$ if and only if $\alpha \in \mathcal{F}_k(\emptyset)$).

If α is an argument, namely $\alpha \in \tilde{\mathcal{X}} \cap \mathcal{X}$, it follows that $q = argst(\alpha) \in T_{k+1} \setminus T_k$. If α is an attack, namely $\alpha \in \tilde{\mathcal{X}} \cap \mathcal{R}$, it follows that $q = repst(\alpha) \in T_{k+1} \setminus T_k$.

In both cases, it holds that q is marked as $\mathbf{in}(k+1)$ by l. 7, hence q is unattacked in $\mathcal{M}_{k+1}$ while it is not unattacked in $\mathcal{M}_k$. This means that any $p \in state - out(q)$ has already been marked as $\mathbf{out}$. Moreover if α is an attack, also any argument state t such that $reparg(t) \in indirdef(\alpha)$ has already been marked as $\mathbf{out}$.

The $\mathbf{out}$ marking can be carried out at l. 9 or l. 12 of Alg. 1. In the case of l. 9 p is marked as $\mathbf{out}$ since a state $q' \in state - out(p)$ has been marked as $\mathbf{in}(i)$ with $i \leq k$. This means that for any $\beta \in dirdef(\alpha)$ (with $\beta\alpha \in AttL(p)$ for some $p \in state - out(q)$ marked as $\mathbf{out}$ at l. 9) $\exists \gamma \in dirdef(\beta\alpha)$ such that $repst(\gamma\beta\alpha) = q'$ is marked as $\mathbf{in}(i)$ with $i \leq k$. By the inductive hypothesis, we have that a (direct) defeater γ of the attack $\beta\alpha$ is in $\mathcal{F}_k(\emptyset)$, hence α is defended by $\mathcal{F}_k(\emptyset)$ with respect to any $\beta \in dirdef(\alpha)$ (with $\beta\alpha \in AttL(p)$ for some $p \in state - out(q)$ marked as $\mathbf{out}$ at l. 9). With a similar reasoning, in the case α is an attack, we may also conclude that for any $\beta \in indirdef(\alpha)$ (with $\beta = reparg(t)$ for some argument state t marked as $\mathbf{out}$ at l. 9) $\exists \gamma \in dirdef(\beta)$ such that $repst(\gamma\beta) = q'$ is marked as $\mathbf{in}(i)$ with $i \leq k$ and hence α is defended by $\mathcal{F}_k(\emptyset)$ with respect to any $\beta \in indirdef(\alpha)$ (with $\beta = reparg(t)$ for some argument state t marked as $\mathbf{out}$ at l. 9).

In the case of l. 12 p is marked out since $q' = argst(\beta)$ has been marked out with β the (only) element of $sym - in(p)$. It can be observed that any argument state can be marked as $\mathbf{out}$ only at l. 9 (to satisfy the condition for marking at l. 12 an argument state should be already marked as $\mathbf{out}$ according to l. 11). This means that $\exists q'' \in state - out(q')$ with q'' marked as $\mathbf{in}(i)$ with $i \leq k$. By the inductive hypothesis $\exists \gamma \in dirdef(\beta)$ such that $\gamma\beta \in AttL(q'')$ and $\gamma\beta \in \mathcal{F}_k(\emptyset)$. This means that an (indirect) defeater of all elements of $AttL(p)$ belongs to $\mathcal{F}_k(\emptyset)$, hence α is defended by $\mathcal{F}_k(\emptyset)$ with respect to any attack in $AttL(p)$.

Summing up, it follows that $\mathcal{F}_k(\emptyset)$ defends α against any $\beta \in dirdef(\alpha)$ (either $\beta\alpha$ is attacked, case of l. 9, or β is attacked, case of l. 12) and, if α is an attack, $\mathcal{F}_k(\emptyset)$ defends α against any $\beta \in indirdef(\alpha)$ (β is attacked, case of l. 9). It ensues $\alpha \in \mathcal{F}_{k+1}(\emptyset)$.

Turning to the other side of the proof of the inductive step, assume now $\alpha \in \mathcal{F}_{k+1}(\emptyset)$. Again, without loss of generality, we may consider only the case $\alpha \in \mathcal{F}_{k+1}(\emptyset) \setminus \mathcal{F}_k(\emptyset)$.

If α is an argument, from case 3. of Prop. 4 it follows that $\forall\beta \in dirdef(\alpha)$ $(totdef(\beta\alpha) \cap \mathcal{F}^k(\emptyset)) \neq \emptyset \wedge \exists\beta \in dirdef(\alpha) \mid (totdef(\beta\alpha) \cap (\mathcal{F}^k(\emptyset) \setminus \mathcal{F}^{k-1}(\emptyset))) \neq \emptyset$. This implies that $\forall\beta \in dirdef(\alpha)$ α is defended by $\mathcal{F}^k(\emptyset)$ against β, namely there is an argument γ such that $\gamma\beta\alpha$ or $\gamma\beta$ belongs to $\mathcal{F}^k(\emptyset)$ (in both cases it must also hold $\gamma \in \mathcal{F}^k(\emptyset)$). Moreover, for one of these elements γ it must hold that either $\gamma\beta\alpha$ or $\gamma\beta$ belongs to $\mathcal{F}^k(\emptyset) \setminus \mathcal{F}^{k-1}(\emptyset)$.

By the inductive hypothesis, if follows that for any such γ, $argst(\gamma)$ is marked as **in**(i) with $i \leq k$ and either $repst(\gamma\beta\alpha)$ or $repst(\gamma\beta)$ is marked as **in**(i) with $i \leq k$ (again, for at least one of these elements, the mark is exactly **in**(k)). It follows that $\forall\beta \in dirdef(\alpha)$ $repst(\beta\alpha)$ is marked out at an iteration $i \leq k$ and one of these $repst(\beta\alpha)$ is marked out exactly at the iteration k. Hence $argst(\alpha)$ becomes unattacked, and hence is marked **in**, exactly at the iteration $k+1$ and $\alpha \in L_{k+1}$ as desired.

If α is an attack, from case 4. of Prop. 4 it follows that $\forall\beta \in totdef(\alpha)$ $(totdef(\beta\alpha) \cap \mathcal{F}^k(\emptyset)) \neq \emptyset \wedge \exists\beta \in totdef(\alpha) \mid (totdef(\beta\alpha) \cap (\mathcal{F}^k(\emptyset) \setminus \mathcal{F}^{k-1}(\emptyset))) \neq \emptyset$. This implies that:

- $\forall\beta \in dirdef(\alpha)$ α is defended by $\mathcal{F}^k(\emptyset)$ against β, namely there is an argument γ such that $\gamma\beta\alpha$ or $\gamma\beta$ belongs to $\mathcal{F}^k(\emptyset)$ (in both cases it must also hold $\gamma \in \mathcal{F}^k(\emptyset)$);
- letting $\epsilon = src(\alpha)$, $\forall\beta \in indirdef(\alpha) = dirdef(\epsilon)$ α is defended by $\mathcal{F}^k(\emptyset)$ against β, namely there is an argument γ such that $\gamma\beta\epsilon$ or $\gamma\beta$ belongs to $\mathcal{F}^k(\emptyset)$ (in other words ϵ is defended by $\mathcal{F}^k(\emptyset)$).

In all cases it must also hold $\gamma \in \mathcal{F}^k(\emptyset)$ and for at least one of these elements γ it must hold that either $\gamma\beta\alpha$ or $\gamma\beta\epsilon$ or $\gamma\beta$ belongs to $\mathcal{F}^k(\emptyset) \setminus \mathcal{F}^{k-1}(\emptyset)$.

By the inductive hypothesis, if follows that for any such γ, $argst(\gamma)$ is marked as **in**(i) with $i \leq k$ and either $repst(\gamma\beta\alpha)$ or $repst(\gamma\beta\epsilon)$ or $repst(\gamma\beta)$ is marked as **in**(i) with $i \leq k$ (again, for at least one of these elements, the mark is exactly **in**(k)). It follows that $\forall\beta \in totdef(\alpha)$ $repst(\beta\alpha)$ is marked out at an iteration $i \leq k$ and one of these $repst(\beta\alpha)$ is marked out exactly at the iteration k. Hence $repst(\alpha)$ becomes unattacked, and hence is marked **in**, exactly at the iteration $k+1$ and $\alpha \in L_{k+1}$ as desired.

On this basis we obtain one of the main results of the paper.

Theorem 2. *Let $\mathcal{M}^+ = \langle \mathcal{X}, Q_{\mathcal{M}^+}, q_0, F_{\mathcal{M}^+}, \delta^+ \rangle$ with $\alpha \in L(\mathcal{M}^+) \Leftrightarrow \alpha \in \mathcal{X} \cup \mathcal{R}$ be a* DFA$^+$ *describing the* AFRA, $\langle \mathcal{X}, \mathcal{R} \rangle$ *with corresponding* AF $\langle \tilde{\mathcal{X}}, \tilde{\mathcal{R}} \rangle$. *It is possible to construct in polynomial time a* DFA $\mathcal{M}_G = \langle \mathcal{X}, Q_G, q_0, F_G, \delta_G \rangle$ *with $\alpha \in L(\mathcal{M}_G) \Leftrightarrow \alpha \in GE(\langle \tilde{\mathcal{X}}, \tilde{\mathcal{R}} \rangle)$*

Proof. Given Prop. 5, we have only to show that Alg. 1 terminates in polynomial time. We have already commented that the *csplit* operation (1. 4) can be carried out in polynomial time and gives rise to a total number of states $\#Q \leq |Q_{\mathcal{M}^+}| * |\mathcal{X}|$. The **repeat** cycle terminates when $\mathcal{M}_i = \mathcal{M}_{i-1}$, which occurs when no unmarked unattacked states are detected at iteration i. Identifying whether a state q is unattacked requires the following checks (check (ii) only applies to attack states): (i) for any state $p \in state - out(q)$ is p in F_i? (ii) for any argument $x \in sym - in(q)$ is any defeater of x in F_i?

Check (i) requires at most $\#Q$ constant time operations for each state q, so its complexity in a single iteration of the **repeat** cycle is $O(\#Q^2)$. Check (ii) requires at most $|\mathcal{X}|^2$ constant time operations for each state q, so its complexity in a single iteration of the **repeat** cycle is $O(\#Q * |\mathcal{X}|^2)$.

Given the identification of unattacked states for granted, in a single iteration of the **repeat** cycle:

- at most $\#Q$ mark operation are executed at l. 7;
- at most $\#Q$ checks on membership to $state - in(q) \cap F_i$ are carried out at l. 8 and at most the same number of marking and removal operations are executed at l. 9;
- the **for** cycle at l. 11 is executed at most $|\mathcal{X}|$ times and for each of these iterations at most $\#Q$ marking and removal operations are executed at l. 12.

Noting that the algorithm never adds accepting states, it follows that the number of removals and, hence, the number of iterations of the **repeat** cycle is bounded by $\#Q$. Finally the **for** cycle at l. 15 is executed at most $\#Q$ times.

Summing up, the order of magnitude of the computational complexity of Alg. 1 is determined by checks (i) and (ii) within the **repeat** cycle, which turn out to be respectively $O(\#Q^3) = O(|Q_{\mathcal{M}^+}|^3 * |\mathcal{X}|^3)$ and $O(\#Q * \#Q * |\mathcal{X}|^2) = O(|Q_{\mathcal{M}^+}|^2 * |\mathcal{X}|^4)$. □

7   Conclusions

This paper proposes a methodology and provides some initial results in the largely unexplored field of computing with infinite argumentation frameworks, using as a starting point the possible existence of infinite attacks in the recently introduced AFRA formalism, exemplified by a case of moral dilemma. While other approaches (for instance, Modgil's EAF [13]) may provide a different formalization of this specific example, from a general point of view it is worth noting that the notion of unlimited recursive attacks, as in the AFRA formalism, may encompass infinite attack sequences even with a finite set of arguments. This can be easily seen as a finite alphabet able to describe infinite attack structures.

In fact, the proposal is built on the main idea of drawing correspondences between the specification of argumentation frameworks and well-known notions and results in formal language theory. While there are cases of infinite attacks which can not be represented with formal grammars, deterministic finite automata provide a convenient way to represent infinite attack relations with potential practical use. In particular we show that, with this representation, the problem of computing the grounded extension, which is tractable in the finite case, preserves its tractability in the infinite case. We are already extending this kind of analysis to other "standard" computational problems in abstract argumentation, like checking whether a set is conflict-free, is admissible or is a stable extension. The representation of special reasoning cases, like dilemmas, is an example of motivation for this kind of studies. In a similar spirit, one might consider the representation of dialogues where the repetition of previous moves is allowed: while this is normally forbidden, in order to ensure dialogue termination, the proposed approach might be used to define a sound semantics for some kinds of non-terminating dialogues, which represent the formal counterpart of situations where dialogue participants decide to keep (some of) their positions forever [10,9].

In the perspective of enlarging its applicability domain, the proposed methodology and techniques could also be applied to other cases of infinite frameworks, either in the context of traditional Dung's AF or in some of its extended versions. In particular, it can be noted that the proposed approach implicitly deals with a family of infinite Dung's AFs since any AFRA with infinite attacks can be translated into a traditional AF with infinite arguments (see Sect. 2). From a more general perspective, one can consider using the DFA representation to specify an infinite set of arguments (so that each accepted word corresponds to an argument) complemented by a compact definition of the attack relation. Just to give an example, one simple option is to state that if both words xw and w are accepted (i.e. both of them represent arguments) then xw attacks w. In this way it is possible, for instance, to represent an infinite chain of attacks with a simple DFA, accepting the words x, xx, xxx, ... A more general option is to specify the attack relations through an expression constructed by a set of operators. A variant of Algorithm 1 could then be devised to compute the grounded extension of this kind of frameworks. A deep investigation of these issues is the subject of ongoing work [3].

References

1. Baroni, P., Cerutti, F., Giacomin, M., Guida, G.: An argumentation-based approach to modeling decision support contexts with what-if capabilities. In: AAAI Fall Symposium. Technical Report SS-09-06. pp. 2–7 (2009)
2. Baroni, P., Cerutti, F., Giacomin, M., Guida, G.: AFRA: argumentation framework with recursive attacks. International Journal of Approximate Reasoning 51(1), 19–37 (2011)

3. Baroni, P., Cerutti, F., Dunne, P.E., Giacomin, M.: Automata for infinite argumentation structures. Technical Report, Department of Computer Science, University of Liverpool, UK (2011)
4. Barringer, H., Gabbay, D., Woods, J.: Temporal Dynamics of Support and Attack Networks: From Argumentation to Zoology. In: Hutter, D., Stephan, W. (eds.) Mechanizing Mathematical Reasoning. LNCS (LNAI), vol. 2605, pp. 59–98. Springer, Heidelberg (2005)
5. Bench-Capon, T.J.M.: Persuasion in practical argument using value based argumentation frameworks. Journal of Logic and Computation 13(3), 429–448 (2003)
6. Dung, P.M.: On the acceptability of arguments and its fundamental role in nonmonotonic reasoning, logic programming, and N-person games. Artificial Intelligence 77, 321–357 (1995)
7. Dunne, P.E.: Computability Theory – concepts and applications. Ellis–Horwood (1991)
8. Dunne, P.E., Modgil, S., Bench-Capon, T.: Computation in extended argumentation frameworks. In: Proceedings of 19th European Conference on Artificial Inteligence (ECAI 2010), Lisbon, pp. 119–124 (2010)
9. Gabbay, D., Woods, J.: More on non-cooperation in dialogue logic. Logic Journal of IGPL 9(1), 305–323 (2001)
10. Gabbay, D., Woods, J.: Non-cooperation in dialogue logic. Synthese 127, 161–186 (2001)
11. Gabbay, D.: Semantics for higher level attacks in extended argumentation frames part 1: Overview. Studia Logica 93, 357–381 (2009)
12. Hopcroft, J.E., Ullman, J.D.: Introduction to Automata Theory, Languages, and Computation. Addison-Wesley (1979)
13. Modgil, S.: Reasoning about preferences in argumentation frameworks. Artificial Intelligence 173(9–10), 901–934 (2009)
14. Walton, D., Reed, C., Macagno, F.: Argumentation Schemes. Cambridge University Press, NY (2008)

Multi-sorted Argumentation

Tjitze Rienstra[1], Alan Perotti[2], Serena Villata[3],
Dov M. Gabbay[4], and Leendert van der Torre[1]

[1] Computer Science and Communication, University of Luxembourg
tjitze.rienstra@uni.lu, leendert@vandertorre.com
[2] Dip. di Informatica, University of Turin
perotti@di.unito.it
[3] Edelweiss, INRIA Sophia Antipolis
serena.villata@inria.fr
[4] Dept. Computer Science, King's College London
dov.gabbay@kcl.ac.uk

Abstract. In the theory of abstract argumentation, the acceptance status of arguments is normally determined for the complete set of arguments at once, under a single semantics. However, this is not always desired. In this paper, we extend the notion of an argumentation framework to a multi-sorted argumentation framework, and we motivate this extension using an example which considers practical and epistemic arguments. In a multi-sorted argumentation framework, the arguments are partitioned into a number of *cells*, where each cell is associated with a semantics under which its arguments are evaluated. We prove the properties of the proposed framework, and we demonstrate our theory with a number of examples. Finally, we relate our theory to the theory of modal fibring of argumentation networks.

1 Introduction

Abstract argumentation frameworks [10] are used to model sets of arguments and the attacks among these arguments. Given an abstract argumentation framework, we can ask the question of which arguments are acceptable, and which arguments are not. This question is answered by what is called an *acceptability semantics*. Different modes of reasoning are possible, each giving rise to a different acceptability semantics. Well-known examples are the *grounded semantics* that minimizes the number of accepted arguments, and the *preferred semantics*, that maximizes the number of accepted arguments. The choice of which semantics is appropriate depends on the kind of arguments, and the attitude towards these arguments. A skeptical attitude, for example, can be modeled with grounded semantics, whereas a credulous attitude can be modeled using preferred semantics [8]. In most literature on acceptability semantics (see for instance [10,3,7,9]), the assumption is made that all arguments of a framework are evaluated under a single semantics.

In this paper, we argue for a generalization, and we answer the research question *how to define an abstract argumentation framework where the arguments*

S. Modgil, N. Oren, and F. Toni (Eds.): TAFA 2011, LNAI 7132, pp. 215–231, 2012.
© Springer-Verlag Berlin Heidelberg 2012

can be evaluated under different semantics? We motivate this through an example about practical and epistemic arguments, and we introduce a system called *multi-sorted argumentation*. The motivating example, which now follows, is taken from Prakken [13], and adapted to an abstract argumentation framework.

Consider a university lecturer (let us call him John) with two conflicting desires. He wants to finish a paper on Friday, but he also promised to give a talk in a town called Faraway on the same day. There are two ways to travel to Faraway: by car and by bus. In neither case will he be able to finish the paper while traveling; he cannot work while driving, and he gets sick when working in a bus. Figure 1 shows an informal instantiation of an abstract framework for this situation: the three arguments a, b and c represent the situation described so far. Note that the arguments are forms of the practical syllogism. For example, argument b is an inference consisting of John's belief that traveling by bus (to Faraway) allows him to give the lecture in Faraway, his desire to give a lecture in Faraway, and the conclusion that he must therefore travel by bus. John has more information: his friend Bob tells him that there is a train connection to Faraway. So if John travels by train, he will be able to finish the paper (argument d, another practical syllogism). Now, John's other friend Mary warns him about a railway strike, which would defeat d (argument e). On the other hand, Bob believes there will be no strike (argument f). John has no reason to trust either of his friends more than the other. To be on the safe side, John does not want to act on the credulous belief that there will not be a train strike, and that there will be a train to Faraway on Friday.

In this example, we have four arguments that pertain to actions (arguments a, b, c, d) and two arguments expressing beliefs about the world (arguments e and f). Prakken [13] calls them *practical arguments* and *epistemic arguments*, and he argues that practical arguments should be evaluated credulously and epistemic arguments skeptically. The reason is that for practical arguments, it is rational for an agent to consider credulously all possible 'action alternatives' that have skeptical support of epistemic arguments. Since a credulous attitude can be modeled using preferred semantics, and a skeptical attitude using grounded semantics, the evaluation of this framework is a combination of preferred semantics for a, b, c and d, and grounded semantics for e and f.

This example motivates the evaluation of different parts of the same framework under both preferred and grounded semantics. We believe that the point made by this example extends to the more general case. Suppose we have a set of frameworks $\{A_1, \ldots, A_n\}$ and for each framework A_i there is an appropriate semantics s_i. Then it is also possible that there is a framework A which is a merge of the frameworks $A_1, \ldots, A_n$, where the n different parts of A may interact mutually through additional attacks. Determining the acceptance status of the arguments in A would amount to the application of the s_i semantics to the part corresponding to A_i, for all parts $1 \ldots n$.

What we need, then, is a method to apply different semantics to different parts of the same framework. To this end, we propose a system called *multi-sorted argumentation*. The system is based on two elements: a regular argumentation

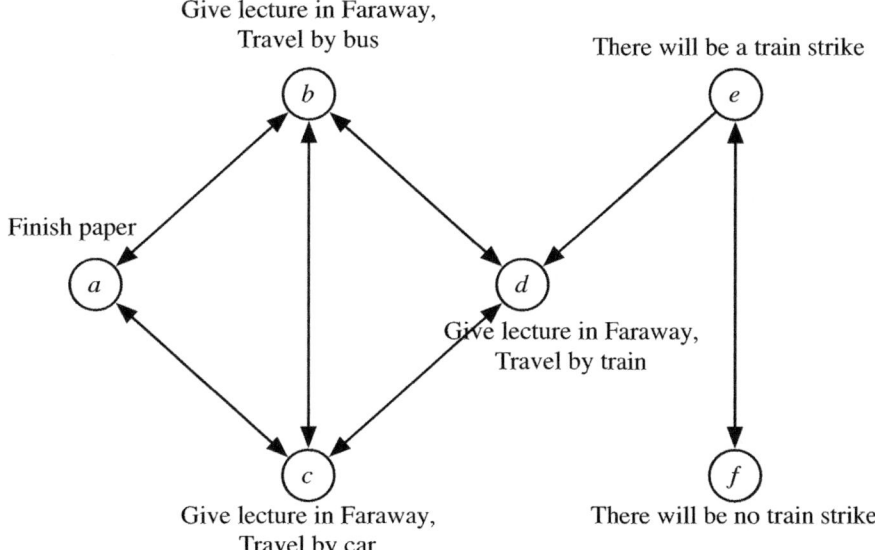

Fig. 1. An example of an argumentation framework with both practical and epistemic arguments

framework and a sorting. A sorting supplements the argumentation framework with information on how the framework is divided into cells, and which cell in the sorting is to be evaluated under which semantics. A sorted extension is a set of arguments that are acceptable with respect to the sorting. We prove a number of desired properties of sorted extensions. For example, sorted extensions should be conflict-free. Moreover, some properties of the semantics associated with each cell are preserved, i.e., if the semantics of all cells are admissible (resp. complete), then the sorted extensions should also be admissible (resp. complete).

Finally, we show how to formalize multi-sorted argumentation using the modal fibring approach. Multi-sorted argumentation is expressed as a special case of the fibring of modal argumentation frameworks. We present this kind of multi-sorted argumentation by means of a number of examples, and we discuss the properties which hold for this method of evaluating the cells under different semantics.

The paper is organized as follows: Section 2 provides the basic concepts of argumentation theory; in Section 3 we introduce the notions of sortings and sorted extension, in Section 4 we study some properties of sorted extensions, and in Section 5 we relate our theory to the theory of modal fibring. Finally, Sections 6 and 7 present related work, conclusions and future work.

2   Preliminaries

The following definitions set forth the basics of Dung's well-known theory of abstract argumentation [10].

Definition 1 (Argumentation Framework). *We assume as given a set $\mathcal{U}$, called the* universe of arguments. *An* argumentation framework $\mathcal{AF}$ *is a pair $\langle A, R \rangle$ with finite $A \subseteq \mathcal{U}$ and a binary relation $R \subseteq A \times A$, called the* attack relation.

Definition 2 (Conflict Free). *Let $\mathcal{AF} = \langle A, R \rangle$ be an argumentation framework. A set $\mathcal{S} \subseteq A$ is* conflict free *iff there are no arguments $a, b \in \mathcal{S}$ such that aRb. If $\mathcal{S}$ is conflict free, we write $cf(\mathcal{S})$.*

We follow Baroni & Giacomin's [3] generalized approach, where the acceptability of arguments is considered with respect to a designated subset of arguments. This set, which we call the set of *qualified arguments*, contains the arguments that an extension may consist of. Intuitively, it is used to filter out arguments that do not qualify for acceptance. This is necessary when we evaluate only a subset of arguments, but at the same time we know that some arguments in this subset cannot be accepted due to attacks from outside the subset.

Definition 3 (Defense). *Let $\mathcal{AF} = \langle A, R \rangle$ be an argumentation framework. A set $\mathcal{S} \subseteq A$* defends a from A *iff $\forall b \in A$ such that bRa, $\exists c \in \mathcal{S}$ such that cRb. Let $D_Q(\mathcal{S}) = \{a \in Q \mid \mathcal{S}$ defends a from $A\}$, where $Q \subseteq A$ is called a* qualified set of arguments.

We will sometimes say that $\mathcal{S} \subseteq A$ defends a, without mentioning the set from which $\mathcal{S}$ defends a. In that case, we mean that $\mathcal{S}$ defends a from all arguments, i.e. from A.

Definition 4 (Acceptance Function). *An* acceptance function

$$\mathcal{E} : 2^{\mathcal{U}} \times 2^{\mathcal{U} \times \mathcal{U}} \times 2^{\mathcal{U}} \to 2^{2^{\mathcal{U}}}$$

is a partial function that associates each argumentation framework $\langle A, R \rangle$ and each set of qualified arguments $Q \subseteq A$, with sets of subsets of A, called extensions: $\mathcal{E}(\langle A, R \rangle, Q) \subseteq 2^A$.

Dung [10] presents several acceptability semantics which produce zero, one, or several sets of accepted arguments. These semantics are grounded on the two main concepts of conflict-freeness and defense. The following definitions are equivalent to those in Dung's original theory, if we set $Q = A$.

Definition 5 (Acceptability Semantics). *Let $\mathcal{AF} = \langle A, R \rangle$ be an argumentation framework and $Q \subseteq A$ a set of qualified arguments. Acceptance functions for conflict free ($\mathcal{E}_{cf}$), admissible ($\mathcal{E}_{ad}$), complete ($\mathcal{E}_{co}$), grounded ($\mathcal{E}_{gr}$) and preferred ($\mathcal{E}_{pr}$) extensions are defined as follows:*

- *$\mathcal{S} \in \mathcal{E}_{cf}(\mathcal{AF}, Q)$ iff $\mathcal{S} \subseteq Q$ and $cf(\mathcal{S})$.*
- *$\mathcal{S} \in \mathcal{E}_{ad}(\mathcal{AF}, Q)$ iff $cf(\mathcal{S})$ and $\mathcal{S} \subseteq D_Q(\mathcal{S})$.*

- $\mathcal{S} \in \mathcal{E}_{co}(\mathcal{AF}, Q)$ iff $cf(\mathcal{S})$ and $\mathcal{S} = D_Q(\mathcal{S})$.
- $\mathcal{S} \in \mathcal{E}_{gr}(\mathcal{AF}, Q)$ iff $\mathcal{S}$ is minimal in $\mathcal{E}_{co}(\mathcal{AF}, Q)$ w.r.t. set inclusion.
- $\mathcal{S} \in \mathcal{E}_{pr}(\mathcal{AF}, Q)$ iff $\mathcal{S}$ is maximal in $\mathcal{E}_{co}(\mathcal{AF}, Q)$ w.r.t. set inclusion.
- $\mathcal{S} \in \mathcal{E}_{st}(\mathcal{AF}, Q)$ iff $\mathcal{S} \in \mathcal{E}_{cf}(\mathcal{AF}, Q)$ and $\forall a \in A, a \notin \mathcal{S} \rightarrow \exists b \in \mathcal{S}$ s.t. bRa.

The definitions above are reformulations of those proposed by Baroni & Giacomin [3].

Example 1 (Admissible extension). Consider the framework of Figure 1. Let $Q = \{a, b, c, d\}$. Given Q, the set $\{a\}$ is an admissible extension, i.e., it is included in $\mathcal{E}_{ad}(\mathcal{AF}, Q)$. This can be seen as follows: we have that $\{a\}$ is conflict free, and $\{a\} \subseteq D_Q(\{a\}) = \{a\}$. However, the set $\{a, d\}$ is not an admissible extension, and is not included in $\mathcal{E}_{ad}(\mathcal{AF}, Q)$. The reason is that the set does not defended itself from the attack by f: $\{a, d\} \not\subseteq D_Q(\{a, d\}) = \{a\}$.

Note, in the running example, that the arguments of an admissible set need to be defended from all arguments in A, so not only from arguments in Q.

Example 2 (Complete extension). Consider the framework of Figure 1. Let $Q = \{a, b, c, d\}$. Given Q, the set $\{b\}$ is a complete extension, i.e., it is included in $\mathcal{E}_{co}(\mathcal{AF}, Q)$. This can be seen as follows: we have that $\{b\}$ is conflict free, and $\{b\} = D_Q(\{b\}) = \{b\}$. However, the set $\{a, d\}$ is not a complete extension, namely for the same reason that it is not an admissible extension (see above). Now, let $Q = \{b, c, d, e, f\}$. The set $\{d, f\}$ is a complete extension, because $\{d, f\} = D_Q(\{d, f\})$. However, the set $\{f\}$ is not, because $\{f\} \neq D_Q(\{f\}) = \{d, f\}$.

Example 3 (Grounded, preferred extensions). Consider the framework of Figure 1. Let $Q = \{b, c, d, e, f\}$. Given Q, the complete extensions are $\{b, e\}$, $\{b, f\}$, $\{b\}$, $\{c, e\}$, $\{c, f\}$, $\{c\}$, $\{d, f\}$, $\{f\}$ and $\emptyset$. The extensions $\{b, e\}$, $\{b, f\}$, $\{c, e\}$, $\{c, f\}$ and $\{d, f\}$ are preferred extensions, since they are maximal with respect to set-inclusion. The extension $\emptyset$ is the grounded extension, since it is minimal with respect to set-inclusion.

3   Multi-sorted Argumentation

We now define the main ingredients of our system: sortings and sorted extensions. A sorting supplements the argumentation framework with information on how the framework is divided into cells, and which cell in the sorting is to be evaluated under which semantics. In the following definitions, we assume a fixed argumentation framework $\mathcal{AF} = \langle A, R \rangle$.

Definition 6 (Sorting). *A sorting $\mathbb{S}$ is a pair $\langle P, T \rangle$, where P is a partition of A and $T : P \rightarrow \{cf, ad, co, gr, pr\}$ a function associating each cell in P to a semantics.*

The following example demonstrates this representation, for the framework shown in Figure 1, and discussed in the introduction.

Example 4. The situation shown in Figure 1 is formally represented by a framework $AF = \langle \{a, b, c, d, e, f\}, R \rangle$, where aRb, bRa, aRc, cRa, bRc, cRb, bRd, dRb, cRd, dRc, eRd, eRf, fRe and sorting $\mathbb{S} = \langle \{C_1, C_2\}, T \rangle$, where $C_1 = \{a, b, c, d\}$, $C_2 = \{e, f\}$, $T(C_1) = pr$ and $T(C_2) = gr$, i.e., arguments a, b, c, d are evaluated under the preferred semantics, and arguments e, f are evaluated under the grounded semantics.

We will shortly give the condition, given a sorting, for a set of arguments to be a multi-sorted extension of an argumentation framework. Before we do so, we introduce the concepts of a subframework and of the set of qualified arguments of a subframework. These concepts define a way of evaluating the arguments in a cell, given an extension $\mathcal{S}$. The intuition behind them is as follows. Given a cell C and extension $\mathcal{S}$, we determine whether $\mathcal{S} \cap C$ is an extension for C, by first restricting C to those arguments that are not defeated by arguments outside C. This set, denoted by C', makes up the arguments of what we call the subframework for C. Next, we further restrict the arguments of C' to those that are defended by $\mathcal{S}$ from attacks outside C. This set, denoted by C'', contains the arguments in C that are qualified for acceptance.

Definition 7 (Subframework). *Let P be a partition of A, $C \in P$ a cell and $\mathcal{S} \subseteq A$ an extension. The* subframework *for C, given $\mathcal{S}$, is the argumentation framework $\langle C', R \downarrow C' \rangle$ where $C' = \{a \in C \mid \nexists b \in \mathcal{S} \setminus C, bRa\}$ and where $R \downarrow C'$ is the attack relation R restricted to the arguments in C', i.e. $R \downarrow C' = \{(a, b) \in R \mid a, b \in C'\}$.*

Definition 8 (Qualified Arguments of a Subframework). *Let $\langle C', R \downarrow C' \rangle$ be a subframework for a cell C and extension $\mathcal{S}$. The* qualified arguments *of $\langle C', R \downarrow C' \rangle$, denoted by C'', are defined as follows.*

$$C'' = \{a \in C' \mid \forall b \in A \setminus C, (bRa \to \exists c \in \mathcal{S}, cRb)\}$$

Given an extension $\mathcal{S}$, we can determine whether it is a sorted extension by checking that for each $C \in P$, we have that $C \cap \mathcal{S}$ is an extension of the subframework for C, given the qualified arguments of the subframework for C. The semantics under which the subframework for C is evaluated, is the semantics associated with C.

Definition 9 (Sorted Extension). *A set $\mathcal{S} \subseteq A$ is a* sorted extension *of $AF = \langle A, R \rangle$ and $\mathbb{S} = \langle P, T \rangle$ iff for all $C \in P$, we have*

$$C \cap \mathcal{S} \in \mathcal{E}_{T(C)}(\langle C', R \downarrow C' \rangle, C'')$$

The sorted acceptance function *$\mathcal{E}_{srt}$ is defined as follows: $\mathcal{S} \in \mathcal{E}_{srt}(AF, \mathbb{S})$ iff $\mathcal{S}$ is a sorted extension of AF and $\mathbb{S}$.*

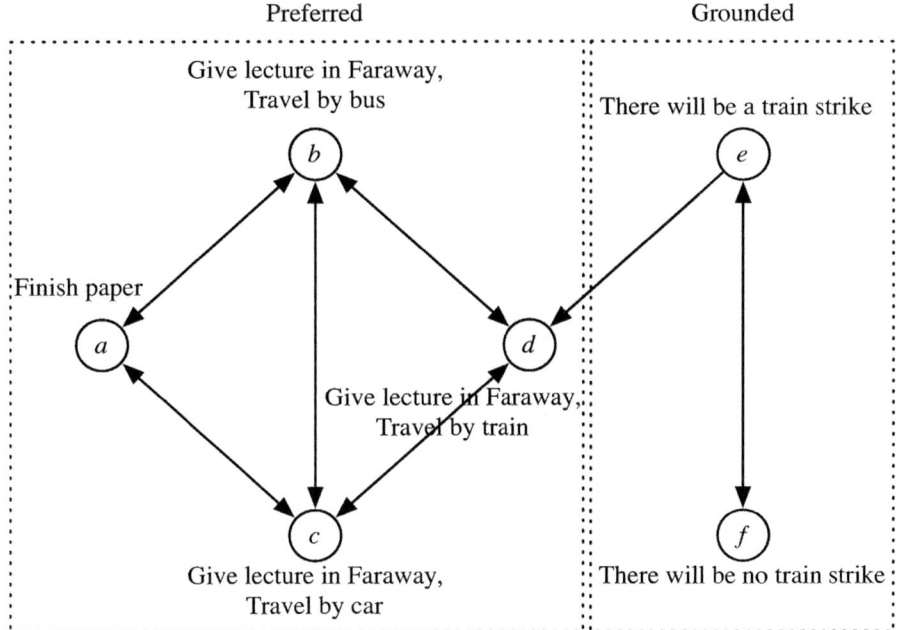

Fig. 2. The multi-sorted argumentation framework of the running example about practical and epistemic arguments

The following example demonstrates the computation of the sorted extensions for the framework used in our running example.

Example 4 (Continued). Consider the following extensions for the framework AF shown in Figure 2: $\mathcal{S}_1 = \emptyset$, which is the grounded extension of AF; $\mathcal{S}_2 = \{a, d, f\}$, which is a preferred extension of AF; and $\mathcal{S}_3 = \{b\}$, which is neither the grounded nor a preferred extension of AF. We determine whether they are multi-sorted extensions of AF, given the sorting $\mathbb{S}$ introduced earlier. That is to say, we determine whether $\mathcal{S}_1, \mathcal{S}_2, \mathcal{S}_3 \in \mathcal{E}_{srt}(AF, \mathbb{S})$.

– Given $\mathcal{S}_1$, we have $C_1' = \{a, b, c, d\}$ and $C_2' = \{e, f\}$ (no argument in C_1 defeats an argument in C_2 or vice versa); and $C_1'' = \{a, b, c\}$ and $C_2'' = \{e, f\}$ (d is not defended from e). We have that $\mathcal{S}_1 \cap C_1 = \emptyset$ and $\emptyset \notin \mathcal{E}_{pr}(\langle C_1', R \downarrow C_1'\rangle, C_1'')$ and $\mathcal{S}_1 \cap C_2 = \emptyset$ and $\emptyset \in \mathcal{E}_{gr}(\langle C_2', R \downarrow C_2'\rangle, C_2'')$. While $\mathcal{S}_1 \cap C_2$ is the grounded extension of $\langle C_2', R \downarrow C_2'\rangle$, $\mathcal{S}_1 \cap C_1$ is not a preferred extension of $\langle C_1', R \downarrow C_1'\rangle$. It follows that $\mathcal{S}_1 \notin \mathcal{E}_{srt}(AF, \mathbb{S})$.

– Given $\mathcal{S}_2$, we have $C_1' = \{a, b, c, d\}$ and $C_2' = \{e, f\}$ (no argument in C_1 defeats an argument in C_2 or vice versa); and $C_1'' = \{a, b, c, d\}$ and $C_2'' = \{e, f\}$ (no argument is undefended from attacks by other cells). We have that $\mathcal{S}_2 \cap C_1 = \{a, d\}$ and $\{a, d\} \in \mathcal{E}_{pr}(\langle C_1', R \downarrow C_1'\rangle, C_1'')$ and $\mathcal{S}_1 \cap C_2 = \{f\}$

and $\{f\} \notin \mathcal{E}_{gr}(\langle C_2', R \downarrow C_2' \rangle, C_2'')$. While $\mathcal{S}_1 \cap C_1$ is a preferred extension of $\langle C_1', R \downarrow C_1' \rangle$, $\mathcal{S}_1 \cap C_2$ is not the grounded extension of $\langle C_2', R \downarrow C_2' \rangle$. It follows that $\mathcal{S}_2 \notin \mathcal{E}_{srt}(AF, \mathbb{S})$.

- Given $\mathcal{S}_3$, we have $C_1' = \{a, b, c, d\}$ and $C_2' = \{e, f\}$; and $C_1'' = \{a, b, c\}$ (d is not defended from e) and $C_2'' = \{e, f\}$. We have that $\mathcal{S}_3 \cap C_1 = \{b\}$ and $\{b\} \in \mathcal{E}_{pr}(\langle C_1', R \downarrow C_1' \rangle, C_1'')$ and $\mathcal{S}_3 \cap C_2 = \emptyset$ and $\emptyset \in \mathcal{E}_{gr}(\langle C_2', R \downarrow C_2' \rangle, C_2'')$. It follows that $\mathcal{S}_3 \in \mathcal{E}_{srt}(AF, \mathbb{S})$.

In conclusion, $\mathcal{S}_3$ is a sorted extension, and $\mathcal{S}_1$ and $\mathcal{S}_2$ are not. The only other sorted extensions are $\{a\}$ and $\{c\}$. Switching back to the language of the running example in the introduction, the three acceptable options are to either finish the paper (extension $\{a\}$), to give the lecture and travel by bus (extension $\{b\}$) or to give the lecture and travel by car (extension $\{c\}$). As desired, the option to travel by train and both give the lecture and finish the paper, which would correspond to the extension $\{a, d, f\}$, is not supported.

4   Properties

In this section, we present some desired properties of sorted extensions. In particular, we aim to show that the sorted extensions presented in Section 3 satisfy conflict-freeness, admissibility, and completeness. We say that a semantics x satisfies conflict-freeness (resp. admissibility, completeness) if, given any framework $\langle A, R \rangle$ and any $Q \subseteq A$, all extensions $\mathcal{E}_x(\langle A, R \rangle, Q)$ are conflict-free (resp. admissible, complete). We then have that all the semantics considered here satisfy conflict-freeness; that all semantics except conflict-free satisfy admissibility; and that all semantics except conflict-free and admissible satisfy completeness.

We first prove the preservation of the conflict-free, admissible and completeness properties of sorted extensions, i.e., whenever the semantics associated with all cells of the partitioning satisfy these properties, then the sorted extensions satisfy them as well.

Proposition 1. *For any AF and* $\mathbb{S} = \langle P, T \rangle$*, if* $\forall C \in P$*,* $T(C)$ *is a conflict-free semantics, then* $\forall \mathcal{S} \in \mathcal{E}_{srt}(AF, \mathbb{S})$*,* $\mathcal{S}$ *is conflict-free.*

Proof. Let $AF = \langle A, R \rangle$, $\mathbb{S} = \langle P, T \rangle$, and $\mathcal{S} \in \mathcal{E}_{srt}(AF, \mathbb{S})$. We know that $\forall C \in P, T(C)$ is a conflict-free semantics. Note that it follows that $\forall C \in P$, $\mathcal{S} \cap C$ is conflict free. Now suppose the contrary, i.e. there are $a, b \in \mathcal{S}$ s.t. bRa. Let $C \in P$ be the cell s.t. $a \in C$. Because $\mathcal{S} \cap C$ is conflict-free, we have $b \in \mathcal{S} \setminus C$. Then by Definition 7, $a \notin C'$. Because we have that $C'' \subseteq C' \subseteq C$ and $\mathcal{S} \cap (C \setminus C'') = \emptyset$, it follows that $a \notin \mathcal{S}$. Contradiction. $\square$

Note that, since all the semantics that we consider satisfy conflict-freeness, we have that every sorted extension is conflict-free.

Proposition 2. *For any AF and* $\mathbb{S} = \langle P, T \rangle$*, if* $\forall C \in P$*,* $T(C)$ *is an admissible semantics, then* $\forall \mathcal{S} \in \mathcal{E}_{srt}(AF, \mathbb{S})$*,* $\mathcal{S}$ *is admissible.*

Proof. Let $AF = \langle A, R \rangle$, $\mathbb{S} = \langle P, T \rangle$, and $\mathcal{S} \in \mathcal{E}_{srt}(AF, \mathbb{S})$. By Proposition 1 we have that $\mathcal{S}$ is conflict-free. We also know that $\forall C \in P, T(C)$ is an admissible semantics. Note that it follows that $\forall C \in P$, $\mathcal{S} \cap C$ is admissible w.r.t. the framework $\langle C', R \downarrow C' \rangle$. Suppose now that $\mathcal{S}$ is not admissible, i.e., there are $a \in \mathcal{S}$, $b \in A$, bRa and $\nexists c \in \mathcal{S}$ s.t. cRb. Because $\mathcal{S}$ is conflict free, we know that $b \notin \mathcal{S}$. Let $C \in P$ be the cell s.t. $a \in C$. Because $\mathcal{S} \cap C$ is admissible w.r.t. the framework $\langle C', R \downarrow C' \rangle$, we have that $b \notin C'$. By Definition 7 we have that $\forall b' \in (C \setminus C')$, $\exists c' \in \mathcal{S}$ s.t. $c'Rb'$. Therefore, $b \notin (C \setminus C')$ and so $b \in A \setminus C$. By Definition 8 it now follows that $a \notin C''$. But then, because we have that $C'' \subseteq C' \subseteq C$ and $\mathcal{S} \cap (C \setminus C'') = \emptyset$, it follows that $a \notin \mathcal{S}$. Contradiction. $\square$

Proposition 3. *For any AF and $\mathbb{S} = \langle P, T \rangle$, if $\forall C \in P$, $T(C)$ satisfies completeness, then $\forall \mathcal{S} \in \mathcal{E}_{srt}(AF, \mathbb{S})$, $\mathcal{S}$ is complete.*

Proof. Let $AF = \langle A, R \rangle$, $\mathbb{S} = \langle P, T \rangle$, and $\mathcal{S} \in \mathcal{E}_{srt}(AF, \mathbb{S})$. Suppose $\mathcal{S}$ defends a from A. We need to show that $a \in \mathcal{S}$ (i.e. that $\mathcal{S}$ is complete). Let $C \in P$ be the cell s.t. $a \in C$. First we show that $\mathcal{S} \cap C'$ defends a from C'. Let $b \in C'$ be an argument from which $\mathcal{S} \cap C'$ needs to defend a, i.e. bRa. Because $\mathcal{S}$ defends a from A, there is a $c \in \mathcal{S}$ s.t. cRb. There are two possibilities: either $c \in C'$ or $c \notin C'$. Suppose $c \notin C'$. Then by definition 9, $c \notin C$ and by definition 7, $b \notin C'$. Contradiction. It follows that $c \in C'$, and that $\mathcal{S} \cap C'$ defends a from b. Because this holds for any $b \in C'$, we have that $\mathcal{S} \cap C'$ defends a from C'. Finally, from definition 9, and from the assumption that $T(C)$ is complete, it follows that (for any $C'' \subseteq C'$) $a \in \mathcal{S}$. $\square$

These properties are highly desirable in a multi-sorted argumentation framework, because they allow us to guarantee that the properties which hold for the standard Dung framework, are preserved in the multi-sorted one.

Consider now the case where the sorting associates all cells with the same semantics. We call this the *uniform* case. A natural question to ask is whether the set of sorted extensions will then be equivalent to the set of extensions of the framework evaluated under this semantics in the conventional way. We formalize this property as follows.

Definition 10 (Uniform Case Extension Equivalence). *Let $AF = \langle A, R \rangle$ and $\mathbb{S} = \langle \{C_1, \ldots, C_n\}, T \rangle$. Uniform case equivalence holds if and only if*

$$T(C_1) = \ldots = T(C_n) = x \text{ implies } \mathcal{E}_{srt}(AF, \mathbb{S}) = \mathcal{E}_x(AF, A)$$

This property does not hold in all the cases. Consider the following example:

Example 5. Let $AF = \langle \{a, b\}, R \rangle$, where aRb, bRa; and $\mathbb{S} = \langle \{C_1, C_2\}, T \rangle$, where $C_1 = \{a\}$, $C_2 = \{b\}$ and $T(C_1) = T(C_2) = gr$. The grounded extension of AF is $\emptyset$. We now show that $\emptyset \notin \mathcal{E}_{srt}(AF, \mathbb{S})$: we have $C_1' = \{a\}$, and $C_2' = \{b\}$ (no argument is defeated); and $C_1'' = \emptyset$, and $C_2'' = \emptyset$ (both arguments are undefended). We have that $\mathcal{S} \cap C_1 = \emptyset$ and $\emptyset \notin \mathcal{E}_{gr}(\langle C_1', R \downarrow C_1' \rangle, C_1'')$ (the grounded extension of $\langle C_1', R \downarrow C_1' \rangle$ is not a subset of C_1''). It follows that $\emptyset \notin \mathcal{E}_{srt}(AF, \mathbb{S})$.

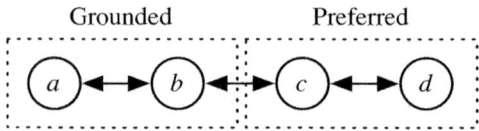

Fig. 3. A sorted argumentation framework

The reason why the uniform case extension equivalence does not hold is the following: every cell is evaluated separately, and the separate evaluation of a cell C under a semantics x may lead to a result that is different from the result of evaluating the complete framework under semantics x. Consider for instance to adopt multi-sorted argumentation to model the merging of the argumentation frameworks of single agents. Even if these agents adopt the same semantics, the evaluation of their single frameworks may lead to different extensions of the merged framework. However, if the multi-sorted framework is used in a context where this property is required to hold, then the equivalence can be guaranteed by replacing the cells associated with the same semantics with their union.

Finally, we underline that, given a cell associated with a certain semantics, say grounded, a sorted extension may not represent a grounded evaluation of the arguments in this cell, when we consider this cell in isolation. Consider the following example.

Example 6. The framework shown in Figure 3 is formally represented by $AF = \langle \{a, b, c, d\}, R\rangle$, where aRb, bRa, bRc, cRb, cRd and dRc; and a sorting $\langle \{C_1, C_2\}, T\rangle$, where $C_1 = \{a, b\}$, $C_2 = \{c, d\}$, and $T(C_1) = gr$, and $T(C_2) = pr$.

Consider the extension $\mathcal{S} = \{a, c\}$. Note that a is accepted, while the cell $\{a, b\}$ is associated with the grounded semantics. Let us check if $\mathcal{S}$ satisfies the conditions for being a sorted extension.

- Given $\mathcal{S}$, we have $C_1' = \{a\}$, and $C_2' = \{c, d\}$ (b is defeated by c); and $C_1'' = \{a\}$, and $C_2'' = \{c, d\}$ (no argument is undefended). We have that $\mathcal{S} \cap C_1 = \{a\}$, and $\{a\} \in \mathcal{E}_{gr}(\langle C_1', R \downarrow C_1'\rangle, C_1'')$, and $\mathcal{S} \cap C_2 = \{c\}$, and $\{c\} \in \mathcal{E}_{pr}(\langle C_2', R \downarrow C_2'\rangle, C_2'')$. It follows that $\mathcal{S} \in \mathcal{E}_{srt}(AF, \mathbb{S})$.

In the example above, selecting c to be accepted accords with the preferred evaluation of the cell $\{c, d\}$. But given this selection, the only complete extension is $\{a, c\}$. Note also that the extension $\{a\}$ is in fact a grounded extension for the subframework $\langle C_1', R \downarrow C_1'\rangle$, where $C_1' = \{a\}$. Consider again the informal example concerning the merging of the single frameworks of the agents. We can note that the merging of the single frameworks may lead to an evaluation such that the arguments accepted under a particular semantics, are then not accepted into the merged framework in the same semantics.

If the multi-sorted framework is used in a context where this behavior needs to be avoided, then a possible way to deal with this behavior is to apply a selection criteria for extensions based on a notion of preference. For example, another sorted extension of the framework described above is $\{d\}$. If actual groundedness for the cell C_1 is important, then this extension would be preferred over $\{a, c\}$.

5   The Modal Fibring Approach

A different representation for multi-sorted argumentation is to express it as a special case of fibring of modal argumentation frameworks. In this perspective, we exploit connections with modal logic, interpreting inter-cell attacks in terms of accessibility relations. Following Barringer and Gabbay [4], we represent argumentation subframeworks as possible worlds in a Kripke structure; moreover, we compute sorted extensions as models of the Kripke structure. We can also apply semantic-based criteria to select desired extensions over the set of possible ones.

Definition 11 (Modal Argumentation Framework). *Let* $\mathcal{AF} = \langle A, R \rangle$ *be an argumentation framework. A modal argumentation framework MAF is a tuple* $\langle A, R, MA, MR, S \rangle$ *where MA is a set of meta-arguments, $MR \subseteq MA \times MA \cup MA \times A$, and $S \in \{cf, ad, co, gr, pr\}$.*

We enrich each sub-framework with meta-arguments, which represent some properties of the original set of arguments (such as the property of *being attacked*). We will refer to the original set of arguments as *actual arguments*.

   We adapt Kripke possible world semantics to the case where possible worlds are argumentation frameworks related by an accessibility relation: argumentation frameworks are possible worlds in a Kripke structure, and modalities are applied to arguments. Thus, $\Diamond \alpha$ in a framework/world w is interpreted as a *possible attack* in the sense that there is a framework/world w' accessible from w, in which α is a justified argument.

Definition 12 (Distributed Argumentation Framework). *A distributed argumentation framework DAF is a tuple* $\langle W, AR \rangle$ *where W is a set of modal argumentation frameworks and $AR \subseteq W \times W$.*

The core idea here is to use modal meta-arguments [11] as pointers between arguments in different worlds, so that we do not loose information about attacks between arguments in different subframeworks; modal relations act as consistency constraints between modal meta-arguments and actual arguments and can be used to ensure global consistency over the justification statuses of arguments in different worlds.

Definition 13 (Sorting-based DAF). *Let* $\mathcal{AF} = \langle A, R \rangle$ *and* $\mathbb{S} = \langle P, T \rangle$, *with* $P = \{C_1, \ldots, C_n\}$. *For each $i = 1 \ldots n$, let $M_i = \langle C_i, R \downarrow C_i, MA_i, MR_i, T(C_i) \rangle$. A DAF $\langle W, AR \rangle$ is a $\mathbb{S}$-based DAF if and only if the following conditions are satisfied:*

1. $W = \{M_1, .., M_n\}$
2. *For any C_i, C_j s.t. $i \neq j$, $\exists a \in C_i, b \in C_j$ s.t. aRb if and only if*
   (a) $\{\Diamond a, x_a\} \subseteq MA_j$, *and* $\{(\Diamond a, x_a), (x_a, \Diamond a), (\Diamond a, b)\} \subseteq MR_j$, *and*
   (b) $(b, a) \in AR$

In the following examples, we show how to construct a sorting-based DAF from the framework of Example 4 and we provide an example of the computation of sorted extensions using a sorting-based DAF created from Example 6.

Example 7. Let $\mathcal{AF} = \langle A, R \rangle$ and $\mathbb{S} = \{P, T\}$ be the framework and sorting of Example 6 (shown in Figure 3). The corresponding sorting-based $DAF = \langle W, AR \rangle$, visualized in Figure 4. We have $W = \{M_1, M_2\}$ with $M_1 = \{A, R, MA_1, MR_1, pr\}$, $MA_1 = \{a, b, c, d, x_e, \Diamond e\}$ and $MR_1 = \{(a, b), (b, a), (a, c), (c, a), (b, c), (c, b), (b, d),$ $(d, b), (c, d), (d, c), (x_e, \Diamond e), (\Diamond e, x_e), (\Diamond e, d)\}$. $M_2 = \{A, R, MA_2, MR_2, gr\}$, $MA_2 = \{e, f\}$ and $MR_2 = \{(e, f), (f, e)\}$. $AR = \{(M_1, M_2)\}$. Note how, in Figure 4, attacks across cells are not part of the DAF; the fact that e attacks d is conveyed by the modal meta-arguments $\Diamond e$ attacking d.

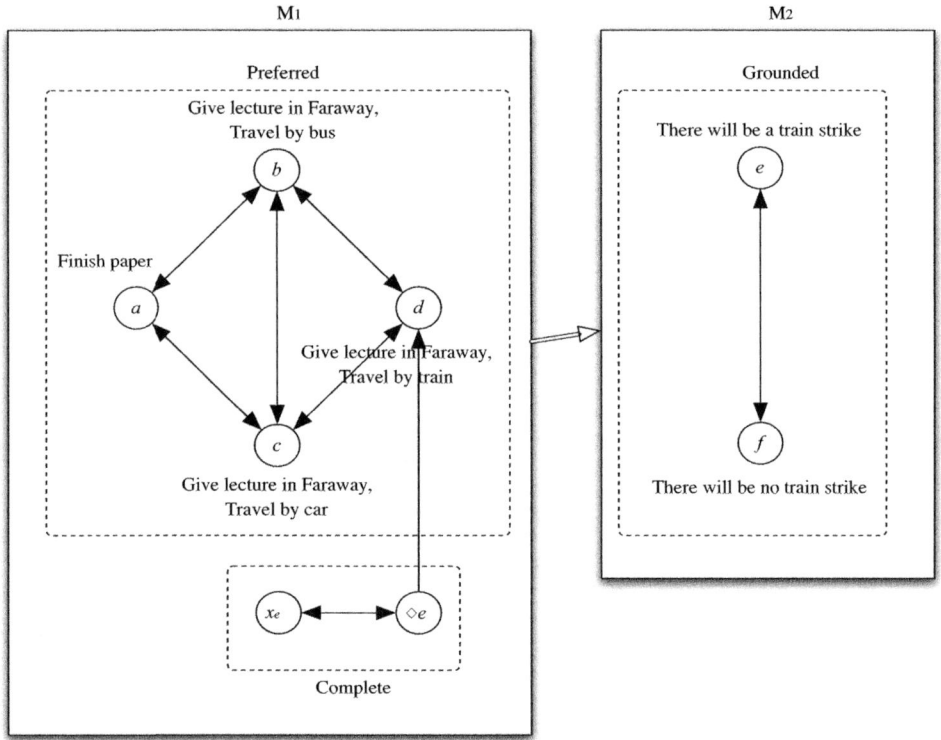

Fig. 4. The sorting-based DAF corresponding to the framework in Figure 1

We now demonstrate the computation of sorted extensions using the sorting-based DAF. Whenever there is a requirement about a subframework to have a specific semantics-based extension we convert it into a requirement of extensions on the actual arguments in the model world; complete extensions are computed on the meta-arguments for the sake of generality.

Every complete multi-sorted extension on the whole framework is the union over the extensions of the subframeworks.

1. For each $M_i \in \{M_1, M_2\}$ (with $M_i = \langle C_i, R \downarrow C_i, MA_i, MR_i, T(C_i) \rangle$).
   (a) Compute the complete extensions of $\langle MA_i, MR_i \rangle$ (i.e., of the meta-arguments of the subframework/world)
   (b) For each admissible extension, compute the possible extensions of the actual arguments of the subframework/world according to the semantics specified by $T(C_i)$.
2. For each M_i, we thus obtain a set of extensions. Each union of these extensions on the condition that it satisfies the *consistency check*, is a sorted extension of the modal framework. Projecting these extensions over the actual arguments (i.e., removing the meta-arguments) gives an actual sorted extension.

The consistency check of an extension involves checking whether the information conveyed by the modal meta-arguments is consistent with the justification statuses of the actual arguments to which they correspond. For example, let e_1 and e_2 be extensions of M_1 and M_2, obtained as described above. We then have that the union $e_1 \cup e_2$ satisfies the consistency check if and only if,

$$\Diamond \alpha \in e_i \leftrightarrow \alpha \in e_j \wedge \Diamond \beta \in e_j \leftrightarrow \beta \in e_i$$

We now exemplify the computation of the multi-sorted extension on an easier framework, visualized in in Figure 5.

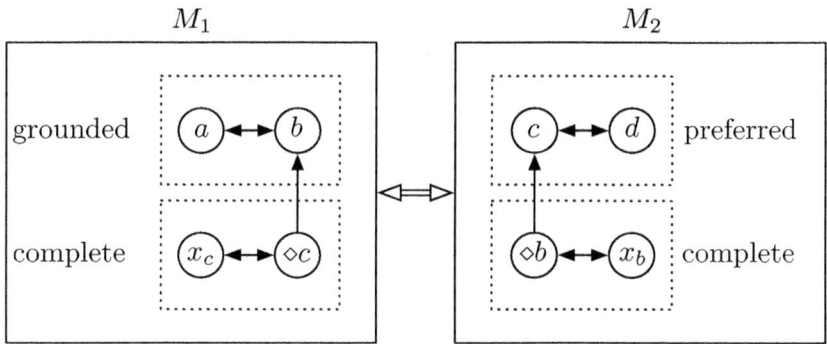

Fig. 5. The sorting-based DAF corresponding to the framework in Figure 3

We do the following:

- The possible complete extensions of the meta-arguments in the subframework/world M_1 are $\{\emptyset, \{x_c\}, \{\Diamond c\}\}$. Then, starting from each one of them, we compute the grounded extension over the actual arguments of M_1. These extensions are: $\emptyset$, $\{x_c\}$ and $\{\Diamond c, a\}$.

- Similarly, we compute the complete extensions over the meta-arguments in the subframework/world M_2 and then, for each one of them, the preferred extensions over the actual arguments of M_2: from $\emptyset$ we get $\{c\}$ and $\{d\}$, from $\{\Diamond b\}$ we get $\{\Diamond b, d\}$ and from $\{x_b\}$ we get $\{x_b, c\}$ and $\{x_b, d\}$.
- We compute the sorted extension by computing the cartesian product over the two sets of extensions (a set of possible extensions for each subframework/world) and then selecting the ones that satisfy the consistency check. So, for instance, the $\{\emptyset\}$ extension from the first world can not be merged with the $\{c\}$ extension from the second world since $\Diamond c$ does not belong to the first one. The resulting set of consistent extensions consists of $\{d\}$, $\{x_b, d\}$, $\{x_c, d\}$, $\{x_c, x_b, d\}$, $\{\Diamond c, a, c\}$ and $\{\Diamond c, a, x_b, c\}$. Projecting over the actual arguments, we get $\{\{d\}, \{c, a\}\}$.

Another property is that, in uniform cases (every world is associated with the same semantics) we can impose a preference criteria over the admissible multi-sorted extensions in order to get the same semantics we would have by evaluating the whole framework using that semantics. For instance, in Example 2, the possible multi-sorted extensions we compute by means of modal fibring are $\{\emptyset, \{a\}, \{b\}\}$. Note that we do not allow $\{a, b\}$ to be a multi-sorted extension: this is because we do not have both a and $\Diamond b$ in the possible extensions of the first world nor we have b and $\Diamond a$ in the possible extensions of the second world: basically, by means of modal pointers, we know that those two arguments attack each other and we prevent them from being part of the same multi-sorted extension. We can impose a semantics-based selection criteria: since all worlds are associated with grounded semantics, we may select the minimal extension over the set of multi-sorted extensions $\{\emptyset, \{a\}, \{b\}\}$: this is $\emptyset$, which is exactly the grounded extension of the whole framework. Selecting the maximal sets w.r.t. set inclusion we get $\{\{a\}, \{b\}\}$, which are the preferred extensions of the whole framework.

Modal fibring is another way to deal with multi-sorted argumentation frameworks. Compared with the main approach discussed in this paper, the difference is that it does not require an extension S as a parameter, but the framework has to undergo a precise transformation process: adding meta-arguments, removing cross-cell attacks, introducing accessibility relations. The property of uniform case extension equivalence holds in this approach. In multi-agent systems, the modal meta-argument is intended as a call to a remote procedure or a generic communication process with another agent, the owner of the referenced argument whose justification status the caller wants to know.

6   Related Work

The proposal that is most related to ours comes from Prakken [13], from which we also took our running example. He proposes an argument-based semantics that combines grounded and preferred semantics. The motivation, as we discussed in the introduction, is that reasoning about beliefs should be skeptical, while reasoning about actions should be credulous. Prakken's formalism can be seen as a special case of ours: there are just two cells: a preferred cell, containing practical arguments; and a grounded cell, containing epistemic arguments. Moreover, arguments in the preferred cell do not attack arguments in the grounded cell. This reflects the principle that no *Is* should be derived from an *Ought*. Other than that, the formalism takes an approach similar to ours: an extension of a framework $\mathcal{AF}$ is the preferred extension of the framework obtained by removing all arguments not defended by the epistemic part of the grounded extension of $\mathcal{AF}$ (of course, this includes the grounded extension of $\mathcal{AF}$ itself).

An interesting feature of Prakken's formalism is a dialectical proof procedure which is sound and complete with respect to the 2-sorted semantics. This proof procedure combines previously developed proof procedures for the skeptical grounded and credulous preferred semantics (see e.g., [12]). It would be interesting to see whether a generalized dialectical proof procedure could be developed for our semantics. This would, of course, depend on the existence of dialectical proof procedures for the semantics associated with the individual cells.

Another related formalism comes from Brewka & Eiter [6]. They propose a framework for group argumentation, which they call argument context systems. It allows a collection of abstract argument systems to interact via mediators, where a mediator consists of so called bridge rules that associate arguments from one framework with a context to another framework. A context for a framework consists of a set of expressions that determine certain properties of that framework. One framework may then decide on these properties for another framework through the acceptance status of the arguments that appear in the body of the bridge rules. Among the properties controlled by the context are values and preferences, i.e., the framework supports value based and preference based argumentation [1,5]. Another property is the acceptance status of an argument. This is effectuated through an extra argument *def*, that may invalidate or validate an argument, by attacking it or attacking its attackers. The resulting framework allows the interaction between different frameworks in the argument context system, where one framework may decide about values, preferences and argument acceptance status of other frameworks.

Like in our work, different frameworks may be evaluated under different semantics. Moreover, the semantics under which a framework is evaluated, is also part of the framework's context. This means that, in addition to values, preferences and acceptance status, the semantics under which a framework is evaluated is also a property about which another framework may decide. Of course, this goes beyond the expressivity of our system. On the other hand, different cells of a sorting in our system are part of the same framework, and may interact through attacks. Brewka & Eiter's system does not allow different frameworks in the same argument context system to this. This may be simulated by bridge rules that validate and invalidate arguments, but only partially. Their approach does not account for the distinction between defeated and undefended arguments.

Other related work includes Amgoud & Prade [2], who introduce explanatory, rewards and threats arguments for negotiation dialogues. In practical reasoning, Rotstein et al. [14] propose different types of arguments to represent categorized domain information, like belief, goals or plans. These works, however, do not explicitly apply different semantics to the different types of arguments they define.

7 Conclusion and Future Work

We have presented a theory of multi-sorted argumentation, that generalizes Dung's theory of abstract argumentation in that it allows different parts of a framework to be evaluated under different semantics. We have proven some basic properties, namely the preservation of conflict-freeness, admissibility and completeness. Moreover, we have analyzed the behavior of the multi-sorted framework in the cases where the same semantics is used to evaluate all the cells of the framework, or where the arguments are not accepted in the framework using the same semantics applied to evaluate the cells.

We justify the introduction of a multi-sorted argumentation framework by using a running example from Prakken [13]. In this example, some arguments pertain to actions, and some others pertain to beliefs about the world. As argued by Prakken [13] practical arguments and epistemic arguments have to be evaluated in a different way. We propose to perform this evaluation using a multi-sorted framework.

The modal fibring approach adds another interesting angle to our theory. The fact that multi-sorted argumentation is expressible in modal argumentation frameworks demonstrates the generality of modal argumentation. We expect that modal argumentation will be a useful framework to investigate more sophisticated forms of multi-sorted argumentation.

There is much work still to be done, on all the aspects described above. First of all, a further generalization is possible if we make some of the assumptions that we made optional. For example, instead of a strict partitioning of the framework, we could allow overlapping subsets. This is natural, because the same argument may be put forward by different agents, each associated with a different semantics.

Secondly, we have applied our theory only to some small examples. It will be interesting to apply it to real-world examples, and to compare it with other approaches to multi-agent argumentation and reasoning about trust.

Third, we are applying our theory to different challenges in multiagent systems. One of the possible applications is bounded reasoning in multi-agent systems: dividing a framework into different sets could facilitate a stepwise evaluation of smaller parts of a larger framework. In addition, arguments that are not the focus of a particular issue, could be evaluated using a computationally cheaper semantics. For example, a 'don't care' attitude towards a set of arguments could result in only requiring conflict-freeness for this set. Another application is, as mentioned in the paper, the merging of the argumentation frameworks of single agents into a common framework, in order to allow an easier collaboration among the agents.

Forth, we aim to redefine multi-sorted argumentation in terms of argument labelling [8], instead of argument semantics. The labelling approach is widely adopted in the argumentation community, and it may allow a simpler representation of the sorted extension.

References

1. Amgoud, L., Cayrol, C.: On the acceptability of arguments in preference-based argumentation. In: Cooper, G.F., Moral, S. (eds.) UAI 1998: Proceedings of the Fourteenth Conference on Uncertainty in Artificial Intelligence, pp. 1–7. Morgan Kaufmann (1998)
2. Amgoud, L., Prade, H.: Handling threats, rewards, and explanatory arguments in a unified setting. Int. J. Intell. Syst. 20(12), 1195–1218 (2005)
3. Baroni, P., Giacomin, M., Guida, G.: SCC-recursiveness: a general schema for argumentation semantics. Artificial Intelligence 168(1-2), 162–210 (2005)
4. Barringer, H., Gabbay, D.M.: Modal and Temporal Argumentation Networks. In: Manna, Z., Peled, D.A. (eds.) Time for Verification. LNCS, vol. 6200, pp. 1–25. Springer, Heidelberg (2010)
5. Bench-Capon, T.J.M.: Value-based argumentation frameworks. In: Benferhat, S., Giunchiglia, E. (eds.) NMR, pp. 443–454 (2002)
6. Brewka, G., Eiter, T.: Argumentation Context Systems: A Framework for Abstract Group Argumentation. In: Erdem, E., Lin, F., Schaub, T. (eds.) LPNMR 2009. LNCS, vol. 5753, pp. 44–57. Springer, Heidelberg (2009)
7. Caminada, M.: Semi-stable semantics. In: Computational Models of Argument; Proceedings of COMMA, pp. 121–130 (2006)
8. Caminada, M.: On the Issue of Reinstatement in Argumentation. In: Fisher, M., van der Hoek, W., Konev, B., Lisitsa, A. (eds.) JELIA 2006. LNCS (LNAI), vol. 4160, pp. 111–123. Springer, Heidelberg (2006)
9. Coste-Marquis, S., Devred, C., Marquis, P.: Prudent semantics for argumentation frameworks (2005)
10. Dung, P.M.: On the acceptability of arguments and its fundamental role in non-monotonic reasoning, logic programming and n-person games. Artif. Intell. 77(2), 321–358 (1995)
11. Gabbay, D.M.: Fibring argumentation frames. Studia Logica 93(2-3), 231–295 (2009)
12. Modgil, S., Caminada, M.: Proof theories and algorithms for abstract argumentation frameworks. In: Argumentation in Artificial Intelligence, pp. 105–129 (2009)
13. Prakken, H.: Combining sceptical epistemic reasoning with credulous practical reasoning. In: Dunne, P.E., Bench-Capon, T.J.M. (eds.) COMMA. Frontiers in Artificial Intelligence and Applications, vol. 144, pp. 311–322. IOS Press (2006)
14. Rotstein, N.D., García, A.J., Simari, G.R.: Reasoning from desires to intentions: A dialectical framework. In: Proceedings of the Twenty-Second AAAI Conference on Artificial Intelligence (AAAI 2007), pp. 136–141. AAAI Press (2007)

Conditional Labelling
for Abstract Argumentation

Guido Boella[1], Dov M. Gabbay[2], Alan Perotti[1],
Leendert van der Torre[3], and Serena Villata[4]

[1] Dipartimento di Informatica, Università di Torino
{guido,perotti}@di.unito.it
[2] King's College London
dov.gabbay@kcl.ac.uk
[3] ICR, University of Luxembourg
leon.vandertorre@uni.lu
[4] INRIA, Sophia Antipolis
serena.villata@inria.fr

Abstract. Agents engage in dialogues having as goals to make some arguments acceptable or unacceptable. To do so they may put forward arguments, adding them to the argumentation framework. Argumentation semantics can relate a change in the framework to the resulting extensions but it is not clear, given an argumentation framework and a desired acceptance state for a given set of arguments, which further arguments should be added in order to achieve those justification statuses. Our methodology, called *conditional labelling*, is based on argument labelling and assigns to each argument three propositional formulae. These formulae describe which arguments should be attacked by the agent in order to get a particular argument *in*, *out*, or *undecided*, respectively. Given a conditional labelling, the agents have a full knowledge about the consequences of the attacks they may raise on the acceptability of each argument without having to recompute the overall labelling of the framework for each possible set of attack they may raise.

1 Introduction

Agents engage in dialogues having as goals to make some arguments acceptable or unacceptable: for instance, *agent A wins the auction* or *agent B is proven guilty*. At each turn, an agent owns a set of possible arguments she can add to the framework: each addition of further arguments to the framework is called a *move*. Argumentation semantics allow us to relate the introduction of a new argument (a *move*) to the resulting justification status of an argument (the *goal*): for instance, *if you defeat argument α then argument β will be labeled undec*. What is missing is a mechanism for making inferences from goals to moves: suppose an agent wants to make an argument β *undec*. How can she compute which arguments to add in order to achieve this goal? What she can do is to try and simulate the introduction of every possible argument she owns to the

S. Modgil, N. Oren, and F. Toni (Eds.): TAFA 2011, LNAI 7132, pp. 232–248, 2012.
© Springer-Verlag Berlin Heidelberg 2012

framework and then compute β's resulting label, comparing it to her goal. Beside this exhaustive approach there is no way, so far, for an agent to know which move to make in order to achieve her goal. Since reaching a goal may require the insertion of several arguments, the complexity of the exhaustive approach is exponential (cardinality of the powerset) over the number of arguments an agent can add to the framework.

The research question of the paper is:

– How to change an abstract argumentation framework, by introducing new arguments and their associated attacks, in order to have one or more arguments accepted or rejected?

Suppose that two agents, Ag_1 and Ag_2, initiate a dialogue. Ag_1 proposes argument a, as depicted in Figure 1.1. Assume that Ag_2 wants to *defeat* Ag_1's argument but we have that argument a is *in*, and the only way to have it labelled *out* is to attack it. Thus, Ag_2 attacks a with her new argument b, *defeating* it. At this turn, as shown in Figure 1.2, it is up to Ag_1 to decide how to proceed in the dialogue. She wants to have her argument a accepted, so she puts forward argument c which attacks b, obtaining the framework in Figure 1.3. In this basic framework, it is straightforward to see which arguments the agents should attack in order to get their arguments accepted. In more complex argumentation frameworks, where also cycles are involved, it is less simple to detect these arguments. Consider the framework depicted in Figure 2: it contains loops and multiple attacks. Suppose that an agent wants to defend argument i: it is not intuitive at all to see which potential modifications of the framework allow her to do that. Moreover, if she has a set A^{ag} of arguments she may add to the framework, she may have to run $2^{|A^{ag}|}$ tests in order to find out whether she can defend i, thus making this process' complexity dependent on the number of possible moves she has.

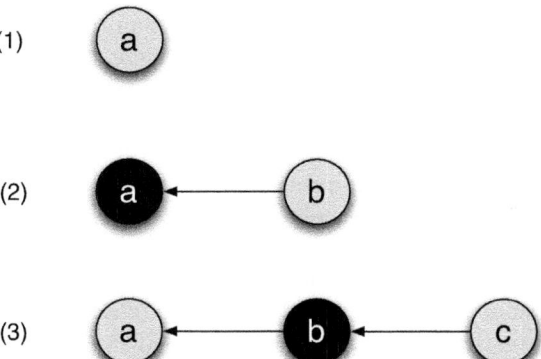

(1)

(2)

(3)

Fig. 1. An argumentation framework with a basic reinstatement. Arguments labelled *in* are depicted as grey nodes, arguments with label *out* are depicted as black nodes.

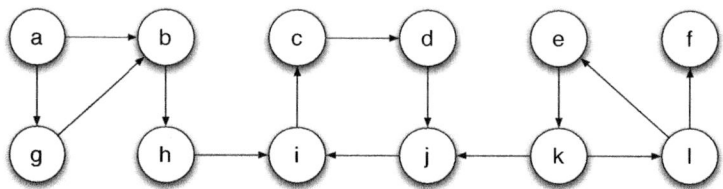

Fig. 2. A more complex argumentation framework

Thus, the research question breaks down into the following subquestions:

1. What kind of information can we associate to each argument concerning its possible justification statuses depending on the acceptability of other arguments in the framework?
2. How to compute this information in an efficient way?

We deal with abstract argumentation frameworks [4], where the internal structure of the arguments is left unspecified. We are inspired by Caminada's labelling [3], which assigns to each argument a label *in, out, undec*, and we extend this idea by assigning a triple of propositional formulae, called *conditional lables*, to every argument in the framework. These formulae are a guide in the dialogic process and suggest which move should be made next. Note that these formulae (and the algorithmic process to compute them) are in no way related to the number of agents: our approach does not depend on the number of argumenting agents and we apply it to a two-agent scenario for the sake of explanation. Consider the framework of Figure 1.3, the conditional label of *a* for making it accepted is the emptyset because *a* is already *in* and no "move" is needed to get it accepted. The conditional label, instead, for making *a* unaccepted is $a \vee c$, because *a* can be *defeat*ed by *defeat*ing *a* itself or *c*. Depending on the further arguments at her disposal, an agent may not be able to directly *defeat* an argument and therefore giving all alternatives is required. Conditional labelling assigns a conditional label to each abstract argument in the framework, even if the framework involves one or more cycles.

The implementation of the algorithm of conditional labelling deals with a number of complexity issues, mostly due to loops in the argumentation frameworks: some preprocessing techniques allow to speed up the performances displayed by a straightforward implementation of the conditional labels' theoretical definition.

In this paper, we are interested in introducing the basic ideas of the conditional labelling and explain it using a number of examples. We do not treat belief revision, and we restrict our examples to grounded semantics.

The paper is organized as follows: Section 2 provides the basic concepts of argumentation theory, Section 3 introduces the conditional evaluation of arguments, Section 4 discusses an algorithmical definition of the conditional labelling and some possible optimizations for the implementation. Finally, some conclusions are drawn.

2   Background

We provide the basic concepts and insights of Dung's abstract argumentation [4].

Definition 1. *(Abstract argumentation framework) An abstract argumentation framework is a pair $\langle \mathcal{A}, \rightarrow \rangle$. $\mathcal{A}$ is a set of elements called arguments and $\rightarrow \subseteq \mathcal{A} \times \mathcal{A}$ is a binary relation called attack. We say that an argument A_i attacks an argument A_j if and only if $(A_i, A_j) \in \rightarrow$.*

Definition 2. *(Conflict-free, Defence) Let $C \subseteq \mathcal{A}$. A set C is conflict-free if and only if there exist no $A_i, A_j \in C$ such that $A_i \rightarrow A_j$. A set C defends an argument A_i if and only if for each argument $A_j \in A$ if A_j attacks A_i then there exists $A_k \in C$ such that A_k attacks A_j.*

Definition 3. *(Acceptability semantics) Let C be a conflict-free set of arguments, and let $\mathcal{D} : 2^{\mathcal{A}} \mapsto 2^{\mathcal{A}}$ be a function such that $\mathcal{D}(C) = \{A | C \text{ defends } A\}$.*

- *C is admissible if and only if $C \subseteq \mathcal{D}(C)$.*
- *C is a complete extension if and only if $C = \mathcal{D}(C)$.*
- *C is a grounded extension if and only if it is the smallest (w.r.t. set inclusion) complete extension.*
- *C is a preferred extension if and only if it is a maximal (w.r.t. set inclusion) complete extension.*
- *C is a stable extension if and only if it is a preferred extension that attacks all arguments in $\mathcal{A} \setminus C$.*

The concepts of admissibility, as well as those of Dung's semantics are originally stated in terms of sets of arguments. It is equal to express these concepts using argument *labeling*. This approach has been proposed firstly by Jakobovits and Vermeir [5] and then by Caminada [3] with the aim to provide quality postulates for dealing with the reinstatement of arguments. The simplest example of reinstatement is: argument A_1 attacks argument A_2 and argument A_2 attacks argument A_3. We have that argument A_1 reinstates argument A_3, i.e., it makes argument A_3 accepted by attacking the attacker of A_3. In a reinstatement labeling [3], an argument is labeled *in* if all its attackers are labeled *out* and it is labeled *out* if it has at least an attacker which is labeled *in*.

Definition 4. *(AF-labeling) Let $\langle \mathcal{A}, \rightarrow \rangle$ be an abstract argumentation framework. An AF-labeling is a total function $lab : \mathcal{A} \rightarrow \{in, out, undec\}$. We define $in(lab) = \{A_i \in \mathcal{A} | lab(A_i) = in\}$, $out(lab) = \{A_i \in \mathcal{A} | lab(A_i) = out\}$, $undec(lab) = \{A_i \in \mathcal{A} | lab(A_i) = undec\}$.*

Definition 5. *(Reinstatement labeling) Let lab be an AF-labeling. We say that lab is a reinstatement labeling if and only if it satisfies the following:*

- *$\forall A_i \in \mathcal{A} : (lab(A_i) = out \equiv \exists A_j \in \mathcal{A} : (A_j \rightarrow A_i \wedge lab(A_j) = in))$ and*
- *$\forall A_i \in \mathcal{A} : (lab(A_i) = in \equiv \forall A_j \in \mathcal{A} : (A_j \rightarrow A_i \supset lab(A_j) = out))$ and*
- *$\forall A_i \in \mathcal{A} : (lab(A_i) = undec \equiv \exists A_j \in \mathcal{A} : (A_j \rightarrow A_i \wedge \neg(lab(A_j) = out)) \wedge \nexists A_k \in \mathcal{A} : (A_k \rightarrow A_i \wedge lab(A_k) = in).$*

3   Conditional Labels

Our goal is to enrich each argument with some information about his *vulnerability*, i.e., we want to know how this argument could be successfully (even if indirectly) attacked, defended or made undecided. We purposely restrict our attention to argument defeating, due to two considerations: first of all, attacks are not resources but consequences of the insertion of the arguments and given a couple of arguments the existence of attacks between them is determined and not subject to strategic moves of agents. In second place, the building of an argumentation framework is a monotonic process and arguments can be defeated with new arguments rather than removed from the framework. Hence our proposal is to attach **three** formulae to each argument, meaning respectively.

- Which arguments should be attacked in order to have this argument labelled *in*?
- Which arguments should be attacked in order to have this argument labelled *out*?
- Which arguments should be attacked in order to have this argument labelled *undec*?

Given an argumentation framework $\langle A, R \rangle$, we associate to each argument α three formulae: $\alpha^+, \alpha^-, \alpha^?$. We indicate a generic formula associated to argument α as α^*. The language of the formulae is the same:

Definition 6. *(Language of conditional labels)*

- *if $\beta \in A$, β° is a formula.*
- *$\top$ and $\bot$ are formulae*
- *if α_1^* and α_2^* are formulae, also $\alpha_1^* \wedge \alpha_2^*$ and $\alpha_1^* \vee \alpha_2^*$ are.*

We will refer to α^+ (respectively: α^-, $\alpha^?$) formulae as *green* (*red, grey*) formulae.

    The interpretation of the formulae is: a *green* formula α^+, if satisfied, guarantees that the related argument α is accepted (labelled *in*). The same holds for *red* formulae for *out* labels and *grey* formulae for *undec* labels respectively. The atoms of those formulae are argument names β° or the special values $\top, \bot$.

- β° means *you have to defeat argument β* (to reach your goal)
- $\top$ means *you do not need to do anything* (to reach your goal)
- $\bot$ means *you can not do anything* (to reach your goal)

Figure 3 provides a simple example of a framework with conditional labels.

- Figure 3.1: There is no need to modify the framework in order to achieve a's acceptability (it is already labelled *in*) ($a^+ : \top$); to *defeat* a you have to *defeat* a ($a^- : a^\circ$), you can not make a undecidable by *defeat*ing any combination of the arguments of the framework ($a^? : \bot$).

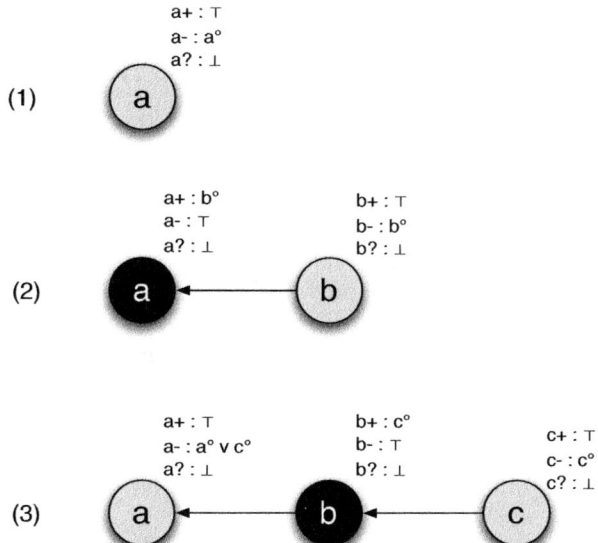

Fig. 3. An argumentation framework with a basic reinstatement and conditional labels

- Figure 3.2: a can be reinstated *defeating* b ($a^+ : b^\circ$), a is already *out* ($a^- : \top$); b is already *in* ($b^+ : \top$) and can only be *defeated* by being directly *defeated* ($b^- : b^\circ$); no argument can be made undecidable by *defeating* any combination of the arguments of the framework ($a^?, b^? : \bot$).
- Figure 3.3: a is *in* ($a^+ : \top$) and can be *defeated* by *defeating* a itself **or** c ($a^- : a^\circ \vee c^\circ$); b is *out* ($b^- : \top$) and can be reinstated *defeating* c ($b^+ : c^\circ$); c is *in* ($c^+ : \top$) and can only be *defeated* by direct (and successful) attack ($c^- : c^\circ$); no argument can be made undecidable by *defeating* any combination of the arguments of the framework ($a^?, b^?, c^? : \bot$).

Now we can introduce a more formal definition of what conditional labels are and what can they be used for.

Definition 7. *(Disjunctive Normal Form, makeset)*
Let Γ be a propositional formula. $dnf(\Gamma)$ is the normalization of Γ in Disjunctive Normal Form.
Let $makeset(\bigvee_i \bigwedge_j \alpha_j^i) = \{\{\alpha_1^1, \alpha_2^1, .., \alpha_p^1\}, \{\alpha_1^2, \alpha_2^2, .., \alpha_q^2\}, .., \{\alpha_1^n, \alpha_2^n, .., \alpha_m^n\}\}$.

The *makeset* function translates a dnf-formula in a set of sets of atoms, where each set corresponds to a conjunctive subformula of the dnf-formula in input. For instance, $dnf(a^\circ \wedge (b^\circ \vee c^\circ)) = (a^\circ \wedge b^\circ) \vee (a^\circ \wedge c^\circ)$ and $makeset((a^\circ \wedge b^\circ) \vee (a^\circ \wedge c^\circ)) = \{\{a, b\}, \{a, c\}\}$.

Definition 8. *(Label of an argument in a framework)*
Let $\mathbb{L}(\alpha, \langle A, R \rangle)$ be the label of argument α in the framework $\langle A, R \rangle$.

Definition 9. *(Defeat)*
Let U be the universe of arguments and $A \subset U$, let $AF = \langle A, R \rangle$ be an abstract argumentation framework and let $\alpha \in U \backslash A$. defeat$(\alpha) = \{\beta \mid \beta \in A, \mathbb{L}(\beta, \langle A \cup \{\alpha\}, R \rangle) = out\}$.

The *defeat* function gives information about which arguments β of a framework are defeated inserting a new argument α into it. For instance, considering the framework in Figure 1.c with $A = \{a\}$, defeat$(b) = \{a\}$, defeat$(c) = \{\emptyset\}$. The definition of *defeat* can be easily extended for sets of arguments: it will point out which arguments of a framework are defeated by inserting a set of new arguments. For instance, considering the framework in Figure 1.c with $A = \{a\}$, defeat$(\{b, c\}) = \{\emptyset\}$

A *move M* is the insertion of a set of arguments into the framework: in the previous example, $\emptyset$, $\{b\}$, $\{c\}$ and $\{b, c\}$ are (possible) moves. Applying a move $M = \{\alpha_1, .., \alpha_n\}$ to a framework $AF = \langle A, R \rangle$ transforms it into a new framework $AF^M = \langle \{A \cup M\}, R \rangle$.

Definition 10. *(Conditional labels' structure)*
A conditional label $\alpha^i : body^i_\alpha$ (where α is an argument, $i \in \{+, -, ?\}$ and $body^i_\alpha$ is a propositional formula) is a relation between a justification status and a set of targets.

Let js (for *justification status*) be this function: $js(+) : in$, $js(-) : out$, $js(?) : undec$. js maps label symbols to the acceptability of arguments. The justification status is expressed by the head of the label: α^i means that α is labelled $js(i)$. The set of targets is expressed by the body of the label, and it consists in a set of sets of argument to defeat.

Definition 11. *(Conditional labels' use)*
Given a framework $AF = \langle A, R \rangle$, an argument $\alpha \in A$ with label $\alpha^i : body^i_\alpha$ and a move M,

$$(defeat(M) \in makeset(dnf(body^i_\alpha))) \Rightarrow \mathbb{L}(\alpha, AF^M) = js(i)$$

This means: when we modify a framework via a move M we can defeat a set of arguments $defeat(M)$. If this set is one of the allowed target sets for the conditional label of an argument α (that is, if this set belongs to $makeset(dnf(body^i_\alpha))$ for some α, i), then the labelling of α in the resulting framework will be the one expressed by the head of the label α^i (that is, $js(i)$).

In the next sections we will explain how to associate labels to arguments. Problems arise when cycles (loops) are introduced in the framework, since they introduce *undecided* labels and the same argument could be given different labels according to different semantics. In this paper we focus on the *grounded* semantics, since it always allows to compute one single labelling. Our approach can be extended to deal with different semantics, but semantics with multiple or no extensions must be handled with care, in particular when investigating about credulous approaches to multiple extensions semantics.

4   Creating Conditional Labels

The formal definition of conditional labels we gave is not constructive and therefore the issue about how to actually compute conditional labels has to be addressed. One of the key aspects of argumentation frameworks is the possibility for arguments to influence their own justification status through loops: this is a *global* property of the framework which is hard to instantiate on a single argument. A first approach could be considering the unfolding of the graph (that is, building a tree rooted in a node of the graph such that each path in the tree corresponds to a (possibly cyclic) path stating from the root node in the graph), but this can not be done for two main reasons: first of all, *breaking* the loops causes an irreparable loss of information (and therefore one could end up computing conditional labels for a completely different framework); secondly, the number of unfolding could be exponential over the number of arguments: in this case, the overall complexity is the same of the exhaustive approach (*try all combinations of attacks and see what is the result*), thus making the whole process pointless.

Our approach consists in assigning to each argument a triple of **local** labels (that is, labels created by only taking into account the attackers of the argument) and then using a substitution mechanism to generate the final labels. The local labels correspond to:

$$a^+ = \bigwedge_{b \; s.t. \; (b,a) \in R} b^-$$

The meaning of this formula is: *in order to ensure a's acceptance, all of a's attackers must be out.*

$$a^- = a^\circ \vee \bigvee_{b \; s.t. \; (b,a) \in R} b^+$$

The meaning of this formula is: *in order to ensure a's rejection, either a is defeated or one of a's attackers is accepted.*

$$a^? = \left(\bigvee_{b \; s.t. \; (b,a) \in R} b^? \right) \wedge \left(\bigwedge_{b \; s.t. \; (b,a) \in R} b^- \vee b^? \right)$$

The meaning of this formula is: *in order to have an argument a undecided, at least one of a's attackers has to be undecided and all of a's attackers must be out or undecided.*

Note that this definition of grounded semantics mirrors Dung's original formulation.

The a° in the second formula means *a has to be defeated* and no substitution is required; b^+, b^- and $b^?$ refer to other formulae and have to be substituted to the actual formulae they refer to.

After this initial definition, the substitution process takes place. It consists in substituting the references to other labels to those labels' actual values.

Simplifications need to be specified:

- $\top \vee \alpha \rightsquigarrow \top$ (you either do nothing or do α: doing nothing is more convenient)
- $\bot \vee \alpha \rightsquigarrow \alpha$ (you can either fail or do α: in order to succeed you have to do α)
- $\top \wedge \alpha \rightsquigarrow \alpha$ (you have to both do nothing and α, therefore α)
- $\bot \wedge \alpha \rightsquigarrow \bot$ (you fail and you have to do α: you still fail)
- $\alpha \wedge \alpha \rightsquigarrow \alpha$
- $\alpha \vee \alpha \rightsquigarrow \alpha$
- $\alpha \vee (\alpha \wedge \beta) \rightsquigarrow \alpha$
- $\alpha \wedge (\alpha \vee \beta) \rightsquigarrow \alpha$

Consider again the framework in Figure 1.3 (reinstatement a-b-c). The initial conditional labels are:

- a^+: $\top$, a^-: a°, $a^?$: $\bot$
- b^+: a^-, b^-: $a^+ \vee b^\circ$, $b^?$: $a^? \wedge (a^? \vee a^-)$
- c^+: b^-, c^-: $b^+ \vee c^\circ$, $c^?$: $b^? \wedge (b^? \vee b^-)$

Substituting in b^* we get:

- b^+: a°, b^-: $\top \vee b^\circ$, $b^?$: $\bot \wedge (\bot \vee a^\circ)$

And after simplifying:

- b^+: a°, b^-: $\top$, $b^?$: $\bot$

Doing the same for c^* we get the conditional labels:

- a^+: $\top$, a^-: a°, $a^?$: $\bot$
- b^+: a°, b^-: $\top$, $b^?$: $\bot$
- c^+: $\top$, c^-: $a^\circ \vee c^\circ$, $c^?$: $\bot$

The conditional labels give us information about the 'static' Caminada labelling of the arguments and also provide us information about what minimal set of arguments we should *defeat* in order to assign a certain label to a certain argument.

In case of loops, new problems arise: the substitution mechanism can end up visiting the same node multiple times, so some termination techniques have to be addressed. Consider, for instance, the framework $AF = \langle \{a\}, \{(a,a)\} \rangle$. Computing the conditional labels without termination techniques we obtain:

- $a^+ : a^- = \underline{a^+ \vee a^\circ} = a^- \vee a^\circ = a^+ \vee a^\circ \vee a^\circ \rightsquigarrow \underline{a^+ \vee a^\circ} = \dots$
- $a^? : \underline{a^? \wedge (a^? \vee a^-)} \rightsquigarrow a^? = \underline{a^? \wedge (a^? \vee a^-)} \rightsquigarrow \dots$

Simplification rules keep the size of formulae under control, but both in a^+ and $a^?$ we end up cycling among the same set of labels without termination. The main consideration is that, according to the definition we have given so far, the substitution process goes on until it reaches unattacked arguments (the only ones which do not require further substitution). But if a framework's component is a loop with no ingoing arcs, this will never happen. Moreover, considering the a^+ label in the previous example, one could notice that both a^+ and a^- appear in the label: this is, intuitively, an unsatisfiable request. Therefore, termination rules have to be applied.

Let $i, j \in \{+, -, ?\}$. If α^i appears in the body of α^j:

- if $i = j =?$, $\alpha^i \rightsquigarrow \top$
- else, $\alpha^i \rightsquigarrow \bot$

We express our termination conditions as simplification rules. The meaning is the following: if, substituting in the body of a conditional formula for an argument α, a conditional formula over the same argument is reached, the argument α belongs to a loop. So in this case the $a^?$ label is satisfied while a^+, a^- are not: if there is no way to give this argument an *in-out* label navigating the whole loop, it is pointless to go through the whole loop again.

Applying these rules to the previous example we get:

- $a^+ : a^- \rightsquigarrow \bot$
- $a^- : a^+ \vee a^\circ \rightsquigarrow \bot \vee a^\circ \rightsquigarrow a^\circ$
- $a^? : a^? \wedge (a^? \vee a^-) \rightsquigarrow a^? \rightsquigarrow \top$

which is exactly what we want to obtain: there is no way to make a *in* ($a^+ : \bot$), a can be directly defeated ($a^- : a^\circ$), a is already *undec* so there is no need to do anything in order to make it *undec* ($a^? : \top$).

We now present some examples of conditional labelling.

Consider the example visualized in Figure 4.1. The basic labels are:

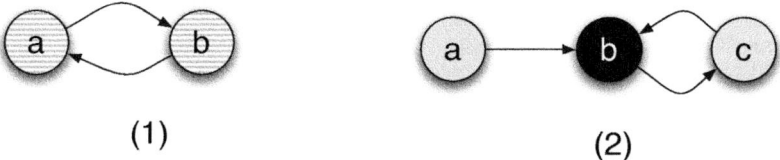

(1)   **(2)**

Fig. 4. Basic frameworks. As in the previous figures, plain grey nodes represent *in* arguments and black nodes represent *out* arguments. *undec* arguments are depicted as dashed grey nodes.

- $a^+ : b^-$, $a^- : b^+ \vee a^\circ$, $a^? : b^?$
- $b^+ : a^-$, $b^- : a^+ \vee b^\circ$, $b^? : a^?$

Solving the labels, for a we get $a^+ : b^\circ$, $a^- : a^\circ$, $a^? : \top$, and this is exactly what we want to obtain.

Consider the example visualized in Figure 4.2. The basic labels are:

- $a^+ : \top$, $a^- : a^\circ$, $a^? : \bot$
- $b^+ : a^- \wedge c^-$, $b^- : a^+ \vee c^+ \vee b^\circ$, $b^? : (a^? \vee c^?) \wedge (a^- \vee a^?) \wedge (c^- \vee c^?)$
- $c^+ : b^-$, $c^- : b^+ \vee c^\circ$, $c^? : b^?$

Consider argument b: it is *out*, but can be labelled *in* if we attack both a and c or *undec* if we attack a (thus activating the $b-c$ loop). We compute the conditional labels in the following way:

b^+ : $a^- \wedge c^-$
$\quad = a \wedge (b^+ \vee c^\circ)$
$\quad = a^\circ \wedge (\bot \vee c^\circ)$
$\quad \rightsquigarrow a^\circ \wedge c^\circ$ (b can be labelled *in* by defeating a and c)
b^- : $a^+ \vee c^+ \vee b^\circ$
$\quad = \top \vee b^- \vee b^\circ$
$\quad = \top \vee \bot \vee b^\circ$
$\quad \rightsquigarrow \top$ (no move is required in order to label b *out*)
$b^?$: $(a^? \vee c^?) \wedge (a^- \vee a^?) \wedge (c^- \vee c^?)$
$\quad = (\bot \vee b^?) \wedge (a^\circ \vee \bot) \wedge ((b^+ \vee c^\circ) \vee b^?)$
$\quad \rightsquigarrow (b^?) \wedge (a^\circ) \wedge ((b^+ \vee c^\circ) \vee b^?))$
$\quad = (b^?) \wedge (a^\circ) \wedge ((\bot \vee c^\circ) \vee \top)$
$\quad \rightsquigarrow (b^?) \wedge (a^\circ) \wedge (\top)$
$\quad = (\top) \wedge (a^\circ) \wedge (\top)$
$\quad \rightsquigarrow a^\circ$ (b can be labelled *undec* by defeating a)

Our approach can be decomposed in four phases:

1. associate each argument to three base labels,
2. compute conditional labels by substitution,
3. find target sets (for instance, by dnf-normalizing the formulae),
4. find a move such that it satisfies a target set of the goal formula.

The biggest challenge lies in step (2), because the substitution process for each formula has the size of the framework as upper bound and the same substitutions take place several times, especially in highly connected frameworks. A support for implementation can be a preprocessing phase of loop detection: loops are the main cause of complexity in label substitution, and knowing which loops an argument belongs to can help propagating activation-deactivation conditions. We call *active* a loop of arguments such that all arguments are labelled *undec* under grounded semantics, *not active* otherwise. Attacking some argument in order to make the arguments of the loop switch from *undec* to *in* or *out* is what we call *deactivating* the loop; we call the opposite process *activating* the loop. For instance, in the framework in Figure 5.1, the *b-c-e-f* loop is active.

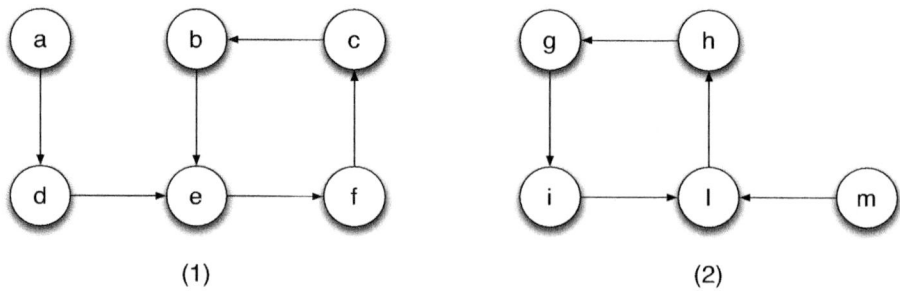

(1)   (2)

Fig. 5. Two argumentation frameworks with even cycles

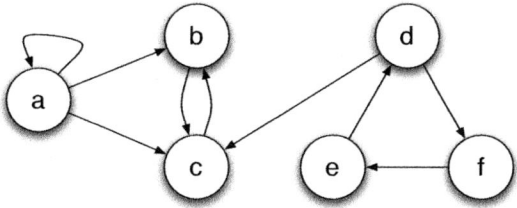

Fig. 6. Frameworks with loops

Some conditional labels are:

- $b^+ : c^\circ \vee e^\circ \vee a^\circ = f^+ = c^- = e^-$
- $b^- : b^\circ \vee f^\circ = f^- = c^+ = e^+$
- $b^? : \top = f^? = c^? = e^?$

According to the definition and the substitution algorithm, all those labels would be computed sequentially. But they just mirror the possible deactivations of the cycle, splitted in two sets according to the position (even/odd) of arguments along the cycle. So detecting the cycle one could compute the conditional labels of a single argument and then copy them (alternating from green to red formulae according to even/odd path) for each argument in the loop. This also holds for activation conditions for not active loops (like the one in Figure 5.2) and can be easily extended to odd-length cycles.

Therefore, loop detection can be a major improvement for our algorithm's performances.

We developed a methodology based on computing the powers of the adjacency matrix of the graph. Consider the framework visualized in Figure 6: there is a single loop (a) and two bigger ones (b-c and d-e-f), plus edges which do not belong to loops. The adjacency matrix of the framework is represented in Table 1 (let it be m^1).

Table 1. Adjacency matrix for the framework in Figure 6

	a	b	c	d	e	f
a	1	1	1	0	0	0
b	0	0	1	0	0	0
c	0	1	0	0	0	0
d	0	0	1	0	0	1
e	0	0	0	1	0	0
f	0	0	0	0	1	0

In fact, m^1 gives us information about self-loops in the framework: the arguments attacking themselves correspond to the 1 on the main diagonal of m^1: in our example, a. We express this property as $a \in diagonal(m^1)$.

But what happens if we compute $m^2 = m^1 * m^1$? On the n-th element of the main diagonal of m^2 we will have a 1 iff the n-th argument is able to reach itself in n steps: that is, if it belongs to a n-deep loop. This is easy to constructively prove by showing how a multiplication between matrices is made. So, for a framework $AF = \langle A, R \rangle$ with $\mid A \mid = n$ we can just compute $m^2, m^3, \ldots m^n$ to detect all loops.

Table 2. Powers of adjacency matrix

m^2	a	b	c	d	e	f
a	1	1	1	0	0	0
b	0	1	0	0	0	0
c	0	0	1	0	0	0
d	0	1	0	0	1	0
e	0	0	1	0	0	1
f	0	0	0	1	0	0

m^3	a	b	c	d	e	f
a	1	1	1	0	0	0
b	0	0	1	0	0	0
c	0	1	0	0	0	0
d	0	0	1	1	0	0
e	0	1	0	0	1	0
f	0	0	1	0	0	1

m^4	a	b	c	d	e	f
a	1	1	1	0	0	0
b	0	1	0	0	0	0
c	0	0	1	0	0	0
d	0	1	1	0	0	1
e	0	0	1	1	0	0
f	0	1	0	0	1	0

Note that in m^1 we detect a, in m^2 b and c, in m^3 d,e,f and no new loop is detected in m^4. Notice that $\alpha \in diagonal(m^p) \Rightarrow \alpha \in diagonal(m^{k*p}), \forall k \in \mathbb{N}$. For instance, in the previous example, $a \in diagonal(m^p)\forall p > 0$ and $b, c \in diagonal(m^2), diagonal(m^4)$. This redundancy of information can be overcome by cross-checking or by removing the elements on the diagonal of the adjacency matrix before multiplying it again.

The number of arguments is the upper bound for loop depths but can be narrowed down in several ways, for instance by detecting connected components or pruning siphons and traps: in the first case, the deepest loop consists of the maximal values over the number of arguments of each connected component; in the second one, siphons and traps can not be part of loops, thus allowing the lowering of the upper bound.

This paper mainly deals with the first two phases listed (the association of each argument to three base labels and the computation of conditional labels by substitution); some observations about the last two phases (finding the target sets and evaluating and comparing the moves) need to be addressed.

Concerning phase (3), we suggested DNF-normalization as a shortcut to find the solutions of the formulae (labels). This is basically just a working hypothesis, as a DNF-normalized formula can be fed to a SAT-solver, thus allowing us to rely on an external tool. Other approaches could be investigated, mainly due to the particular structure of the conditional labels: they only include conjunctions and disjunctions as connectives, and all atoms are positive. We are currently exploring new solutions techniques, mostly based on the connections and similarities with BDDs and AND-OR graphs.

Concerning phase (4), many evaluation criteria may be based on preferences of the agents and could depend on their inner parameters: for instance, an agent may not be able to attack a given argument a_1, or she may prefer argument a_2 over argument a_3 and thus decide to attack the latter rather than the previous, and so on.

Although belief revision is beyond the scope of this paper, one of its main assumption, namely *minimal change*, could be borrowed and used as a criteria for evaluating and comparing possible moves. According to the minimal change principle, the knowledge before and after the change should be as similar as possible. This principle enforces as much information as possible to be preserved by the change. One could wonder whether we could measure the 'impact' of a move, that is, how much a framework changes after a move. A target set alone is not sufficient to provide such information: consider the framework in Figure 7.1.

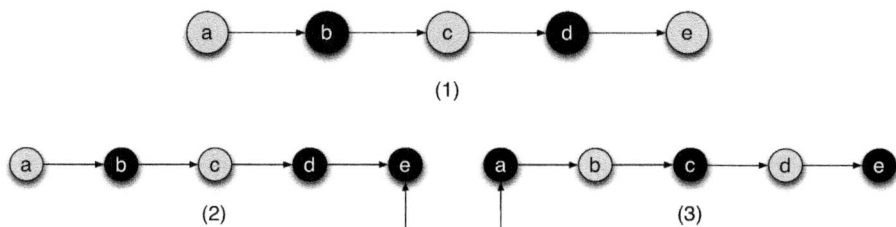

Fig. 7. Reinstatement framework

Since e^- : $a^\circ \vee c^\circ \vee e^\circ$, in order to defeat argument e our target sets are $\{\{a\}, \{c\}, \{e\}\}$. But on the one hand, directly attacking e causes only e itself to flip from *in* to *out* (Figure 7.2), while on the other hand (Figure 7.3) defeating a causes all arguments to change justification status. In order to measure the 'impact' of a target set, the first concept which has to be taken into account is how to define the 'distance' between two labellings: roughly speaking, we want to compare the labelling of the framework before and after the modification and, considering how many arguments changed justification status, measure how 'close' the two labellings are. Intuitively, considering the example in Figure 7, the framework obtained attacking e (Figure 7.2) is similar to the previous one (Figure 7.1, only one argument changes justification status, minimal distance) while the framework obtained attacking a (Figure 7.3) is totally different (every argument changed justification status, long distance). This topic has been recently analyzed by Booth et al. [2] and we may, in fact, use their definition of *distance*. On the other hand, integrating this measure with the existing definitions is not trivial: suppose, for the sake of explanation, that we are just interested in enriching each target set with the *number of arguments* that change justification status if that target set is defeated. Considering the example in Figure 7 again, our enriched target sets are $\{\{a\} : 5, \{c\} : 3, \{e\} : 1\}$: this means that defeating the target set $\{a\}$ five arguments will change justification status, and so on. We could take into account the cardinality of the target set, but this is not relevant: in our example, we have three target sets of one element each, and they correspond to different distances. Consider the frameworks in Figure 8.

For each framework, let t be the only argument in the target set and h the head of the label (that is, the argument whose justification status we want to modify). What we want to show is how intuitive approaches do not correspond to distances or reachability-related concepts.

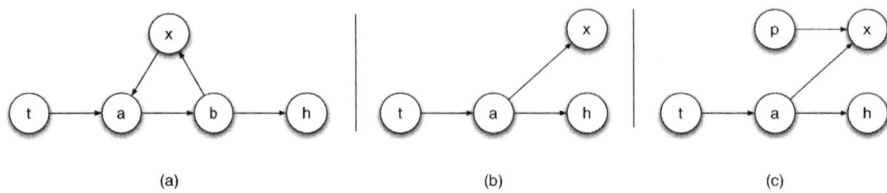

Fig. 8. Impact and graphs

The first framework (Figure 8.a) shows that considering the distance between t and h is not enough: in this case we are ignoring argument x because it is not on the shortest path from t to h, but defeating t would modify x's justification status too.

We can relax the definition and just consider the arguments from which h is reachable (in the substitution process, starting from h, the graph is navigated backwards). But Figure 8.b shows that in this case we would ignore argument x, from which h is not reachable but that would be affected defeating t.

On the other hand, considering only the arguments reachable from t is misleading again: Figure 8.c shows how we would consider argument x (since it is reachable from t), whose justification status would not change defeating t (since it depends on p too).

Therefore simple graph-based considerations are not enough to capture the impact of a target set, and the problem has to be analyzed in its specificity.

5   Related Work

Conditional labelling is closely related to the dialogues games [6,1]. In argumentation theory, such games regulate dialogues where two parties argue about the tenability of one or more claims or arguments, each trying to persuade the other participant to adopt their point of view. Such dialogues are often called persuasion dialogues. Among others, Prakken [6] presents a formal framework for a class of argumentation dialogues, where each dialogue move either attacks or surrenders to a preceding move of the other participant. For instance, each claim, why and since move is viewed as an attacking reply and each concede move is a surrendering reply.

Amgoud and Hameurlain [1] argue that a strategy is a two steps decision process: i) to select the type of act to utter at a given step of a dialogue, and ii) to select the content which will accompany the act. The first step consists of selecting among all the acts allowed by the protocol, the best option which according to some strategic beliefs of the agent will at least satisfy the most important strategic goals of the agent. The second step consists of selecting among different alternatives, the best one which, according to some basic beliefs of the agent, will satisfy the functional goals of the agent.

Roth et al. [7] start from two principles: i) the outcome of a dispute depends on the strategies actually adopted by parties, but ii) this does not mean that the outcome can never be predicted because by using game theoretical solution concepts, the actions themselves can often be found. They use defeasible logic in combination with standard probability calculus in order to prove that a defeasible proof holds, on the basis of the probabilities assigned to the premises. This probability of a claim was then interpreted in the game theoretical sense as the payoff for the proponent of the claim.

In comparison with this kind of frameworks, we share the idea that the first step consists in choosing the next move depending on the strategies of the agents. The differences are that we are not interested in providing a complete framework for argumentation dialogues games, we aim at providing a tool which can be used in those systems and which can be integrated with strategies. We do not restrict our framework to deal with two agents, and we extend the well-known argumentation labelling in order to provide a complete information about the argumentation framework on which it is applied.

6   Summary

In this paper, we present a new kind of argument labelling, called conditional labelling. Conditional labelling allows to associate to each argument the information concerning its possible justification statuses, depending on the changes in the framework. In particular, we express this information by means of propositional formulae which express which arguments should be attacked in order to get the desired argument accepted, not accepted, or undecided. While it is quite straightforward to assign those conditional labels in argumentation frameworks without cycles and multiple attacks, it is rather complicated in the general case. When an argumentation framework with cycles is considered, it is possible to have in the conditional label α^* of an argument another α^* because the conditional labelling algorithm, using substitution, looks for all the attackers of the node until it finds the node itself. The conditional labelling allows the agents to avoid the exhaustive search of all the possible combinations in adding new arguments, and decreases the exponential complexity this search requires. Loop detection via powers of the adjacency matrix is proposed as a preprocessing mechanism to compute common labels among the arguments in a loop.

Future work addresses several issues: first of all, a deeper investigation on the complexity results related to the computation of the new labellings is necessary. From a purely argumentative perspective, it would be nice to find out how conditional labels can be useful *after* a move: that is, if the previous information can be used to compute new conditional labels after the framework has been modified. Associating a cost concept to moves, our labelling lets agents link action costs to goals' outcomes, and can therefore be used as an underlying mechanism to develop strategies in a game theoretical context.

References

1. Amgoud, L., Hameurlain, N.: A formal model for designing dialogue strategies. In: Nakashima, H., Wellman, M.P., Weiss, G., Stone, P. (eds.) Autonomous Agents and Multiagent Systems (AAMAS), pp. 414–416. ACM (2006)

2. Booth, R., Caminada, M., Podlaszewski, M., Rahwan, I.: Quantifying disagreement in argument-based reasoning. In: International Workshop on the Theory and Applications of Formal Argumentation (TAFA), Barcelona, Spain (2011)
3. Caminada, M.: On the Issue of Reinstatement in Argumentation. In: Fisher, M., van der Hoek, W., Konev, B., Lisitsa, A. (eds.) JELIA 2006. LNCS (LNAI), vol. 4160, pp. 111–123. Springer, Heidelberg (2006)
4. Dung, P.M.: On the acceptability of arguments and its fundamental role in non-monotonic reasoning, logic programming and n-person games. Artif. Intell. 77(2), 321–358 (1995)
5. Jakobovits, H., Vermeir, D.: Robust semantics for argumentation frameworks. J. Log. Comput. 9(2), 215–261 (1999)
6. Prakken, H.: Coherence and flexibility in dialogue games for argumentation. J. Log. Comput. 15(6), 1009–1040 (2005)
7. Roth, B., Riveret, R., Rotolo, A., Governatori, G.: Strategic argumentation: a game theoretical investigation. In: International Conference on AI and Law (ICAIL), pp. 81–90. ACM (2007)

Bottom-Up Argumentation

Francesca Toni[1] and Paolo Torroni[2]

[1] Department of Computing - Imperial College London
London, UK
ft@imperial.ac.uk
[2] DEIS - University of Bologna
V.le Risorgimento, 2, 40136, Bologna - Italy
paolo.torroni@unibo.it

Abstract. Online social platforms, e-commerce sites and technical fora support the unfolding of informal exchanges, e.g. debates or discussions, that may be topic-driven or serendipitous. We outline a methodology for analysing these exchanges in computational argumentation terms, thus allowing a formal assessment of the dialectical validity of the positions debated in or emerging from the exchanges. Our methodology allows users to be engaged in this formal analysis and the assessment, within a dynamic process where comments, opinions, objections, as well as links connecting them, can all be contributed by users.

1   Introduction

Online social platforms, such as Facebook[1], e-commerce sites, such as Amazon[2], and technical fora, such as TechSupport Forum[3] support the unfolding of informal exchanges, in the form of debates or discussions, amongst several users. Some of these exchanges may be topic-driven (e.g. is a particular holiday destination worth visiting? Which book by Umberto Eco is best? How can a software bug be fixed?). Others may be serendipitous (e.g. while discussing the recent tsunami in Japan one may end up debating pros and cons of nuclear power stations).

While it is acknowledged (e.g. in [11]) that computational argumentation could benefit these online systems by supporting a formal analysis of the exchanges taking place therein, virtually all of the existing work considering online systems and argumentation focuses on extracting argumentation frameworks of one form or another manually or semi-automatically from these exchanges. For example, Heras et al. [11] suggest the use of argument schemes as a way to understand the contributions in these exchanges, while Rahwan et al. [13] suggest to map these contributions onto the AIF (Argument Interchange Format), again using argument schemes as well as semantic web technology for editing and querying arguments. These works implicitly assume that the extraction of argumentation frameworks is down to "argumentation engineers" external

---

[1] http://www.facebook.com/
[2] http://www.amazon.com/
[3] http://www.techsupportforum.com/forums/

S. Modgil, N. Oren, and F. Toni (Eds.): TAFA 2011, LNAI 7132, pp. 249–262, 2012.

to/passively engaged in the exchanges, and "fluent" in (one form or another of) computational argumentation.

On the other hand, work in computational argumentation predominantly focuses on determining the dialectical validity of a set of arguments, a single argument, or a claim, supported by arguments, with respect to a given, statically defined argumentation framework. Several notions of dialectical validity have been defined (e.g. see [5,8,1]) and several systems, for some or several of these notions, are available (e.g. see [10,9]).

We propose a methodology linking these two lines of work. Rather than assuming the intervention of "argumentation engineers" observing the exchanges, we envisage that the active participants in the exchanges are annotating them. In order for ordinary (rather than computational-argumentation fluent) users to be engaged in these annotations, we keep them very simple and graphical: annotations indicate that pieces of text in natural language are either *comments* or *opinions*, and *links* can be drawn to indicate source, support or objection. Opinions are expressed about comments, and comments and/or opinions can be linked to links too, very freely and in natural language as in online informal exchanges. We then propose an automated mapping from these annotations to an existing computational argumentation framework, Assumption-Based Argumentation (ABA) [7], paving the way to the automatic computation of the dialectical validity of comments, opinions, and links, and thus topics that these encompass. We envisage that users will add comments, opinions and links dynamically, in the same way exchanges grow over time in existing online systems.

We term our methodology *bottom-up argumentation* because it takes a grass-root approach to the problem of deploying computational argumentation in online systems:

- the argumentation frameworks are obtained bottom-up starting from the users' comments, opinions and suggested links;
- no top-down intervention of or interpretation by "argumentation engineers" is required;
- our automated translation feeds building blocks of arguments and attacks up to an argumentation system for determining computational validity;
- topics emerge, bottom-up, during the underlying process, possibly serendipitously.

We choose ABA as the underlying computational argumentation framework since it is the simplest system we are aware of that i) is well suited to support practical argumentative reasoning [4], ii) can distinguish arguments, support as well as attack amongst them, iii) can support defeasibility of information as the system evolves over time, iv) is equipped with a variety of well-defined semantics and computational counterparts for assessing dialectical validity.

We will focus in this paper on social networks as these allow for the most free kinds of exchanges, and are thus the most general setting in which to show our methodology.

The paper is organised as follows.

Darwin's natural selection rules supreme.

If you have ever been in GB you must have experienced washing your hands with separate taps. You know what I mean.

The picture shows a tap specimen now inhabiting Imperial College restrooms. You can clearly see a minor, but significant, mutation in the DNA of its ancestors. In particular, with respect to the "hardcore separate taps" variety, which used to live there not so long ago, cold and hot water are still separate, but they seem to have developed a form of symbiosis.

Besides, the population of "hardcore separate taps" (the only tap variety accounted for, until recently) seems to be on its way to extinction. Even in my hotel I couldn't find any.

This is quite impressive, considering that we are only in 2011.

Fig. 1. Initial post on Facebook

In section 2 we provide a concrete, motivating example for our methodology, of an exchange in a social network. We also discuss the main motivations for our proposed methodology. In section 3 we provide our basic system of annotations, in the context of the motivating example. In section 4 we give background on ABA. In section 5 we define the automated mapping between exchanges as given in section 3 and ABA, again illustrated for the motivating example. In section 6 we discuss some directions for future work and conclude.

2   Motivation

Let us consider a concrete case[4], where Facebook user Paolo Rossi posts the picture and comment shown in Figure 1.

This post does not have a precisely identified subject or purpose. There is a picture showing two separate taps controlling the water flow of a single faucet,

---

[4] This is a real discussion that took place in Facebook. The comments have not been edited. We instead modified the users' names for reasons of privacy. As a disclaimer, this paper does not intend to take any position regarding the opinions in this illustration.

🖉 Darwin's natural selection rules supreme.

If you have ever been in GB you must have experienced washing your hands with separate taps. You know what I mean.

The picture shows a tap specimen now inhabiting Imperial College restrooms. You can clearly see a minor, but significant, mutation in the DNA of its ancestors. In particular, with respect to the "hardcore separate taps" variety, which used to live there not so long ago, cold and hot water are still separate, but they seem to have developed a form of symbiosis.

Besides, the population of "hardcore separate taps" (the only tap variety accounted for, until recently) seems to be on its way to extinction. Even in my hotel I couldn't find any.

This is quite impressive, considering that we are only in 2011.

Added 14 March · Like · Comment

👍 Leyla Gencer, Benjamin Franklin, Enrico Fermi and 3 others like this.

Isabella Rossellini this is what my kitchen tap used to be only up until 5 years ago :-)
Monday at 11:10 · Unlike · 👍 1 person

Benjamin Franklin I remember not knowing whether I freeze or burn my hands under the tap. Took me a while to realise that there were 2 seperate streams coming out of the tap...
Monday at 11:54 · Unlike · 👍 1 person

Robert Axelrod i was also fascinated by the "hardcore separate taps" design until someone explained to me that brits regarded a "mixer" (miscelatore) needed to produce a single stream "unsanitary" and a "health hazard". after all, we know that mixing the hot and cold sources in the tapped basin to produce a tolerable temperature is absolutely sanitary! just give them another 10–20 more generations (with lots of mutations) and the two distinct streams will merge into a single one.
Monday at 12:58 · Unlike · 👍 2 people

Isabella Rossellini I actually think the health problem was a real issue: the water for the hot tap used to come from a cistern, while the water for the cold tap comes directly from the mains, so the cold water is suitable for drinking, the water coming from the hot tap is not, that's why mixing was not advisable. Nowadays I think both taps get water from the mains, though for older homes this might not be the case (and therefore the two streams of water in the picture are a solution to the problem, if you want to drink tap water you can safely just turn the cold tap on, safe in the knowledge water won't be 'contaminated'
Monday at 13:22 · Unlike · 👍 1 person

Fig. 2. Separate Taps discussion: comments (a)

from which two separate streams of water flow. The *comment* is intended to be humourous, but it does not say whether separate taps are inconvenient, or antiquate, although that may be implied. Then, as more Facebook users comment on this post (see Figures 2 and 3, where comments are labelled $C1, C2, \ldots, C14$), some *opinions* start to emerge between the lines, grounded in the comments, and people start discussing them, to express their agreement and bring additional *support* to comments/opinions of other users, or else to show disagreement and bring up *objections*. For example, the first three comments seem to agree,

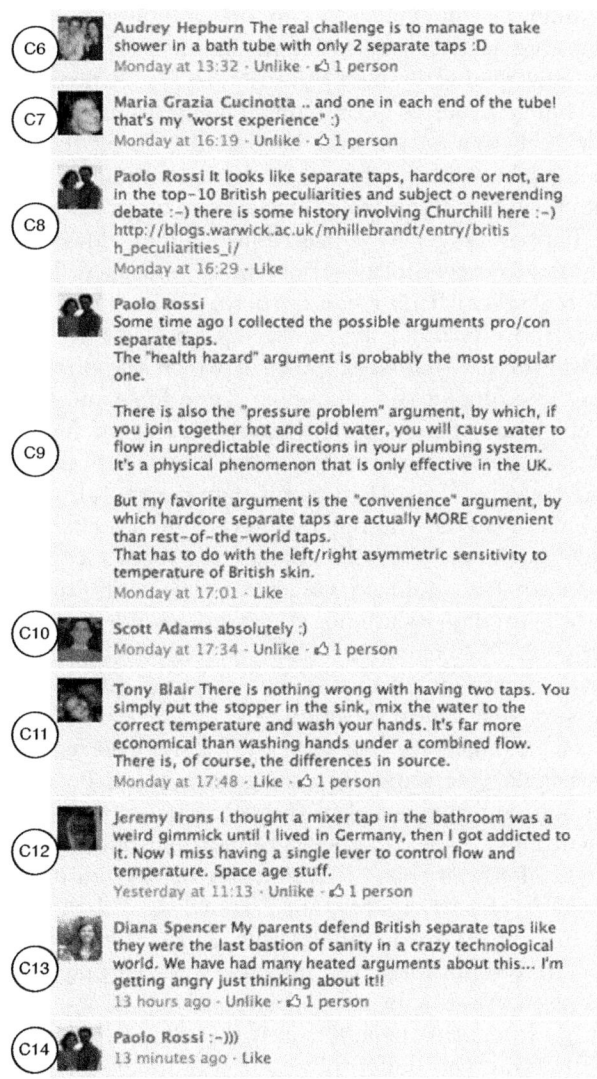

Fig. 3. Separate Taps discussion: comments (b)

directly or indirectly, on the opinion (let us call it $O1$) that "separate taps are common in GB". $C1$ also seems to convey another opinion: "separate taps are antiquate" ($O3$). $O1$ and $O3$ together may support the further opinion that "GB is a backward country" ($O2$), although in a somehow implicit way. These relationships between comments and emerging opinions are of a positive nature, i.e., comments support certain opinions. However, there are also comments representing objections to opinions or to other comments. For example, $C3$ may support the opinion that "separate taps are inconvenient because they freeze/burn hands"

(*O*4) whereas a different comment *C*11 supports a different, conflicting opinion, that "separate taps are not inconvenient as the basin solves temperature problems" (*O*17), hence we may read *O*17 as an objection to *O*4. A possible annotation of the Facebook exchange in terms of opinions, objections and links is given at the end of the paper. This annotation may be contributed by the users engaged in the exchange or by other users, external to the exchange.

It has been often said that the Web 2.0 is a place for grassroots. Actually, this is exactly what happens here. New contributions and ideas are produced and shared in an exquisitely serendipitous, bottom-up approach. In general, debates in the social Web start with no clear purpose. If the one who posts the first comment has a purpose in mind, he or she does not usually state it. Different is the case of structured debates, or polls in which the objective is clear, for instance choosing one among three possible dates for a meeting. Here instead we are looking at chains of pseudo-random posts, like we find in Facebook, in Amazon or at the bottom of an online newspaper's article. Sometimes such chains of posts converge to some topic, then again they may totally diverge and focus on some other topic. They may happen to never find a focus.

Despite these features, we can still abstract away and recognize, within these exchanges, *arguments*. But, unlike arguments in the computational argumentation literature, these arguments are not structured or relevant to any predefined topic, opinion or goal. They emerge, *bottom-up*, from the grassroots. From these arguments, a few mainstream opinions may emerge as the result of many comments, as if in a sort of "natural selection".

In this form of exchange we can identify a "struggle for existence" of arguments. The struggle determines what arguments will be most appreciated, upheld, agreed upon, and influential in the definition of the forthcoming generations of arguments, if we stick to the metaphor. But what are the forces that govern the struggle for existence of arguments in bottom-up argumentation? The rhetoric abilities of participants, their knowledge, their logical and social skills, all contribute greatly to the final result. But, since this is enabled by the presence of a social Web platform, the medium is also a player.

In this paper we outline a methodology for bringing computational argumentation (with its evaluative benefits) into these kinds of unstructured online exchanges while keeping the philosophy and style (simplicity, fun, freedom of expression) of the existing medium for social network. Indeed, we envisage that users can add further annotations, in the form of opinions grounded in/based on comments, objections, as well as (directed) links connecting them. The opinions are in the same format (free text) as the comments. Links are just graphical.

3   Annotations

We will use the following terminology:

- *comment* stands for a "base-level" user comment, i.e. a comment posted in an online debate by a user; *comments* will be denoted $C1, C2, \ldots$;

- *opinion* stands for a "meta-level" comment, containing information extracted or digested from part of one or more *comments* or other *opinions*, again by a user; *opinions* will be denoted $O1, O2, \ldots$;
- *links* are of two types:
  - continuous lines, connecting a target with one or more starting points marked by solid circles. These circles indicate either that an item at the starting point is the source for the information held at the end of the connecting line, if the starting point item is a *comment*, or that the starting point item provides support for the end point, if the starting point item is not a *comment*. These connecting lines can be seen as expressing a *basedOn* relation in the first case, and a *supportedBy* relation in the second;
  - dashed lines, again connecting a target with one or more starting points marked by solid circles. These lines indicate objections from the starting points (typically opinions) to the end point, and can be seen as expressing an *objection* relation.

In our motivating example, the *basedOn* relation is used to model that the source of $O1$ is $C1$; the *supportedBy* relation is used to model that $O1, O3$, together, support $O2$; the *objection* relation is used to model that $O17$ disagrees with $O4$. In general, *basedOn*, *supportedBy* and *objection* relations can also hold between a *comment* or *opinion* and another *basedOn*, *supportedBy* or *objection* relation. Indeed, in our motivating example, the *objection* to $O4$ originating from $O17$ is *basedOn* another *comment*, $C11$.

We will see how to map *comments*, *opinions*, and *links* onto a computational argumentation framework. The idea (and expected benefit) is to determine which *opinions* are acceptable given the current state of the discussion, in relation with other *comments/opinions*. As the exchange proceeds, different views will emerge and become more or less acceptable.

The dialectical process we are considering is full of implicit user assumptions. For example, if a user agrees on some *opinion* supported by some *comments*, we could say that the user "assumes" that such *comments* make sense, unless there are reasons not to do so. Likewise, if such *opinion* is subject to some *objections*, we could say that the user does not "assume" that such *objections* make sense, unless there are reasons to do so.

These considerations (as well as the reasons put forward in the introduction) make us believe that Assumption-Based Argumentation [7] is a very natural candidate framework for modeling bottom-up argumentation.

4   Assumption-Based Argumentation

Assumption-Based Argumentation (ABA) is a general-purpose argumentation framework where arguments and attacks between them are built from *ABA frameworks*, which are tuples $\langle \mathcal{L}, \mathcal{R}, \mathcal{A}, {}^{\overline{}} \rangle$ where

- $(\mathcal{L}, \mathcal{R})$ is a *deductive system*, with $\mathcal{L}$ a language and $\mathcal{R}$ a set of inference rules,

- $\mathcal{A} \subseteq \mathcal{L}$, referred to as the set of *assumptions*,
- $\bar{\ }$ is a (total) mapping from $\mathcal{A}$ into $\mathcal{L}$, where $\bar{x}$ is referred to as the *contrary* of x.

In this paper, we assume that inference rules have the syntax $s_0 \leftarrow s_1, \ldots, s_n$ (for $n \geq 0$) where $s_i \in \mathcal{L}$. We refer to $s_1, \ldots, s_n$ as the *premises* and to s_0 as the *head* of the rule. If $n = 0$, we represent a rule simply by its head and we call the rule a *fact*. As in [6], we restrict attention to *flat* ABA frameworks, such that no assumption occurs in the head of a rule.

Rules may be domain-dependent or not, and some of the premises of rules may be assumptions. These can be used to render the rules defeasible. In this setting, contraries of assumptions can be seen as representing "defeaters".

An *(ABA) argument* in favour of a sentence $c \in \mathcal{L}$ supported by a set of assumptions $A \subseteq \mathcal{A}$ is a proof of c from A and (some of) the rules in $\mathcal{R}$. This proof can be understood as a tree (with root the claim and leaves the assumptions), as in [7], as a backward deduction, as in [6,8], or as a forward deduction, as in [2], equivalently. For the purposes of this paper, we will use the notation $A \vdash_R c$ to stand for an argument for c supported by A by means of rules $R \subseteq \mathcal{R}$. When the rules can be ignored, we write an argument $A \vdash_R c$ simply as $A \vdash c$.

An argument $A \vdash c$ *attacks* an argument $A' \vdash c'$ if and only if $c = \overline{\alpha}$ for some $\alpha \in A'$.

Several "semantics" for ABA have been defined in terms of sets of assumptions fulfilling a number of conditions. These are expressed in terms of a notion of attack between sets of assumptions, where $A \subseteq \mathcal{A}$ attacks $A' \subseteq \mathcal{A}$ if and only if there is an argument $B \vdash c$, with $B \subseteq A$, attacking and argument $B' \vdash c'$, with $B' \subseteq A'$.

In this paper we will focus on the following notions:

- $A \subseteq \mathcal{A}$ is *conflict-free* if and only if A does not attack itself;
- $A \subseteq \mathcal{A}$ is *admissible* if and only if A is conflict-free and attacks every $B \subseteq \mathcal{A}$ that attacks A;
- $A \subseteq \mathcal{A}$ is *preferred* if and only if A is (subset) maximally admissible.

Note that these notions can be equivalently expressed in terms of arguments, rather than assumptions, as shown in [8].

Given an ABA framework $\mathcal{F} = \langle \mathcal{L}, \mathcal{R}, \mathcal{A}, \bar{\ } \rangle$ and a (conflict-free or admissible) set of assumptions $A \subseteq \mathcal{A}$ in $\mathcal{F}$, the (conflict-free or admissible) *extension* (respectively) $\mathcal{E}_{\mathcal{F}}(A)$ is the set of all sentences supported by arguments with support a set of assumptions $B \subseteq A$:

$$\mathcal{E}_{\mathcal{F}}(A) = \{s \in \mathcal{L} | \exists B \vdash s \text{ with } B \subseteq A\}.$$

In the remainder of this section, we will use the following conventions. Uppercase letters denote variables that are implicitly universally quantified. Variables O, C, L are used to represent opinions, comments and links between them, respectively. Variables X, Y are used to represent items that can be either opinions or comments. Variable Z is used to represent items that can be either opinions or links. The rules/assumptions/contraries are to be intended as schemata, standing for all their ground instances over appropriate universes (for comments, objections and links). Assumptions are always of the form $asm(\_)$, where asm is either α (for assumptions about opinions), χ (for assumptions about comments), or λ/λ^a (for assumptions about continuous/dashed links). The contrary of assumption $asm(a)$ is of the form $c\_asm(a)$, for any a, formally: $\overline{asm(a)} = c\_asm(a)$.

5   An ABA Mapping for Bottom-Up Argumentation

In this section we show how comments, opinions and links, as envisaged in section 3, can be translated onto an ABA framework. This translation from an annotated exchange of views on the social Web into ABA can be performed automatically. The resulting ABA framework can then be fed into an ABA system, such as CaSAPI [10], to determine which items (opinions, links etc) can be accepted dialectically.

The ABA framework resulting from this translation consists of a domain-dependent part (facts and rules), directly obtained from the annotated exchanges, and a domain-independent part (facts, rules, assumptions and contraries) which is generic, but to be used in conjunction with the domain-dependent part.

QUI

Domain-Dependent Facts and Rules. For each *comment*, the ABA model contains a fact *comment*(C), where C is the comment's label. In our illustration, we have 14 facts: *comment*$(c1)$, *comment*$(c2)$, ..., *comment*$(c14)$.

For each *opinion*, the ABA model contains a fact *opinion*(O), where O is the opinion's label. In our illustration, we have 19 opinions: *opinion*$(o1)$, *opinion*$(o2)$, ..., *opinion*$(o19)$.

For each continuous link, the ABA model contains a fact *link*(L, Y, X), where L is the link's label (chosen to determine it univocally), Y the starting point item and X the target item. For our example, these links are listed in Table 1.

Table 1. Continuous links

link(l_1_1,o1,c1).	*link*(m_2_1_3,o2,o1).	*link*(m_2_1_3,o2,o3).
link(l_2_19,o2,o19).	*link*(l_3_1,o3,c1).	*link*(l_3_2,o3,c2).
link(l_3_3,o3,c3).	*link*(l_4_3,o4,c3).	*link*(l_5_4,o5,c4).
link(l_6_5,o6,c5).	*link*(m_7_6_7,o7,c6).	*link*(m_7_6_7,o7,c7).
link(l_8_8,o8,c8).	*link*(l_8_9,o8,c9).	*link*(l_9_9,o9,c9).
link(l_10_9,o10,c9).	*link*(l_11_9,o11,c9).	*link*(l_12_9,o12,c9).
link(l_13_11,o13,c11).	*link*(l_14_12,o14,c12).	*link*(l_14_18,o14,o18).
link(l_15_9,o15,c9).	*link*(l_15_10,o15,c10).	*link*(l_16_9,o16,c9).
link(l_17_11,o17,c11).	*link*(l_18_12,o18,c12).	*link*(l_19_13,o19,c13).
link(l_19_14,o19,c14).	*link*(l_1_4_17_11,l_4_17,c11).	
link(l_1_11_12_9,l_11_12,c9).	*link*(l_1_16_15_9,l_16_15,c9).	

Darwin's natural selection rules supreme.

If you have ever been in GB you must have experienced washing your hands with separate taps. You know what I mean.

The picture shows a tap specimen now inhabiting Imperial College restrooms. You can clearly see a minor, but significant, mutation in the DNA of its ancestors. In particular, with respect to the 'hardcore separate taps' variety, which used to live there not so long ago, cold and hot water are still separate, but they seem to have developed a form of symbiosis.

Besides, the population of 'hardcore separate taps' (the only tap variety accounted for, until recently) seems to be on its way to extinction. Even in my hotel I couldn't find any.

This is quite impressive, considering that we are only in 2011.

Added 14 March · Like · Comment
👍 Leyla Gencer, Benjamin Franklin, Enrico Fermi and 3 others like this.

Isabella Rossellini this is what my kitchen tap used to be only up until 5 years ago :~)
Monday at 11:10 · Unlike · ♡1 person

Benjamin Franklin I remember not knowing whether I freeze or burn my hands under the tap. Took me a while to realise that there were 2 separate streams coming out of the tap...
Monday at 11:54 · Unlike · ♡1 person

Robert Axelrod i was also fascinated by the 'hardcore separate taps' design until someone explained to me that brits regarded a 'mixer' (miscelatore) needed to produce a single stream 'unsanitary' and a 'health hazard', after all, we know that mixing the hot and cold sources in the tapped basin to produce a tolerable temperature is absolutely sanitary! just give them another 10-20 more generations (with lots of mutations) and the two distinct streams will merge into a single one.
Monday at 12:58 · Unlike · ♡2 people

Isabella Rossellini I actually think the the health problem was a real issue: the water for the hot tap used to come from a cistern, while the cold water for the cold tap comes directly from the mains, so the cold water is suitable for drinking, the water coming from the hot tap is not, that's why mixing was not advisable. Nowadays I think both taps get water from the mains, though for older homes this might not be the case (and therefore the two streams of water in the picture are a solution to the problem, if you want to drink tap water you can safely just turn the cold tap on, safe in the knowledge water won't be 'contaminated'
Monday at 13:22 · Unlike · ♡1 person

Audrey Hepburn The real challenge is to manage to take shower in a bath tube with only 2 separate taps :D
Monday at 13:32 · Unlike · ♡1 person

Maria Grazia Cucinotta ... and one in each end of the tube! that's my 'worst experience' :)
Monday at 16:19 · Unlike · ♡1 person

Paolo Rossi It looks like separate taps, hardcore or not, are in the top~10 British peculiarities and subject o neverending debate :~) there is some history involving Churchill here :~)
http://blogs.warwick.ac.uk/mhillebrandt/entry/britis h_peculiarities.1/
Monday at 16:29 · Like

Paolo Rossi Some time ago I collected the possible arguments pro/con separate taps.
The 'health hazard' argument is probably the most popular one.
Monday at 17:01 · Like

There is also the 'pressure problem' argument, by which, if you join together hot and cold water, you will cause water to flow in unpredictable directions in your plumbing system. It's a physical phenomenon that is only effective in the UK.

But my favorite argument is the 'convenience' argument, by which hardcore separate taps are actually MORE convenient than rest-of-the-world taps. That has to do with the left/right asymmetric sensitivity to temperature of British skin.
Monday at 17:01 · Like

Scott Adams absolutely :)
Monday at 17:34 · Unlike · ♡1 person

Tony Blair There is nothing wrong with having two taps. You simply put the stopper in the sink, mix the water to the correct temperature and wash your hands. It's far more economical than washing hands under a combined flow. There is, of course, the differences in source.
Monday at 17:48 · Like · ♡1 person

Jeremy Irons I thought a mixer tap in the bathroom was a weird gimmick until I lived in Germany, then I got addicted to it. Now I miss having a single lever to control flow and temperature. Space age stuff.
Yesterday at 11:13 · Unlike · ♡1 person

Diana Spencer My parents defend British separate taps like they were the last bastion of sanity in a crazy technological world. We have had many heated arguments about this.... I'm getting angry just thinking about it!!
13 hours ago · Unlike · ♡1 person

Paolo Rossi :-)))
13 minutes ago · Like

Argument nodes / boxes:

- O1: Separate taps are antiquate
- O3: Separate taps are common in GB
- O4: Separate taps are inconvenient because they freeze/burn hands
- O5: Separate taps are unsanitary as they require mixing in the basin
- O6: Mixer taps are unsanitary if hot water comes from a cistern (as cistern water not drinkable)
- (O2: GB is a backward country)
- O7: Separate taps are impractical/inconvenient for showers ... especially if taps at opposite ends- it has happened in GB
- O8: GB is a peculiar country
- O9: Mixer taps give pressure problems
- O10: Pressure problems are not scientifically justified
- O11: Separate taps are more convenient
- O12: A11 only holds for GB skins
- O16: Convenience argument justifies separate taps
- (O15: Convenience argument is ridiculous)
- O13: Separate taps are more economical as far as water consumption
- O17: Separate taps are not inconvenient as basin solves temperature problem
- O18: Mixer taps are more convenient
- O14: Mixer taps are modern comfort
- O19: Arguments justifying separate taps are backward/typical of older generation

C1 C2 C3 C4 C5 C6 C7 C8 C9 C10 C11 C12 C13 C14

All these links are from *comments* to *opinions* or from *opinions* to *opinions*, except for the last three that are from *comments* to *links*.

For each dashed link (objection), the ABA model contains a sentence *alink*(*L*, *O*, *X*), where *L* is the link's label, *O* the attacked *opinion*, and *X* the objecting item. For our example, dashed links are listed in Table 2.

Table 2. Dashed links

alink(l_4_17,o4,o17).	*alink*(l_9_10,o9,o10).	*alink*(l_11_12,o11,o12).
alink(l_11_18,o11,o18).	*alink*(l_18_11,o18,o11).	*alink*(l_16_15,o16,o15).
alink(l_17_7,o17,o7).	*alink*(l_17_19,o17,o19).	

Opinions can be supported by comments, or by other opinions, or by both. For each *opinion*, the ABA framework contains:

- one or more rules *basedOn*(*Z*) ← *C*, . . . , *L*, . . . indicating the *links* and *comments*on which item *Z* is based;
- one or more rules *supportedBy*(*Z*) ← *O*, . . . , *L*, . . . indicating the *links* and *opinions*supporting item *Z*.

For our example, the *basedOn* relations are listed in Table 3 and the *supportedBy* relations are listed in Table 4. Note that links from multiple starting points (such as that between $c6, c7$ and $o7$) are modeled by a single rule, whereas multiple independent links (such as that between $c1$ and $o3$, or between $c2$ and $o3$) are modeled by multiple rules. Absence of links is modelled by rules with an empty body (facts).

Table 3. *basedOn* relations

basedOn(o1)	← c1,l_1_1.	*basedOn*(o2)	← .
basedOn(o3)	← c1,l_3_1.	*basedOn*(o3)	← c2,l_3_2.
basedOn(o3)	← c3,l_3_3.	*basedOn*(o4)	← c3,l_4_3.
basedOn(o5)	← c4,l_5_4.	*basedOn*(o6)	← c5,l_6_5.
basedOn(o7)	← c6,c7,m_7_6_7.	*basedOn*(o8)	← c8,l_8_8.
basedOn(o8)	← c9,l_8_9.	*basedOn*(o9)	← c9,l_9_9.
basedOn(o10)	← c9,l_10_9.	*basedOn*(o11)	← c9,l_11_9.
basedOn(o12)	← c9,l_12_9.	*basedOn*(o13)	← c11,l_13_11.
basedOn(o14)	← c12,l_14_12.	*basedOn*(o15)	← c9,l_15_9.
basedOn(o15)	← c10,l_15_10.	*basedOn*(o16)	← c9,l_16_9.
basedOn(o17)	← c11,l_17_11.	*basedOn*(o18)	← c12,l_18_12.
basedOn(o19)	← c13,l_19_13.	*basedOn*(o19)	← c14,l_19_14.
basedOn(l_4_17)	← c11,l_l_4_17_11.	*basedOn*(l_11_12)	← c9,l_l_11_12_9.
basedOn(l_16_15)	← c9,l_l_16_15_9.		

Table 4. *supportedBy* relations

$supportedBy(o1) \leftarrow .$	$supportedBy(o2) \leftarrow o1,o3,m\_2\_1\_3.$
$supportedBy(o2) \leftarrow o19,l\_2\_19.$	$supportedBy(o3) \leftarrow .$
$supportedBy(o4) \leftarrow .$	$supportedBy(o5) \leftarrow .$
$supportedBy(o6) \leftarrow .$	$supportedBy(o7) \leftarrow .$
$supportedBy(o8) \leftarrow .$	$supportedBy(o9) \leftarrow .$
$supportedBy(o10) \leftarrow .$	$supportedBy(o11) \leftarrow .$
$supportedBy(o12) \leftarrow .$	$supportedBy(o13) \leftarrow .$
$supportedBy(o14) \leftarrow o18,l\_14\_18.$	$supportedBy(o15) \leftarrow .$
$supportedBy(o16) \leftarrow .$	$supportedBy(o17) \leftarrow .$
$supportedBy(o18) \leftarrow .$	$supportedBy(o19) \leftarrow .$

Domain-Independent Facts, Rules, Assumptions, Contraries. Domain-independent facts and rules are used as follows.

- We rely on an *opinion* O if we can rely on other *comments* on which O is based (if any) and/or on *opinions* that support O (if any), and if it is legitimate to assume O. Therefore the following ABA rule is used, for all *opinions* O:

$$O \leftarrow basedOn(O), supportedBy(O), \alpha(O), opinion(O).$$

The defeasibility of an *opinion* O is modeled by the assumption $\alpha(O)$.
- We rely on a *comment* C if it is legitimate to assume C. Therefore the following ABA rule is used, for all *comments* C:

$$C \leftarrow \chi(C), comment(C).$$

The defeasibility of a *comment* C is modeled by the assumption $\chi(C)$.
- We rely on a continuous *link* L if it is legitimate to assume L. Therefore the following ABA rule is used, for all continuous *links* L:

$$L \leftarrow \lambda(L), link(L, \_, \_).$$

The defeasibility of a continuous *link* L is modeled by the assumption $\lambda(L)$.
- We rely on a dashed *link* L to provide an attack against X given Y if it is legitimate to assume L and if the attacker Y holds. The following rule is then used:

$$c\_\alpha(X) \leftarrow Y, \lambda^a(L), alink(L, X, Y).$$

The defeasibility of a dashed *link* L is modeled by the assumption $\lambda^a(L)$.
- All *opinions, comments, links* are in principle legitimate. Therefore the sentences $\alpha(O), \chi(C), \lambda(L), \lambda^a(L')$ are possible assumptions for all O, C, L, L' in our universe of symbols such that $opinion(O)$, $comment(C)$, $link(L)$, $alink(L')$ hold.
- Finally, the following contraries are given:

$$\overline{\alpha(O)} = c\_\alpha(O). \quad \overline{\chi(C)} = c\_\chi(C). \quad \overline{\lambda(L)} = c\_\lambda(L). \quad \overline{\lambda^a(L)} = c\_\lambda^a(L).$$

6   Conclusions

We have outlined a generic methodology to benefit exchanges of views in social networks (but also e-commerce systems or technical fora) by deploying computational argumentation. We have taken the view to modify only minimally the existing style for social networks, and allow users to unearth opinions and links. We have supported our proposal by means of a concrete illustration on top of Facebook. Our methodology consists of 1) allowing users to comment on exchanges, thus adding to and refining them; 2) applying a formal mapping from these augmented exchanges onto an (assumption-based) argumentation framework; 3) use standard argumentation semantics to provide an informed view to users as to the dialectical validity of the positions debated.

There are several directions for future work. We mention just a few here.

We have ignored the possibility of feedback by users, e.g. using the Like button in Facebook. These need to be incorporated within our methodology.

We have introduced a separation between "base-level" (the comments as in existing social net sites) and a "meta-level" (our opinions, links etc). We envisage that these will need to blend eventually, and that, for example, opinions may feed back into comments.

We have proposed an annotation for enriched exchanges in social networks, that we believe has the right level of simplicity and ease of use for ordinary users while at the same time being easily translatable into ABA. It would be interesting to see whether existing annotations used in sense-making tools, such as Cohere [3], would be suitable and/or would lend themselves to be mapped onto ABA format.

We envisage to use ABA as the underlying mechanism for computational argumentation. A novel bottom-up tool for computing extensions will be required for ABA to support a query-independent evaluation of arguments.

We have glossed over the choice of argumentation semantics: experimental psychology may be able to provide us with hints as to which semantics is the most suited. It may be possible that none of the existing semantics for argumentatin may be appealing or suitable, as indicated, in a different setting, in [12].

We also need to design effective methods and incentives that encourage users to annotate their discussions. For example, it will be important to understand how bottom-up argumentation may increase users satisfaction and engagement in online conversations. To this end, we will need to run empirical and theoretical investigations.

The theoretical implications of bottom-up argumentation will also be subject of further research. For example, if we asked several different users to independently mark up the same discussion, we would obtain different results. Would this be a problem? How would different mark-ups relate with each other?

We did not elaborate on concrete ways to exploit bottom-up argumentation in existing or future social networks. Clearly, if we want to use it as a run-time support for users on a large scale, some further analysis needs to be done to understand the computational complexity of the underlying reasoning. Suitable user testing and benchmarking tasks will also have to be designed and carried out. A more thorough study must also be done in order to make our methodology better defined and structured. With this article we mean to describe the general ideas that, if successful, may underlie a groundbreaking use of computational argumentation, for the benefit of communities of non-argumentation-savvy individuals.

Acknowledgments. The realization of this work was made possible by a Short-Term Scientific Mission grant, kindly provided by the "Agreement Technologies" COST Action IC0801. We also thank the anonymous TAFA reviewers for valuable feedback and Henry Prakken for helpful comments at the workshop.

References

1. Besnard, P., Hunter, A.: Elements of Argumentation. MIT Press (2008)
2. Bondarenko, A., Dung, P., Kowalski, R., Toni, F.: An abstract, argumentation-theoretic approach to default reasoning. Artif. Intell. 93(1-2), 63–101 (1997)
3. Buckingham Shum, S.: Cohere: Towards Web 2.0 argumentation. In: Hunter, A. (ed.) Proceedings of the Second International Conference on Computational Models of Argument (COMMA 2008). IOS Press (2008)
4. Dung, P.M., Toni, F., Mancarella, P.: Some design guidelines for practical argumentation systems. In: Baroni, P., Cerutti, F., Giacomin, M., Simari, G. (eds.) Proceedings of the Third International Conference on Computational Models of Argument (COMMA 2010), vol. 216, pp. 183–194. IOS Press (2010)
5. Dung, P.: On the acceptability of arguments and its fundamental role in non-monotonic reasoning, logic programming and n-person games. Artif. Intell. 77, 321–357 (1995)
6. Dung, P., Kowalski, R., Toni, F.: Dialectic proof procedures for assumption-based, admissible argumentation. Artif. Intell. 170, 114–159 (2006)
7. Dung, P., Kowalski, R., Toni, F.: Assumption-based argumentation. In: Rahwan, I., Simari, G. (eds.) Argumentation in AI, pp. 199–218. Springer, Heidelberg (2009)
8. Dung, P., Mancarella, P., Toni, F.: Computing ideal sceptical argumentation. Artif. Intell. 171(10-15), 642–674 (2007)
9. Egly, U., Gaggl, S.A., Woltran, S.: ASPARTIX: Implementing Argumentation Frameworks Using Answer-Set Programming. In: Garcia de la Banda, M., Pontelli, E. (eds.) ICLP 2008. LNCS, vol. 5366, pp. 734–738. Springer, Heidelberg (2008)
10. Gaertner, D., Toni, F.: On computing arguments and attacks in assumption-based argumentation. IEEE Intelligent Systems, Special Issue on Argumentation Technology 22(6), 24–33 (2007)
11. Heras, S., Atkinson, K., Botti, V., Grasso, F., Julian, V., McBurney, P.: How argumentation can enhance dialogues in social networks. In: Baroni, P., Cerutti, F., Giacomin, M., Simari, G. (eds.) Proceedings of the Third International Conference on Computational Models of Argument (COMMA 2010), vol. 216, pp. 267–274. IOS Press (2010)
12. Rahwan, I., Madakkatel, M.I., Bonnefon, J.F., Awan, R.N., Abdallah, S.: Behavioural experiments for assessing the abstract argumentation semantics of reinstatement. Cognitive Science 34(8), 1483–1502 (2010)
13. Rahwan, I., Zablith, F., Reed, C.: Laying the foundations for a world wide argument web. Artificial Intelligence 171, 897–921 (2007)

A First Step towards Argumentation Dialogues for Discovery

Xiuyi Fan and Francesca Toni

Imperial College London, London, SW7 2AZ, UK
{xf309,ft}@imperial.ac.uk

Abstract. We present a formal model for two-agent discovery dialogues. The model allows agents to collectively discover a realization for a shared goal, using argumentation dialogues to exchange information. This information is in the form of rules, assumptions, and contraries of assumptions as in Assumption-based Argumentation (ABA). With dialogues, agents jointly build arguments and construct shared ABA frameworks. We define successful discovery dialogues as those giving admissible arguments that realize the shared goal. The main novelty of this paper is the modelling of the *buttom-up* relation between utterances. This new relation helps building "higher level" arguments from existing "lower level" supports, which we deem essential for discovery.

1 Introduction

Argumentation dialogues have been studied by a number of researchers [1,8,12]. Walton & Krabbe [14] introduce a dialogue taxonomy that categorizes dialogues into six types: *persuasion, inquiry, information seeking, negotiation, deliberation,* and *eristic*. McBurney and Parsons [8] introduce *discovery* as an additional type of dialogue, different from other types in that "[discovery dialogues] discover something not previously known; the question whose truth is to be ascertained may only emerge in the course of the dialogue." In this paper, we present a two-agent dialogue framework that supports a special type of discovery dialogues.

Most previous work on argumentation dialogues, e.g. [1,11], define dialogue models as dialogues starting from a known proposition; through dialogues, the acceptability of this proposition is examined. However, in discovery dialogues, there may be no such known proposition to start with. Instead, the dialogue participants face an open problem. Inour discovery dialogues, the participants start with an abstract description of the goal of the dialogue and proceed by putting forward information that may contribute to identify the proposition and determine its acceptability. This abstract description is the same for the two participating agents. We call this abstract description the *goal*. None of the two agents have sufficient information to produce an acceptable concrete realization of this goal. The agents' task is then to discover an acceptable concrete goal realization.

In this work, the two agents communicate to each other using Assumption-Based Argumentation (ABA) [3]. ABA is a general-purpose, widely applicable

S. Modgil, N. Oren, and F. Toni (Eds.): TAFA 2011, LNAI 7132, pp. 263–279, 2012.

form of argumentation where arguments are built from *rules* and supported by *assumptions*, and attacks against arguments are directed at the assumptions supporting the arguments, and are provided by arguments for the *contrary* of assumptions. ABA is applicable in several settings, as discussed in [3].

During dialogues, agents communicate with ABA by putting forward rules, assumptions, and contraries of assumptions as their utterances. In order to perform the joint discovery, the dialogue starts with one agent putting forward a goal. Then the two agents can either expand this goal or some sentence in a *top-down* manner to explore its supports and attacks or in a *bottom-up* manner to identify any "higher level" arguments that are supported by it. Utterances jointly form the ABA framework *drawn* from the dialogue. A discovery dialogue is successful if it produces a goal realization. We justify the soundness of our approach by showing that the produced goal realization is admissible with respect to the ABA framework drawn from the dialogue.

This work extends [6], which present a conflict resolution framework in a two-agent system. The two agents in [6] share the same goal but each agent has its own realization to start with. Hence, in [6] the two agents examine each realization solely in (what we call) the *top-down* fashion to determine its admissibility. Since [6] is an application of the dialogue framework in [5], this paper also extends [5]. The main novelty of our work here is the recognition of the *bottom-up* relation, which we believe is essential for modelling discovery dialogues.

The rest of this paper is organized as follows. Section 2 introduces background information on ABA and *abstract dispute trees* [4], that we use to prove our results. Section 3 introduces a motivating example that we use throughout this paper. Section 4 introduces our dialogue model. Section 5 introduces the *debate tree* that we use to refine our dialogue model. Section 6 presents formal results of our model. We describe a few related works in section 7 and conclude in section 8.

2   Background

An ABA framework [3,4] is a tuple $\langle \mathcal{L}, \mathcal{R}, \mathcal{A}, \mathcal{C} \rangle$ where

- $\langle \mathcal{L}, \mathcal{R} \rangle$ is a deductive system, with $\mathcal{L}$ the *language* and $\mathcal{R}$ a set of *rules* of the form $s_0 \leftarrow s_1, \ldots, s_m (m \geq 0)$;
- $\mathcal{A} \subseteq \mathcal{L}$ is a (non-empty) set, referred to as *assumptions*;
- $\mathcal{C}$ is a total mapping from $\mathcal{A}$ into $\mathcal{L}$; $\mathcal{C}(\alpha)$ is referred to as the *contrary* of $\alpha \in \mathcal{A}$.

Given a rule $s_0 \leftarrow s_1, \ldots, s_m$, s_0 is referred to as the *head* and $s_1, \ldots, s_m$ as the *body*. We use the following notation: $Head(s_0 \leftarrow s_1, \ldots, s_m) = s_0$ and $Body(s_0 \leftarrow s_1, \ldots, s_m) = \{s_1, \ldots, s_m\}$. As in [4], we enforce that ABA frameworks are *flat*, namely assumptions do not occur in the head of rules. Moreover, without loss of generality, we enforce that no two assumptions may have the same contrary.[1]

---

[1] Indeed, an ABA framework $\langle \mathcal{L}, \mathcal{R}, \mathcal{A}, \mathcal{C} \rangle$ where $a, b \in \mathcal{A}$ such that $\mathcal{C}(a) = \mathcal{C}(b) = c$ can be equivalently rewritten as an ABA framework $\langle \mathcal{L}', \mathcal{R}', \mathcal{A}, \mathcal{C}' \rangle$ where $\mathcal{L}' = \mathcal{L} \cup \{c_1, c_2\}$ (with $c_1, c_2 \notin \mathcal{L}$), $\mathcal{R}' = \mathcal{R} \cup \{c_1 \leftarrow c, c_2 \leftarrow c\}$, and $\mathcal{C}(a) = c_1$, $\mathcal{C}(b) = c_2$.

In ABA, *arguments* are deductions of claims using rules and supported by assumptions, and *attacks* are directed at assumptions. Informally, following [3]:

- *an argument for (the claim)* $c \in \mathcal{L}$ *supported by* $S \subseteq \mathcal{A}$ ($S \vdash c$ in short) is a (finite) tree with nodes labelled by sentences in $\mathcal{L}$ or by the symbol τ^2, such that the root is labelled by c, leaves are either τ or assumptions in S, and non-leaves s have as many children as elements in the body of a rule with head s, in a one-to-one correspondence with the elements of this body.
- *an argument* $S_1 \vdash c_1$ *attacks an argument* $S_2 \vdash c_2$ iff $c_1 = \mathcal{C}(\alpha)$ for $\alpha \in S_2$.

Attacks between arguments correspond in ABA to attacks between sets of assumptions, where *a set of assumptions* A *attacks a set of assumptions* B iff an argument supported by $A' \subseteq A$ attacks an argument supported by $B' \subseteq B$.

With argument and attack defined, standard argumentation semantics can be applied in ABA [3]. We focus on the admissibility semantics:

- *a set of assumptions is admissible* (w.r.t. $\langle \mathcal{L}, \mathcal{R}, \mathcal{A}, \mathcal{C} \rangle$) iff it does not attack itself and it attacks all $A \subseteq \mathcal{A}$ that attack it;
- *an argument* $S \vdash c$ *belongs to an admissible extension supported by* $\Delta \subseteq \mathcal{A}$ (w.r.t. $\langle \mathcal{L}, \mathcal{R}, \mathcal{A}, \mathcal{C} \rangle$) iff $S \subseteq \Delta$ and Δ is admissible;
- *a sentence is admissible* iff it is the claim of an argument that belongs to an admissible extension supported by some $\Delta \subseteq \mathcal{A}$.

Our main result will be proven using the abstract dispute trees of [4], with an *abstract dispute tree* for an argument γ a (possibly infinite) tree $\mathcal{T}^a$ such that:

1. every node of $\mathcal{T}^a$ is labelled by an argument and is assigned the status of either a *proponent* (P) node or an *opponent* (O) node, but not both;
2. the root of $\mathcal{T}^a$ is a P node labelled by γ;
3. for every P node n labelled by an argument b, and for every argument c that attacks b, there exists a child of n, which is an O node labelled by c;
4. for every O node n labelled by an argument b, there exists exactly one child of n which is a P node labelled by an argument which attacks some assumption α in the set supporting b. α is said to be the *culprit* in b;
5. there are no other nodes in $\mathcal{T}^a$ except those given by 1-4 above.

The set of all assumptions in (the support of arguments of) the proponent nodes in $\mathcal{T}^a$ is called the *defence set* of $\mathcal{T}^a$. An abstract dispute tree is *admissible* iff no culprit in the argument of an opponent node belongs to the defence set of $\mathcal{T}^a$. The defence set of an admissible abstract dispute tree $\mathcal{T}^a$ for an argument γ is admissible (Theorem 5.1 in [4]), and thus γ belongs to an admissible extension and the sentence supported by γ is admissible.

3   Motivating Example

Two agents, Jenny and Amy, are planning a film night. They would like to jointly decide on a movie. Jenny wants to pick a fun movie. She finds action movies fun.

---

^{2} As in [3], $\tau \notin \mathcal{L}$ intuitively stands for "true". It is used to represent the empty body of a rule.

Jenny also worries about going home late so she prefers a movie that finishes by 10 o'clock. Amy knows that *Terminator* is screening and is an action movie. She also knows that *Terminator* finishes by 10 o'clock. Amy does not have any preference in selecting a movie. In order to reach agreement, the two agents may conduct a dialogue as follows[3].

Jenny: Let's find a movie to watch.
Amy: Sure, I know *Terminator* is an action movie.
Jenny: That's interesting. I think action movies are pretty fun.
Amy: We can watch *Terminator*, as long as it has the right screening time.
Jenny: Agreed. I think *Terminator* starts at the right time.
Amy: Are you sure it won't be too late?
Jenny: Why?
Amy: I don't know. I am just afraid so.
Jenny: It won't be too late if it finishes by 10 o'clock.
Amy: I see. Indeed *Terminator* finishes by 10 o'clock.
Jenny: OK.
Amy: OK.

Jenny starts the dialogue by putting forward the goal of determining some movie to watch, but without specifying which one. Then Amy supplies one possibly relevant fact, that *Terminator* is an action movie. This is just a guess, in the sense that the agent does not know whether a goal realization can be found by exploring information related to *Terminator*. From this utterance, agents reason buttom-up until Amy's second utterance. Then they start top-down. From the inital guess, the dialogue constructs arguments both for and against watching *Terminator*. After examining the arguments, the agents decide that *Terminator* is a good movie to watch.

4   Goals and Discovery Dialogues

We define *goal* and *goal realization* w.r.t. a given ABA framework as in [6].

Definition 1. *[6] A goal (w.r.t. $\mathcal{L}$) is of the form $\exists\, X\, \mathcal{G}$ such that*

- *X is a tuple of variables;*
- *there exists $\sigma = \{X/t\}$ for t a tuple of terms such that $\mathcal{G}\sigma \in \mathcal{L}$*[4].

A (goal) realization (w.r.t. an ABA framework $\langle \mathcal{L}, \mathcal{R}, \mathcal{A}, \mathcal{C}\rangle$) is $\mathcal{G}\sigma \in \mathcal{L}$ such that $\sigma = \{X/t\}$ and $\mathcal{G}\sigma$ is admissible (w.r.t. $\langle \mathcal{L}, \mathcal{R}, \mathcal{A}, \mathcal{C}\rangle$).

The (natural language) dialogue in our motivating example can be formalised as starting with a goal watchMovie(X). The dialogue identifies watchMovie(*Ter*) as a realization (for $\{X/Ter\}$, where *Ter* stands for *Terminator*) w.r.t. the ABA framework in Figure 1.

---

[3] A variant of this dialogue example is in [6]. There, however, the dialogue starts with an initial concrete proposal of a movie to watch.

[4] $\mathcal{G}\sigma$ stands for $\mathcal{G}$ where all occurrences of (elements of) X are replaced by (the corresponding elements of) t. We often leave the existential quantifier of goals implicit.

R_δ: watchMovie(Ter) $\leftarrow$ fun(Ter), goodScreenTime(Ter)
  fun(Ter) $\leftarrow$ actionMovie(Ter)
  actionMovie(Ter) $\leftarrow$
  finishByTen(Ter) $\leftarrow$
A_δ: goodScreenTime(Ter)
  late(Ter)
C_δ: C_δ(goodScreenTime(Ter)) = late(Ter)
  C_δ(late(Ter)) = finishByTen(Ter)

Fig. 1. ABA framework $\langle \mathcal{L}, \mathcal{R}_\delta, \mathcal{A}_\delta, \mathcal{C}_\delta \rangle$ ($\mathcal{L}$ is left implicit)

In the remainder we consider two generic agents a_1 and a_2. In our example these are Jenny (J) and Amy (A). We assume that a_1, a_2 share a language $\mathcal{L}$. We do not assume that these agents hold ABA frameworks internally. However, they excgange information in ABA format, w.r.t. the shared $\mathcal{L}$.

We define dialogues as sequences of utterances between a_1 and a_2. Formally:

Definition 2. *An* utterance *from agent a_i to agent a_j ($i, j \in \{1,2\}$, $i \neq j$) w.r.t. $\mathcal{L}$ is a tuple $\langle a_i, a_j, Target, C, ID, R \rangle$, where:*

- *C (the* content*) is of one of the following forms:*
  - *(1) goal($\mathcal{G}$) for some $\mathcal{G}$ such that $\exists X \mathcal{G}$ is a goal;*
  - *(2) rl($s_0 \leftarrow s_1, \ldots, s_m$) for some $s_0, \ldots, s_m \in \mathcal{L}$ (a rule), and if $m > 0$ then $s_i \neq s_j$ for $0 \leq i, j \leq m, i \neq j$;*
  - *(3) asm(a) for some $a \in \mathcal{L}$ (an assumption);*
  - *(4) ctr(a, s) for some $a, s \in \mathcal{L}$ (a contrary);*
  - *(5) a* pass *sentence π, such that $\pi \notin \mathcal{L}$.*
- *ID $\in \mathbb{N} \cup \{0\}$ (the identifier).*
- *Target $\in \mathbb{N} \cup \{0\}$ (the target); Target $\leq ID$.*
- *R is either td (top-down), bu (bottom-up) or nr (not-related).*

We refer to an utterance with content π as a pass-utterance, *to an utterance with content goal(()_) as a* goal-utterance, *and to utterances with content other than π and goal(_) as* regular-utterances.

Intuitively, a pass indicates that the agent does not have or want to contribute information at that point. This definition is based on Definition 1 of [5], but (i) this definition adds the new *related* field (R) to indicate different utterance relations (*td, bu* or *nr*); and (ii) there is no "claim" used in this definition.

Definition 3. *For two utterances $u_i \neq u_j$, u_j is* top-down related *to u_i iff $u_i = \langle \_, \_, \_, C_i, ID, \_ \rangle$, $u_j = \langle \_, \_, ID, C_j, \_, td \rangle$[5], and one of the following holds:*

1. *$C_j = rl(\rho_j)$, $Head(\rho_j) = h$ and either $C_i = rl(\rho_i)$ with $h \in Body(\rho_i)$, or $C_i = ctr(\_, h)$;*
2. *$C_j = asm(a)$ and either $C_i = rl(\rho)$ with $a \in Body(\rho)$, or $C_i = ctr(\_, a)$;*
3. *$C_j = ctr(a, \_)$ and $C_i = asm(a)$.*

---

[5] Throughout, _ stands for the anonymous variable, as in Prolog.

This definition is based on Definition 3 of [5], but without considering claim utterances (as these are not allowed here). Intuitively, an utterance is top-down related to another if its target is the identifier of the latter and it contributes to expanding an argument (cases 1), identifies an assumption in the support of an argument (cases 2) or starts the construction of a counter-argument (case 3).

Definition 4. *For two utterances $u_i \neq u_j$, u_j is* bottom-up related *to u_i iff $u_i = \langle \_, \_, \_, C_i, ID, \_ \rangle$ and $u_j = \langle \_, \_, ID, C_j, \_, bu \rangle$, and one of the following holds:*

1. *$C_i = rl(\rho_i)$, $C_j = rl(\rho_j)$, and $Head(\rho_i) \in Body(\rho_j)$;*
2. *$C_i = asm(a)$, $C_j = rl(\rho)$, and $a \in Body(\rho)$.*

Intuitively, an utterance is buttom-up related to another if its target is the identifier of the latter and it forms a "higher level" argument (partially) supported by its target. We say that an utterance u_j is *related to* an utterance u_i if u_j is either top-down or bottom-up related to u_i. Note that an utterance may be related to an utterance from the same agent to the same agent or not. Also, no pass-utterance can be related to a regular-utterance and no utterance can be related to a pass-utterance. Further, no utterance can be realted to a goal utterance and goal utterances are not related to any utterance.

Definition 5. *A dialogue $D_{a_j}^{a_i}(\mathcal{G})$ (between a_i and a_j, for goal $\mathcal{G}$ w.r.t. $\mathcal{L}$), $i, j \in \{1, 2\}$, $i \neq j$, is a finite sequence $\langle u_0, \ldots, u_n \rangle$, $n \geq 0$, where each u_l, $l = 0, \ldots, n$, is an utterance from a_i or a_j (w.r.t. $\mathcal{L}$), u_0 is an utterance from a_i, and:*

1. *the content of u_l is $goal(\mathcal{G})$ iff $l = 0$;*
2. *the content of u_1 is either $rl(\_)$ or $asm(\_)$;*
3. *u_0 and u_1 are of the form $\langle \_, \_, 0, \_, \_, nr \rangle$;*
4. *the target of pass-utterances is 0;*
5. *each regular-utterance u_l, $l > 1$, is related to its target utterance;*
6. *no two consecutive utterances are pass-utterances, other than possibly the last two utterances, u_{n-1} and u_n;*
7. *the identifier of u_i is i.*

This definition requires dialogues to start with a goal (condition 1), followed by a rule or an assumption (condition 2). Intuitively, agents make this utterance with a "wild guess" in the hope that a goal realization can be found by exploring around this guess. Thus, these first two utterances do not need to be related to any utterance (condition 3). All subsequent regular-utterances must be related to some earlier utterance in the dialogue (condition 5). We impose, for simplicity, that the identifier of utterances is their position in the dialogue minus one (condition 7), and that 0 is the target of all "unrelated" utterances (u_0, u_1 and pass-utterances). This definition is a variant of Definition 3 in [5] with our dialogues starting with a goal rather than a claim, followed by an utterance not related to it. An example dialogue is given in Figure 2. An informal reading of this dialogue is the natural language dialogue given in Section 3.

Below, $\mathcal{U}$ and $\mathcal{D}$ stand for the sets, respectively, of all utterances as in Definition 2 and of all dialogues as in Definition 5.

By means of dialogues, agents exchange information and build a shared framework, as defined in [5]:

$\langle J, A, 0, goal(\text{watchMovie}(X)), 0, nr \rangle$
$\langle A, J, 0, rl(\text{actionMovie}(Ter) \leftarrow), 1, nr \rangle$
$\langle J, A, 1, rl(\text{fun}(Ter) \leftarrow \text{actionMovie}(Ter)), 2, bu \rangle$
$\langle A, J, 2, rl(\text{watchMovie}(Ter) \leftarrow \text{fun}(Ter), \text{goodScreenTime}(Ter)), 3, bu \rangle$
$\langle J, A, 3, asm(\text{goodScreenTime}(Ter)), 4, td \rangle$
$\langle A, J, 4, ctr(\text{goodScreenTime}(Ter), \text{late}(Ter)), 5, td \rangle$
$\langle J, A, 0, \pi, 6, nr \rangle$
$\langle A, J, 5, asm(\text{late}(Ter)), 7, td \rangle$
$\langle J, A, 7, ctr(\text{late}(Ter), \text{finishByTen}(Ter)), 8, td \rangle$
$\langle A, J, 8, rl(\text{finishByTen}(Ter) \leftarrow), 9, td \rangle$
$\langle J, A, 0, \pi, 10, nr \rangle$
$\langle A, J, 0, \pi, 11, nr \rangle$

Fig. 2. Example dialogue between Jenny and Amy

Definition 6. *[5] The framework drawn from a dialogue* $\delta = \langle u_0, \ldots, u_n \rangle$ *is* $\langle \mathcal{L},$ $\mathcal{R}_\delta, \mathcal{A}_\delta, \mathcal{C}_\delta \rangle$ *where*

- $\mathcal{R}_\delta = \{r | rl(\rho) \text{ is the content of some } u_i \text{ in } \delta\}$;
- $\mathcal{A}_\delta = \{a | asm(a) \text{ is the content of some } u_i \text{ in } \delta\}$;
- $\mathcal{C}_\delta$ *is a mapping such that, for any* $a \in \mathcal{A}_\delta$, $\mathcal{C}_\delta(a) = c$ *such that* $ctr(a, c)$ *is the content of some* u_i *in* δ, *if one exists, and is undefined otherwise.*

The framework drawn from our earlier example dialogue in Figure 2 is $F_\delta = \langle \mathcal{L}, \mathcal{R}_\delta, \mathcal{A}_\delta, \mathcal{C}_\delta \rangle$ shown earlier in Figure 1. Note that F_δ in this example is a flat ABA framework but in general, as discussed in [5], the framework drawn from a dialogue may not be an ABA framework, since $\mathcal{C}_\delta$ may not be total. We follow [5] and impose, by using suitable *legal-move functions* and *outcome functions*, that dialogues are such that the frameworks drawn from them are necessarily flat ABA frameworks.

Definition 7. *A* legal-move function *is a mapping* $\lambda : \mathcal{D} \mapsto \mathcal{U}$ *such that, given* $\delta = \langle u_0, \ldots, u_n \rangle \in \mathcal{D}$ *between* a_i, a_j *(for some goal),* $\lambda(\delta) = \langle a_i, a_j, t, C, id, r \rangle$ *and*

1. $\delta \circ \lambda(\delta) = \langle u_0, \ldots, u_n, \lambda(\delta) \rangle$ *is a dialogue;*
2. *if* $C \neq \pi$, *then there exists no* $k, 1 \leq k \leq n$, *such that* $u_k = \langle \_, \_, t, C, k, \_ \rangle$.
3. *if* $C = ctr(a, c)$ *then there exists no* $u_k = \langle \_, \_, \_, ctr(a, c'), k, \_ \rangle$, *for* $1 \leq k \leq n$, *and* $c' \neq c$.

Given $\delta = \langle u_0, \ldots, u_n \rangle$, *if* $\lambda(\langle u_0, \ldots, u_m \rangle) = u_{m+1}$ *for all* m $(0 \leq m < n)$, *we say that* δ *is* constructed with λ. *We use* Λ *to denote the set of all legal-move functions.*

This definition is analogous to Definition 5 in [5] except for the format of utterances (having here an extra field r to indicate how they are related to their target). This definition imposes that any sequence of utterances constructed using a legal-move function forms a dialogue (condition 1); there is no repeated non-pass utterance to the same target (condition 2); assumptions have a single contrary (condition 3). However, the definition of legal-move function does not impose any "mentalistic" requirement on agents, such as that they utter information they hold true.

Definition 8. *A flat legal-move function is such that*
if $\lambda(\langle u_0, \ldots, u_n \rangle) = \langle a_x, a_y, t, C, n+1, r \rangle$ *then*

- $C = asm(a)$ *only if there exists no* $u_i = \langle \_, \_, \_, C_i, i_\_ \rangle$, *for* $1 \leq i \leq n$, *with* $C_i = rl(\rho)$ *and* $Head(\rho) = a$;
- $C = rl(\rho)$ *only if there exists no* $u_i = \langle \_, \_, \_, C_i, i_\_ \rangle$, *for* $1 \leq i \leq n$, *with* $C_i = asm(a)$ *and* $Head(\rho) = a$.

This definition is analogous to Definition 6 in [5], excpet again for the formal of utterances, with the extra r field. Trivially, the framework drawn from a dialogue generated by a flat legal-move function, if an ABA framework, is flat, in the sense of [2].

Definition 9. *A one-way expansion legal-move function is a flat legal-move function such that*
if $\lambda(\langle u_0, \ldots, u_n \rangle) = \langle a_x, a_y, \_, C, n+1, r \rangle$, *then*

- $C = rl(\rho)$ *only if there exists no* $u_i = \langle \_, \_, \_, rl(\rho'), i, \_ \rangle$, *for* $1 \leq i \leq n$, *with* $Head(\rho) = Head(\rho')$.

This definition imposes that in a dialogue there is only one unique way of expanding a rule. This requirement allows to keep the framework simple.

Legal-move functions provide some guidance as to what agents are allowed to utter during dialogues. In order to guarantee that the contrary mapping in the framework drawn from a dialogue is total, and thus that the framework is indeed an ABA framework, as in [5], we use the notion of *outcome function*, checking specific properties in a generated dialogue:

Definition 10. *An outcome function is a mapping* $\omega : \mathcal{D} \times \Lambda \mapsto \{true, false\}$. *The ABA outcome function,* ω_{ABA}, *is such that given a dialogue* δ *and a legal-move function* λ, $\omega_{ABA}(\delta, \lambda) = true$ *iff* δ *is constructed with* λ *and the framework* $\langle \mathcal{L}, \mathcal{R}_\delta, \mathcal{A}_\delta, \mathcal{C}_\delta \rangle$ *drawn from* δ *is such that for all* $\alpha \in \mathcal{A}_\delta$, $\mathcal{C}_\delta(\alpha)$ *is not undefined.*[6]

In the remainder of the paper, we focus on dialogues where each utterance results from applying a one-way expansion legal-move function to the dialogue prior to that utterance, and for which ω_{ABA} is true. We refer to these dialogues as *ABA dialogues*.

We use *debate trees*, defined in the next section, to refine the notions of legal-move and outcome functions to guarantee that dialogues compute arguments in admissible extensions.

---

[6] Definition 10,18, and 20 are variations of Definition 7, 13, and 15, respectively, in [5] with the extra Λ as a domain for outcome functions, to enforce that dialogues are properly constructed.

5 Debate Trees

Nodes of debate trees are labelled either *proponent* or *opponent* as in the abstract dispute trees in [4]. However, differently from [4], in debate trees each node (1) contains one sentence, (2) is *tagged* as either *unmarked (um), marked-rule (mr)* or *marked-assumption (ma)*, and (3) has an *ID* to identify its corresponding utterance in a dialogue.

When constructing a debate tree from a dialogue, we use a subset of utterances presented in the dialogue. This extraction ignores the goal- and pass-utterances, i.e. a debate tree is extracted from the *goal-π-pruned sequence obtained from a dialogue*, consisting of all regular-utterances. Note that, since no regular-utterance has a goal- or pass-utterance as its target (see definition 5), the target of every utterance in a goal-π-pruned sequence is guaranteed to be in this sequence. Furthermore, by Definition 5, for all utterances $u = \langle \_, \_, \_, \_, i, R \rangle$ in a goal-π-pruned sequence, if $i > 1$, then R is either *td* or *bu*.

The sentence of each node in a debate tree represents an argument's claim or an element of its support. A node is tagged *unmarked* if its sentence is only mentioned in the body of a rule or as contrary of an assumption, but without any further examination, *marked-rule* if it is the head of a rule, and *marked-assumption* if it has been explicitly declared as an assumption. Debate trees are special cases of debate graphs, defined as follows:

Definition 11. *A debate graph drawn from a dialogue $\delta = \langle u_0, \ldots, u_n \rangle$ ($n \geq 0$) is a graph $\mathcal{T}(\delta)$ whose nodes are tuples $(S, F : L[U])$ where*

- *S (the sentence) is a sentence in $\mathcal{L}$,*
- *F (the tag) is either um (unmarked), mr (marked-rule) or ma (marked-assumption),*
- *L (the label) is either P (proponent) or O (opponent),*
- *$U \in \mathbb{N}$ (the ID).*

$\mathcal{T}(\delta)$ is $\mathcal{T}^m(\delta)$ in the sequence $\mathcal{T}^0(\delta), \mathcal{T}^1(\delta), \ldots, \mathcal{T}^m(\delta)$ constructed inductively from the goal-π-pruned sequence $\delta' = \langle u'_1 \ldots, u'_m \rangle$ obtained from δ, as follows:

- *$\mathcal{T}^0(\delta)$ is empty;*
- *$\mathcal{T}^1(\delta)$ is constructed as follows:*
  - *if the content of u'_1 is $asm(\alpha)$, then $\mathcal{T}^1(\delta)$ consists only of $(\alpha, ma : P[1])$;*
  - *if the content of u'_1 is $rl(h \leftarrow b_1, \ldots, b_l)$, then $\mathcal{T}^1(\delta)$ consists of $l + 1$ nodes: $(h, mr : P[1])$ with children*

$$(b_1, um : P[1]), \ldots, (b_l, um : P[1]).$$

- *Let $\mathcal{T}^i(\delta)$ be the i-th tree, for $0 < i < m$, let $u'_{i+1} = \langle \_, \_, t, C, id, R \rangle$, and let $u'_t = \langle \_, \_, \_, C_t, t, \_ \rangle$ be the target utterance of u'_{i+1}; then $\mathcal{T}^{i+1}(\delta)$ is obtained according to one of the following cases:*
  - *If $R = td$, then*

* if $C=rl(h \leftarrow b_1, \ldots, b_l)$ then $\mathcal{T}^{i+1}(\delta)$ is $\mathcal{T}^i(\delta)$ with additional l nodes:

$$(b_1, um : \mathrm{L}[id]), \ldots, (b_l, um : \mathrm{L}[id])$$

as children of the existing node $(h, um : \mathrm{L}[t])$, and this node is replaced by $(h, mr : \mathrm{L}[id])$;
* if $C = asm(\alpha)$ then $\mathcal{T}^{i+1}(\delta)$ is $\mathcal{T}^i(\delta)$ with the existing node $(\alpha, um : \mathrm{L}[t])$ replaced by $(\alpha, ma : \mathrm{L}[id])$;
* if $C = ctr(\alpha, c)$ then $\mathcal{T}^{i+1}(\delta)$ is $\mathcal{T}^i(\delta)$ with an additional node: $(c, um : \mathrm{L}[id])$ child of $(\alpha, ma : \mathrm{L}'[t])$, where $L, L' \in \{P, O\}, L \neq L'$.
* If $R = bu$, then
  * $C = rl(h \leftarrow b_1, \ldots, b_l)$ and $\mathcal{T}^{i+1}(\delta)$ is $\mathcal{T}^i(\delta)$ with l additional nodes:
    · $(h, mr : \mathrm{L}[id])$, added as parent of the existing node $(b_t, F : \mathrm{L}[t])$, such that
    if $C_t = rl(h' \leftarrow b'_1, \ldots, b'_k)$, then $b_t = h'$, $F = mr$;
    if $C_t = asm(\alpha)$, then $b_t = \alpha$, $F = ma$;
    · $(b''_1, um : \mathrm{L}[id]), \ldots, (b''_{l-1}, um : \mathrm{L}[id])$, added as children of the existing node $(h, mr : \mathrm{L}[id])$, where $\{b''_1, \ldots, b''_{l-1}\} = \{b_1, \ldots, b_l\} \setminus \{b_t\}$.

Figure 3 gives the construction of the debate graph drawn from the dialogue in our example[7]. Note that this is a tree but in general it may not be. For example, if the dialogue in Figure 2 is modified to δ^* with utterances 10 and 11 replaced by:
$\langle J, A, 1, rl(\mathrm{noisy}(Ter) \leftarrow \mathrm{actionMovie}(Ter)), 10, bu \rangle$
$\langle J, A, 0, \pi, 11, nr \rangle$
$\langle A, J, 0, \pi, 12, nr \rangle$
then the debate graph $\mathcal{T}(\delta^*)$, given in Figure 4, is not a tree. To ensure that a debate graph is a tree, we give the following definition.

Definition 12. *A debate graph $\mathcal{T}(\delta)$ is properly-structured iff it is a tree.*

This definition is needed as we need to rule out the case of related utterances somewhat being a distraction, as in the previous example leading to the graph in Figure 4.

We ensure that debate graphs drawn from our dialogues are properly-structured using a legal-move function, as follows.

Definition 13. *A legal-move function $\lambda : \mathcal{D} \mapsto \mathcal{U}$ is a properly-structured legal-move function iff it is a one-way expansion legal-move function and, for every dialogue $\delta \in \mathcal{D}$ such that $\mathcal{T}(\delta)$ is properly-structured, then $\mathcal{T}(\delta \circ \lambda(\delta))$ is also properly-structured.*

---

[7] In this figure, $\uparrow$ represents expanding a rule towards the construction of an argument, whereas $\Uparrow$ represents the introduction of a contrary and thus the attack relation between arguments. Here, wM, aM, gST, fBT, and T are a shorthand for watch-Movie, actionMovie, goodScreenTime, finishByTen, and Terminator, respectively.

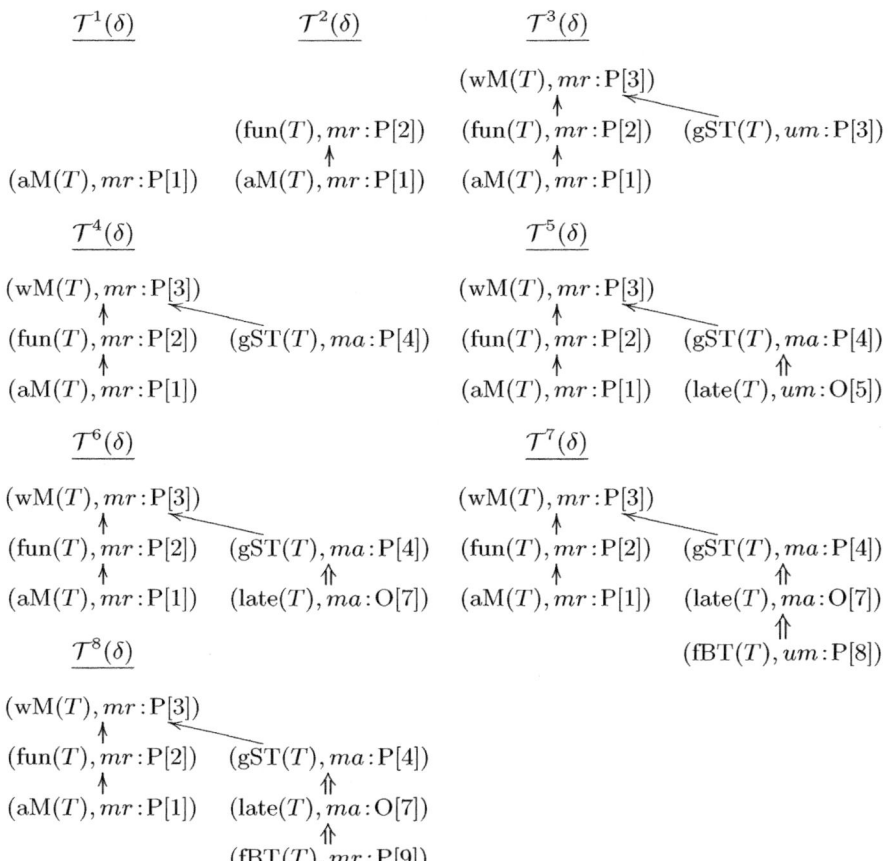

Fig. 3. The construction of the debate graph drawn from the dialogue in Figure 2

Thus, when an agent decides what to utter, it needs to keep the current debate tree into account and make sure that its new utterance keeps the graph properly-structured. As a result, the debate tree can be seen as a *commitment store* for the agents. In the remainder, we assume that all dialogues are constructed with a properly-structured legal-move function.

Definition 14. *Given a debate tree $\mathcal{T}(\delta)$,*

- *the* defence set $\mathcal{D}ef(\mathcal{T}(\delta))$ *is the union of all assumptions s in nodes of the form $(s, ma : \mathrm{P}[\_])$;*
- *the* culprits $\mathcal{C}ul(\mathcal{T}(\delta))$ *are given by the set of all assumptions s in nodes $n = (s, ma : \mathrm{O}[\_])$ such that the child of n in $\mathcal{T}(\delta)$ is $(\_, \_ : \mathrm{P}[\_])$.*[8]

---

[8] Definition 14, 15, 16, and 19 are adaptations to the case of *debate trees* of definitions 10, 11, 12, and 15, respectively, in [5], where they are given for the *dialectical trees* defined therein.

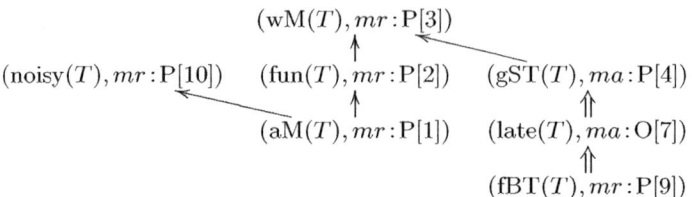

Fig. 4. A debate graph that is not a tree

Arguments can be drawn from a debate tree, as follows:

Definition 15. *An argument drawn from a debate tree $\mathcal{T}(\delta)$ is a subtree $\mathcal{T}$ of $\mathcal{T}(\delta)$ such that:*

- *all nodes in $\mathcal{T}$ have the same label (either P or O),*
- *there exists no node n' in $\mathcal{T}(\delta)$ such that n' is parent or a child of some node n_i in $\mathcal{T}$ and n_i, n' have the same label (either P or O).*

An argument $\mathcal{T}$ is actual if for all nodes $(\_, F: \_[\_])$ in $\mathcal{T}$, F is either mr or ma. If there is at least one node of the form $(\_, um: \_[\_])$ in $\mathcal{T}$, then $\mathcal{T}$ is potential.

The sentence c in the root of $\mathcal{T}$ is the claim of $\mathcal{T}$. The set of all sentences s in nodes of the form $(s, ma: \_[\_])$ is the support of $\mathcal{T}$.

Actual arguments can be mapped to equivalent ABA arguments.

Proposition 1. *Let $\mathcal{T}$ be an actual argument drawn from a debate tree $\mathcal{T}(\delta)$ and let c, S be the claim and support, respectively, of $\mathcal{T}$. Then $S \vdash c$ is an ABA argument (w.r.t. the framework drawn from δ).*

This is trivially true as a node in an actual argument can be mapped to a node in an ABA argument by dropping the tag, the label, and the ID. Then, nodes τ need to be added as children of leaf nodes that do not hold assumptions, as each such node represents a rule with an empty body.

The ABA arguments equivalent to the actual arguments drawn from the dialogue in our running example are: (1) {goodScreenTime(Ter)} $\vdash$ watchMovie(Ter), (2) {late(Ter)} $\vdash$ late(Ter), and (3) {} $\vdash$ finishByTen(Ter). There is no potential argument in our example.

We consider now restricted forms of debate trees, that we then use below to further refine our notion of legal-move function.

Definition 16. *A debate tree $\mathcal{T}(\delta)$ is*

- patient *iff for all nodes $n = (\_, ma : \_[\_])$ in $\mathcal{T}(\delta)$ such that n has a child, then n is in an actual argument drawn from $\mathcal{T}(\delta)$.*
- focused *iff for all arguments Γ drawn from $\mathcal{T}(\delta)$, if Γ contains a node $(\_, ma : O[\_])$, then there is at most one node c of the form $(\_, ma : O[\_])$ in Γ such that c has a child.*

Arguments in a patient tree are fully expanded (cf. actual) before being attacked. In focused trees, no alternative ways to defend claims are considered simultaneously, i.e., an opponent argument is only attacked by a single proponent argument whereas a proponent argument can be attacked in as many ways as the number of its assumptions. The tree in Figure 3 is both patient and focused.

The restricted form of legal-move function we consider is guaranteed to generate patient, focused trees, as follows.

Definition 17. *A legal-move function $\lambda : \mathcal{D} \mapsto \mathcal{U}$ is a* patient and focused legal-move function *iff it is properly-structured and for every dialogue $\delta \in \mathcal{D}$ such that $\mathcal{T}(\delta)$ is patient and focused, $\mathcal{T}(\delta \circ \lambda(\delta))$ is still patient and focused.*

This definition is in the same spirit as Definition 13 presented earlier as it requires agents to consult the debate tree before making utterances. In the remainder we will assume that all dialogues are constructed with patient and focused legal-move functions.

Definition 18. *The* exhaustive outcome function ω_{ex} *is such that, given $\delta \in \mathcal{D}$, $\lambda \in \Lambda$ and $\langle \mathcal{L}, \mathcal{R}_\delta, \mathcal{A}_\delta, \mathcal{C}_\delta \rangle$ the framework drawn from δ, $\omega_{ex}(\delta, \lambda) = true$ iff $\omega_{ABA}(\delta, \lambda) = true$ and there exists no $u' = \lambda(\delta)$ with content either $asm(a)$, for $a \in \mathcal{A}_\delta$, or $rl(\rho)$, for $\rho \in \mathcal{R}_\delta$, or $ctr(a, c)$, for $c = \mathcal{C}_\delta(a)$, such that $\omega_{ABA}(\delta \circ u', \lambda) = true$.*

Note that exhaustiveness does not force agents to contribute to dialogues with all relevant information they hold. Rather, it enforces agents to make all utterances that are compliant with the given λ.

6 Formal Results

In this section we link dialogues and the admissible argumentation semantics. First we refine the outcome function so that if a dialogue has a true outcome then the (fictional) proponent has the last word in the dialogue, namely all leaves in the debate tree are proponent nodes or "dead-end" opponent nodes (not corresponding to any actual argument). Formally:

Definition 19. *The* last word outcome function ω_{lw} *is such that, given $\delta \in \mathcal{D}$, $\lambda \in \Lambda$, and the debate tree drawn from $\mathcal{T}(\delta)$, then $\omega_{lw}(\delta, \lambda) = true$ iff $\omega_{ex}(\delta, \lambda) = true$ and one of the following two cases holds:*

1. *for all leaf nodes n in $\mathcal{T}(\delta)$, n is either $(\_, mr : P[\_])$ or $(\_, ma : P[\_])$;*
2. *if a leaf node n is in the form $(\_, \_ : O[\_])$, then n is in an argument that contains at least one node in the form $(\_, um : O[\_])$.*

We refer to exhaustive dialogues for which ω_{lw} is true as *positive*. The dialogue in our example is positive. Positive dialogues give debate trees corresponding to *abstract dispute tree* with the same defence set and culprits. Formally:

Lemma 1. *Given a positive dialogue δ let $\mathcal{T}(\delta)$ be the debate tree drawn from δ and s be the sentence in the root node of $\mathcal{T}(\delta)$. Then there is an abstract dispute tree $\mathcal{T}^a$ for $S \vdash s$ for some S, such that $\mathcal{D}ef(\mathcal{T}(\delta)) = \mathcal{D}ef(\mathcal{T}^a)$ and $\mathcal{C}ul(\mathcal{T}(\delta)) = \mathcal{C}ul(\mathcal{T}^a)$.*

Proof. We can transform debate trees into abstract dispute trees. Given a debate tree $\mathcal{T}(\delta)$, its equivalent abstract dispute tree $\mathcal{T}^a$ is constructed as follows.

1. Modify $\mathcal{T}(\delta)$ by appending a new *flag* field $Z = \{0, 1\}$ to each node in $\mathcal{T}(\delta)$ and initialize Z to 0 for all nodes, i.e., a node now looks like $(\_, \_ : \_[\_])[0]$.
2. Delete all nodes n from $\mathcal{T}(\delta)$ where n is in a potential argument.
3. $\mathcal{T}^a$ is $\mathcal{T}_m^a$ in the sequence $\mathcal{T}_1^a, \ldots, \mathcal{T}_m^a$ constructed inductively as follows:
   - $\mathcal{T}_0^a$ is empty;
   - obtain the argument Γ that contains the root of $\mathcal{T}(\delta)$, set the flags of all nodes in $\mathcal{T}(\delta)$ that are also in Γ to 1; then $\mathcal{T}_1^a$ contains a single node that is labelled by Γ and is a P node;
   - let $\mathcal{T}_i^a$ be the i-th tree, for $0 < i < m$, then $\mathcal{T}_{i+1}^a$ is $\mathcal{T}_i^a$ with an additional node (Θ, L), where Θ is an argument drawn from $\mathcal{T}(\delta)$, child of Θ', another argument drawn from $\mathcal{T}(\delta)$, such that

     - the flag of nodes in Θ are 0;
     - the root node of Θ has a parent node p in $\mathcal{T}(\delta)$ which has its flag equal to 1 and p is in Θ';
     - L is P if the root node of Θ is a proponent node, otherwise L is O;
     - set the flags of all nodes in $\mathcal{T}(\delta)$ that are also in Θ to 1.

   - $\mathcal{T}_m^a$ is constructed when there is no node in $\mathcal{T}(\delta)$ with its flag equal to 0.

Trivially, $\mathcal{T}^a$ constructed above is an abstract dispute tree. Since $\mathcal{T}^a$ contains the same arguments as $\mathcal{T}(\delta)$ and arguments have the same P/O labelling in both $\mathcal{T}^a$ and $\mathcal{T}(\delta)$, we have $\mathcal{D}ef(\mathcal{T}(\delta)) = \mathcal{D}ef(\mathcal{T}^a)$ and $\mathcal{C}ul(\mathcal{T}(\delta)) = \mathcal{C}ul(\mathcal{T}^a)$.

As in the case of abstract dispute trees, the defence set of a debate tree may not be admissible, as it may attack itself. We refine the notion of outcome function by enforcing that this set does not attack itself, as follows:

Definition 20. *The* admissible outcome function ω_{ADM} *is such that, given $\delta \in \mathcal{D}$, $\lambda \in \Lambda$, $\omega_{ADM}(\delta, \lambda) = true$ iff $\omega_{lw}(\delta, \lambda) = true$ and $\mathcal{D}ef(\mathcal{T}(\delta)) \cap \mathcal{C}ul(\mathcal{T}(\delta)) = \{\}$. If δ is positive and $\omega_{ADM}(\delta, \lambda) = true$, we say that δ is* successful.

Theorem 1. *Given a successful dialogue $D_{a_j}^{a_i}(\mathcal{G}) = \delta$, let s be the sentence in the root node of $\mathcal{T}(\delta)$. Then, there exists an argument $S \vdash s$ that belongs to an admissible extension supported by $\mathcal{D}ef(\mathcal{T}(\delta))$ w.r.t. the ABA framework drawn from δ.*

Proof. If δ is successful, it is positive and, by Lemma 1, there exists an abstract dispute tree $\mathcal{T}^a$ such that $\mathcal{D}ef(\mathcal{T}(\delta)) = \mathcal{D}ef(\mathcal{T}^a)$ and $\mathcal{C}ul(\mathcal{T}(\delta)) = \mathcal{C}ul(\mathcal{T}^a)$. By Theorem 5.1 of [4] (see Section 2), the theorem holds.

Theorem 1 connects dialogues with argumentation semantics. Thus, our dialogues can be seen as a distributed mechanism for computing admissible extensions. With this, we prove a result specifically for discovery dialogues.

Theorem 2. *Given a successful dialogue* $\delta = D_{a_j}^{a_i}(\mathcal{G})$, *if* $(\mathcal{G}\sigma, \_: P[\_])$ *is the root node of* $\mathcal{T}(\delta)$, *where* $\sigma = \{X/t\}$, *then* $\mathcal{G}\sigma$ *is a goal realization for* $\mathcal{G}$ *w.r.t. to the ABA framework drawn from* δ.

Proof. If $\mathcal{G}\sigma$ is the sentence in the root node of $\mathcal{T}(\delta)$ and δ is successful, then $\mathcal{G}\sigma$ is admissible w.r.t. AF_δ, the ABA framework drawn from δ, by Theorem 1. Hence $\mathcal{G}\sigma$ is a goal realization with respect to AF_δ.

7   Related Work

McBurney and Parsons [9] give an overview of dialogue games for argumentation. Our work can be seen as providing a novel dialogue game for ABA.

McBurney and Parsons [8] present a modelling for chance discovery dialogue. The formal system in that work is defined with locutions and rules without linking to any argumentation framework, whereas our work is based on ABA. There is no argumentation semantics used in examining the result of their dialogues, whereas our work makes the connection to the admissibility semantics. Moreover, [8] focuses on chance discovery, whereas our work is applicable to any discoveries as long as the desired outcome can be qualified, essentially, by a predicate.

Rybakov [13] presents a logic modelling of chance discovery. Our work differs from that as is focuses on a dialogue system for discovery whereas his is mainly concerned with constructing a modal/temporal modelling for chance discovery.

Fisher [7] presents a mechanism for concurrent theorem-proving. In his setting, the knowledge base (a set of formulae) is distributed at different objects and each object continuously broadcasts messages about its formulae. Upon receiving messages, an object makes inferences, transforms its formulae and sends out further messages. Even though similarity exists, this work is vastly different from ours as (1) it does not focus on discovering any particular information; (2) it is not concerned with either agents or dialogues; (3) it requires formulae to be represented in Horn Clauses. No formal results have been shown in [7].

Fan and Toni [5] introduce a formal modelling for argumentation dialogues. Though the modelling presented in [5] is generic, it starts a dialogue with a claim and it only uses top-down reasoning, whereas our work here allows agents to start the dialogue with a more generic goal and reason buttom-up. Moreover, the result of [5] is based on the concrete dispute trees of [4] whereas our work here uses the abstract dispute trees of [4]. Our work can be viewed as an extension of [5], apart from the fact that the framework introduced there does not require a dialogue to be built with (restricted) one-way expansion legal-move functions.

Black and Hunter [1] present a formal system for inquiry dialogues based on DeLP as the underlying argumentation framework. Our work differs from theirs as (1) our work does not start the dialogue with a claim; (2) it does not focus on inquiry dialogues; (3) it does not force full disclosure of all relevant knowledge.

Prakken [11] defines a formal system for persuasion. The main differences with our work are: (1) Prakken's work starts the dialogue with a claim (2) proponent and opponent roles are pre-assigned to agents before the dialogue whereas in our work agents can play both roles within the same dialogue; (3) he considers the grounded semantics, whereas we use admissibility; (4) his set of utterances refer to arguments and attacks, as in the case of [10]; (5) he forces the support of arguments to be minimal, whereas we do not.

8   Conclusion

In this paper, we introduce a formal modelling for a simple form of discovery dialogue for two agents, in which the desired outcomes are only partially known when dialogues start. In our setting, the two agents share the same discovery goal but neither of the two agents is capable of discovering a justified goal realization that fulfills the shared goal. A discovery dialogue is successful if a goal realization is found through the dialogue.

In our model, the dialogue effectively starts by one agent putting forward a piece of information that might be related to the goal realization. Through dialogues, more information that is related to the initial utterance is gathered and examined. This process is defined with various legal-move functions with the help of constructing a debate tree. We examine the acceptability of the goal realization with the admissibility semantics.

Future work includes investigating other forms of discovery dialogues, e.g., where agents can change their abstract goals with some form of buttom-up reasoning and studying cases in which results about completeness can be obtained, namely, conditions of a goal realization discovery can be guaranteed. We are also interested in allowing more than one "wild guess" in a dialogue and the possibility of agents to have preferences over realizations. Finally, we plan to consider less restrictive forms of legal-move functions, e.g., not requiring a one-way expansion.

References

1. Black, E., Hunter, A.: An inquiry dialogue system. JAAMAS 19, 173–209 (2009)
2. Bondarenko, A., Dung, P.M., Kowalski, R., Toni, F.: An abstract, argumentation-theoretic approach to default reasoning. AIJ 93(1-2), 63–101 (1997)
3. Dung, P.M., Kowalski, R.A., Toni, F.: Assumption-based argumentation. In: Argumentation in Artificial Intelligence, pp. 25–44. Springer, Heidelberg (2009)
4. Dung, P.M., Kowalski, R.A., Toni, F.: Dialectic proof procedures for assumption-based, admissible argumentation. AIJ 170, 114–159 (2006)
5. Fan, X., Toni, F.: Assumption-based argumentation dialogues. In: Proc. IJCAI (2011)
6. Fan, X., Toni, F.: Conflict resolution with argumentation dialogues – Extended abstract. In: Proc. AAMAS (2011)
7. Fisher, M.: An open approach to concurrent theorem-proving. In: Parallel Processing for Artificial Intelligence, pp. 209–230. Elsevier/North (1997)

8. McBurney, P., Parsons, S.: Chance Discovery Using Dialectical Argumentation. In: Terano, T., Nishida, T., Namatame, A., Tsumoto, S., Ohsawa, Y., Washio, T. (eds.) JSAI-WS 2001. LNCS (LNAI), vol. 2253, pp. 414–424. Springer, Heidelberg (2001)

9. McBurney, P., Parsons, S.: Dialogue games for agent argumentation. In: Argumentation in Artificial Intelligence, pp. 261–280. Springer, Heidelberg (2009)

10. Parsons, S., McBurney, P., Sklar, E., Wooldridge, M.: On the relevance of utterances in formal inter-agent dialogues. In: Proc. AAMAS, pp. 47–62 (2007)

11. Prakken, H.: Coherence and flexibility in dialogue games for argumentation. JLC 15, 1009–1040 (2005)

12. Prakken, H.: Formal systems for persuasion dialogue. Knowledge Eng. Review 21(2), 163–188 (2006)

13. Rybakov, V.: Logic of knowledge and discovery via interacting agents - decision algorithm for true and satisfiable statements. Inf. Science 179, 1608–1614 (2009)

14. Walton, D., Krabbe, E.: Commitment in Dialogue: Basic concept of interpersonal reasoning. State University of New York Press, Albany NY (1995)

Author Index